T0192477

Electromechanical Energy Conversion

Electromechanical Energy Conversion

Dr. Zeki Uğurata KOCABIYIKOĞLU

CRC Press is an imprint of the
Taylor & Francis Group, an **informa** business

First edition published 2020
by CRC Press
6000 Broken Sound Parkway NW, Suite 300, Boca Raton, FL 33487-2742

and by CRC Press
2 Park Square, Milton Park, Abingdon, Oxon, OX14 4RN

CRC Press is an imprint of Taylor & Francis Group, LLC

Library of Congress Cataloging-in-Publication Data
Names: Kocabiyikoğlu, Zeki Uğurata, author.
Title: Electromechanical energy conversion / Zeki Uğurata Kocabiyikoğlu.
Description: First edition. | Boca Raton, FL : CRC Press, 2020. |
Includes bibliographical references and index.
Identifiers: LCCN 2020012068 (print) | LCCN 2020012069 (ebook) |
ISBN 9780367322670 (hbk) | ISBN 9780429317637 (ebk)
Subjects: LCSH: Electric power production—Textbooks. | Electromechanical devices—Textbooks.
Classification: LCC TK1001 .K58 2020 (print) | LCC TK1001 (ebook) |
DDC 621.31/21—dc23
LC record available at https://lccn.loc.gov/2020012068
LC ebook record available at https://lccn.loc.gov/2020012069

ISBN: 978-0-367-32267-0 (hbk)
ISBN: 978-0-367-52402-9 (pbk)
ISBN: 978-0-429-31763-7 (ebk)

Typeset in Times
by codeMantra

To my dearest wife Ayşe ....with my heartiest feelings for her patience and love

January 2020

Contents

Preface

This book is intended to be a textbook on "*Electromechanical Energy Conversion*". The title, "*Electromechanical Energy Conversion*" is often the title of a course given in almost every electrical engineering department of the universities, colleges, and polytechnics. Electromechanical energy conversion is one of the essential topics for a license in electrical engineering. A new approach to this rather old and classical subject will be presented in this book.

There are quite a few books published on the subject. This new approach, however, will provide modern, lean, and up-to-date information on electromechanical energy conversion as well as a review of energy forms with an emphasis on renewable energy.

Electromechanical energy conversion is a vast subject and can be difficult to grasp if went into details. What is more a right amount of this knowledge is a necessity for every electronic and electrical engineer. Information given in this book will be brief, precise and suitable for a university course of one single semester. Only the necessary information on the subject will be conveyed to the students in a simple and precise manner. Electromechanical Energy Conversion is an important course which is a must for every electrical and electronic engineer. This book is aimed at filling this gap in universities and colleges.

Introduction

The purpose of this book is to convey to the reader the basics of electromechanical energy conversion which can be summarised as conversion of energy from mechanical to electrical form or vice versa. Here we will discuss about the principles of equipment and systems involved in such conversion process. The electrical or magnetic fields in the conversion equipment act as the media for the conversion of electromechanical energy to take place. In this book, we shall be dealing with electromechanical conversion through magnetic media.

During this conversion, magnetic energy behaves like a buffer or an interface between mechanical and electrical energies (Figure 1). Therefore, an extensive knowledge of magnetic circuits is a necessity, and this information will be covered in Chapter 2 of this book.

The relay shown in Figure 2 is an example of such converters. In fact, it is an electrical relay which has found a great deal of application area in the past in telephone exchanges. However, recently, it has been replaced by electronic relays which don't have moving contacts.

But before everything, there will be an extensive chapter on energy, power, energy resources, and generation of electrical energy with an emphasis on renewable energy resources. An emphasis on the alarming subject of "*CO_2 emission*" which is dangerously threatening the climate of the world has also been covered in this chapter.

After magnetic circuits and magnetic materials, we shall move onto conversion of magnetic energy to mechanical energy in a general manner. In this book, we shall

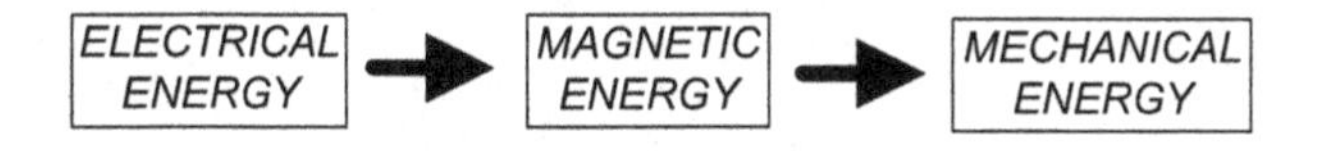

FIGURE 1 Electromechanical energy conversion.

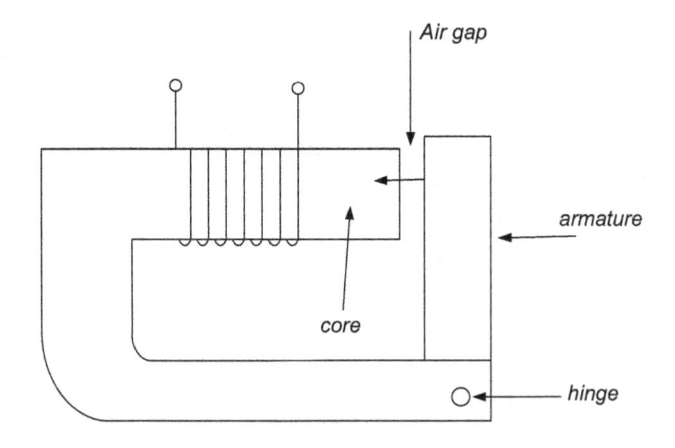

FIGURE 2 An electromechanical energy converter (relay).

be more concerned with conversion of energy from electrical to mechanical form. However, information on the conversion of mechanical energy to electrical energy (i.e., electrical power generators) will in a general sense stay within the scope of this book too. In fact, electromechanical energy conversion is a reversible process.

Then, general working principles of electric motors shall be looked into. Advancements in electronics have made "reluctance and step motors" more important than in the past. These subjects and the general concept of rotating machinery are covered in Chapter 4. Finally, we will move onto DC and AC machines. Main electromechanical energy converters are the rotational ones which are named "electrical machines." Chapters 5 through 9 are extensively allocated to this ever important electrical engineering subject.

Author

Zeki Uğurata Kocabiyikoğlu is a graduate of the University of Sheffield, England in Electrical and Electronic Engineering department. After completing his PhD studies in the same department, he continued his professional career in industry. He has several papers published in several respected citations and is the co-author of a paper presented by professor G.S. Hobson in "8th International Conference on Microwaves and Optical Generation and Amplification" Kluwer-Deventer, the Netherlands.

After 4 years of industrial experience, he then joined in the Turkish defence industry, namely ASELSAN Inc. In the 28 years of his career with this company, he fulfilled several positions starting from R&D engineer and finally he retired as vice president responsible for the division of communication systems and equipment. He then joined various universities and currently is teaching in the Electronic and Electrical Engineering department of the TOBB University of Economics and Technology in Ankara, Turkey.

Author

1 Basic Principles of Energy, Power and Electrical Power Systems

1.1 PHYSICAL FOUNDATIONS OF ENERGY AND POWER

1.1.1 ENERGY

Here, we shall provide a reminder about

- what energy is
- how it changes
- how electrical energy is obtained.

Energy is defined as the "capacity of a system to do work". The most common unit used for energy is joule, abbreviated J. Electric utilities bill the energy in kilowatt hours (kWh). A kWh represents "one kW being used for one hour", and it equals 3.6 MJ of energy. Energy can exist in many forms like heat, light, sound, mechanical, and electrical. It cannot be consumed but can be converted from one form to another. This is the law of "conservation of energy" described in the "first law of thermodynamics". Whenever energy changes from one form to another, some of it is lost as heat. This is because energy transformations are not energywise reversible. Heat is a disorderly form of energy. On the other hand, electrical energy, kinetic energy, and potential energy are orderly forms of energy.

Orderly forms of energy can easily be converted to heat. For example, electrical energy can be easily converted to heat by making a current flow in a resistor or by dropping a stone from a hill onto a hard surface and converting partially its kinetic energy to heat. On the other hand, transforming heat into electrical energy is much more difficult.

The second law of thermodynamics says that "nature has a tendency towards disordered forms of energy". This means that converting heat to electricity will not be as efficient as converting electricity to heat. About 60% of the heat energy input to power plants is lost as waste. Energy can be in the form of "kinetic energy" as in the case of a truck travelling down the hill, or it can be in the form of "potential energy" as in the case of a skier standing at the top of a hill (Figure 1.1).

As it can be stored for future use, potential energy bears a special importance. The pond of water could be drained through a turbine at the bottom of a hill, which in turn could drive a generator to produce electricity (Figure 1.2).

A turbine is a rotary engine that extracts energy from a fluid flow and converts it into useful work.

The simplest turbines have a rotor assembly, which is a shaft or drum with blades attached. Moving fluid acts on the blades, so that they move and impart rotational energy to the rotor. Early turbine examples are windmills and water wheels (Figure 1.3).

FIGURE 1.1 Potential energy of a skier and kinetic energy of a truck.

FIGURE 1.2 A pond of water with potential energy.

FIGURE 1.3 Water wheel and windmill.

1.1.1.1 Potential Energy

In more general terms, we can define energy as the work done in moving an object to a certain distance x under a certain force F. The product of force and distance would give the amount of work done or the energy spent during this process. An important thing to remember about gravitational potential energy "mgh", where m is the mass, g is the gravitational constant, and h is the height, is that the amount of energy stored by moving a mass to a higher altitude is independent of how (through which path) the mass got to its final position (Figure 1.4).

Exercise 1.1:

A crane is made to lift a load of 4,000 kg to an altitude of 300 m. How much energy must the motor provide (neglect losses in the hoist mechanism)?

Solution:

$$\text{Energy} = 4000 \times 9.81 \times 300 = 11.77 \text{ MJ}$$

1.1.1.2 Kinetic Energy

The kinetic energy stored in a moving mass is given by

For linear motion $\rightarrow E = mv^2/2$ (Figure 1.5a)
For rotational motion $\rightarrow E = I\omega^2/2$ (Figure 1.5b)

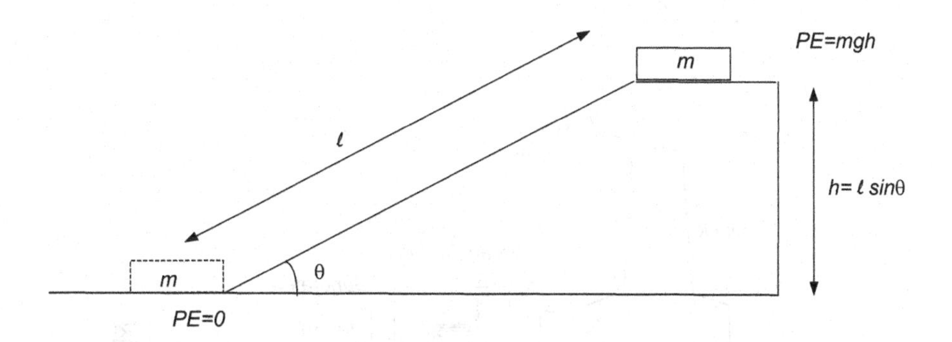

FIGURE 1.4 Potential energy = mgh.

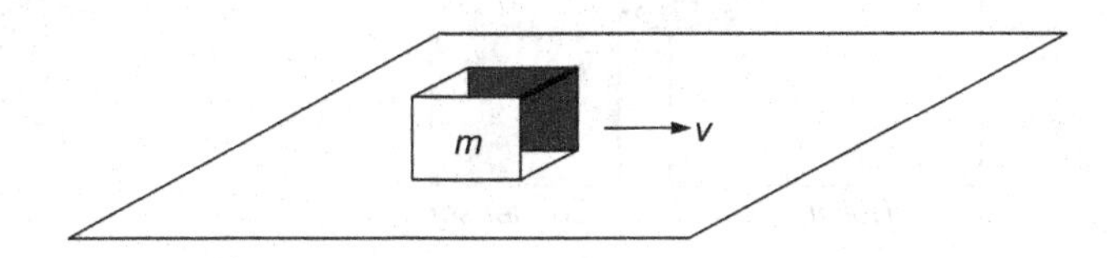

FIGURE 1.5a Linear motion.

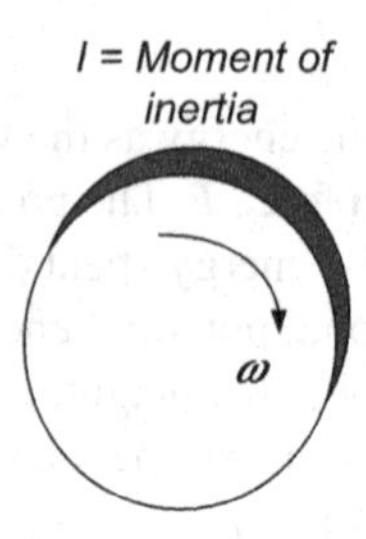

FIGURE 1.5b Rotational motion.

Here, m is the mass of the object, v is its velocity, I is its moment of inertia, and ω is its angular velocity in radians/second.

The potential energy of a pond of water at a hill is much more useful than the kinetic energy of a truck. Rotational kinetic energy is an important actor when load changes take place in the electrical power systems.

When customers demand more power, the rotors of the generators slow down, thus reducing their stored kinetic energy (Figure 1.6). This drop in kinetic energy is converted to electricity to meet the increased load. Eventually, the control system reacts, releasing more steam or water into the turbines driving the generators, which then return to their normal speed. The opposite happens when loads are removed from the system.

Sometimes, a flywheel on the shaft of the generator is intentionally included in the power systems to store energy and meet the sudden load changes without losing much speed (Figure 1.7).

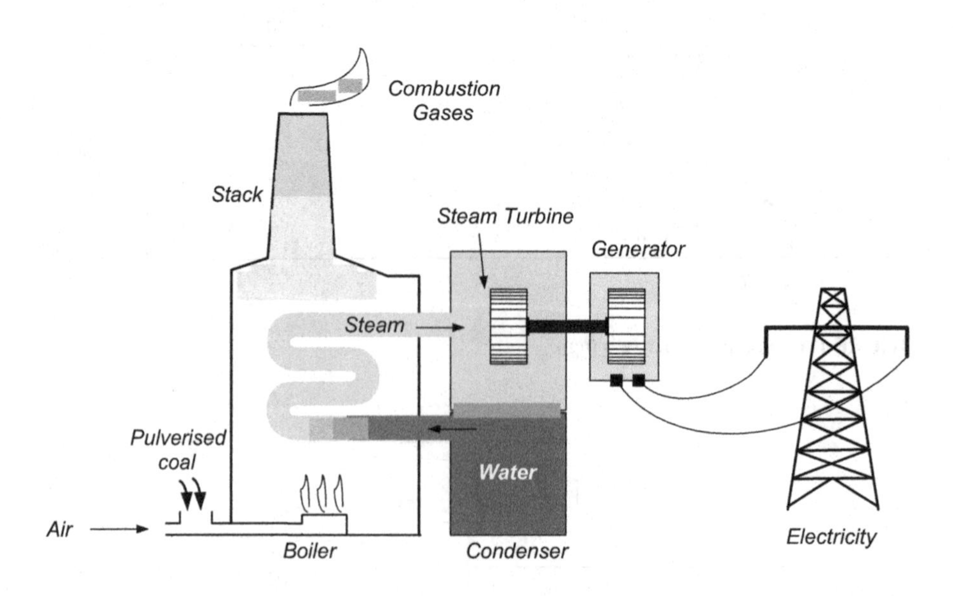

FIGURE 1.6 Power station.

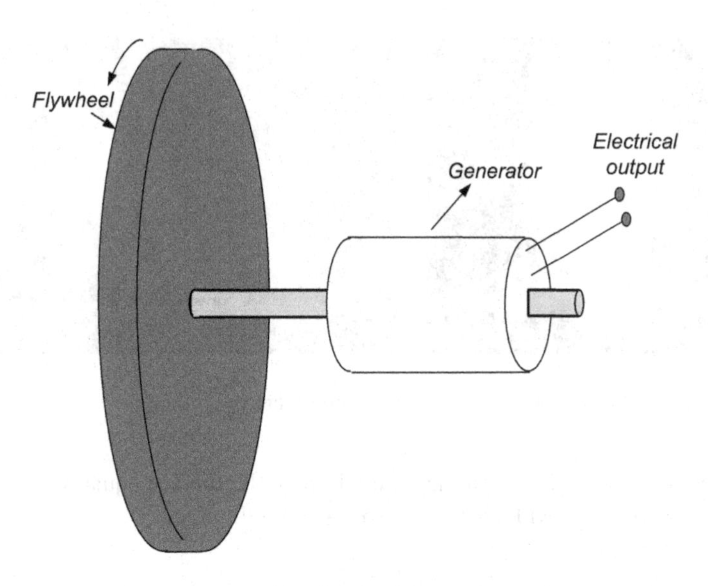

FIGURE 1.7 Electrical generator with a flywheel for nonfluctuating output.

Exercise 1.2:

A synchronous machine is used for power factor correction and has a rotor that is 1 m in diameter and 4 m long. Assuming the rotor is solid steel, with a density of 7.65 g/cm^3, calculate the kinetic energy of the rotor if the machine runs at 500 rpm (revolutions per minute).

Solution:

$$\text{Angular velocity} \rightarrow \omega = 2\pi f = 2\pi\left[\frac{500}{60}\right] = 52.36\,\text{rad/s}$$

$$\text{Volume} = V = \pi r^2.l = \pi(0.5\,\text{m})^2 \times (4\,\text{m}) = 3.14\,\text{m}^3 = 3.14\times10^6\,\text{cm}^3$$

Rotor mass is $m = \rho V = 3.14\times10^6\times7.65 = 24\times10^3\,\text{kg}$
Moment of inertia of the cylindrical rotor is given by

$$I = \frac{1}{2}mr^2 = (0.5)\times(24{,}000)\times(0.5)^2 = 3{,}000\,\text{kg} - m$$

$$\text{Energy stored, } W = \frac{I\omega^2}{2} = 0.5\times3{,}000\times52.36^2 = 4.1\,\text{MJ}$$

1.1.1.3 Nuclear Energy

Nuclear energy is the massive energy hidden in the nucleus (core) of an atom. Every object in the universe is made of atoms. There is enormous energy in the bonds that hold atoms together (Figure 1.8). If nuclear energy is released, it can be used

FIGURE 1.8 Nuclear explosion and nuclear power plant.

to generate electricity. From physics, we all know the famous equation of Einstein which relates mass (m) with energy (E) and speed of light (c):

$$E = mc^2$$

This equation describes that if mass can be destroyed, immense amount of energy can be released. When heavy atoms like uranium 235 split under neutron bombardment (process is named fission), the mass of the fragments is less than the mass of the original atom. This way energy is created by the destruction of some of the original mass. This is the principle of operation of all current nuclear power plants (Figure 1.8).

Exercise 1.3:

How much energy would be released by the destruction of 1.0 μg of matter?

Solution:

$$\text{Energy} = mc^2 = 10^{-9} \times (3 \times 10^8)^2 = 9 \times 10^7 \text{ J} \text{ or } 90\,\text{MJ}$$

1.1.2 Power

Power is the rate of flow of energy per second, and its unit is watts, which is given by

$$P(\text{Watts}) = \frac{W(\text{joules})}{t(\text{secs})}$$

1.1.3 Torque

Force is when you push or pull on an object, whereas torque involves the rotation of the object. Torque is the application of force on an object at a distance so that the object turns (Figure 1.9):

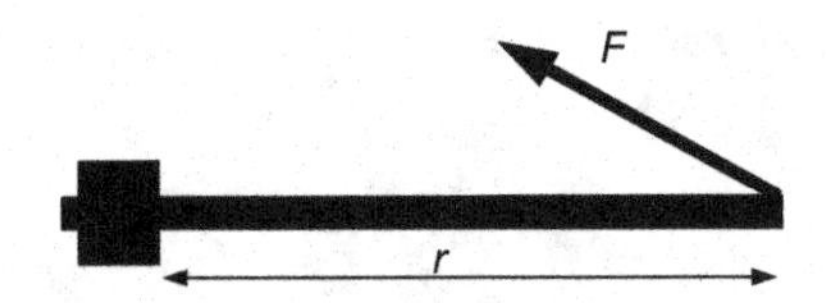

FIGURE 1.9 Torque.

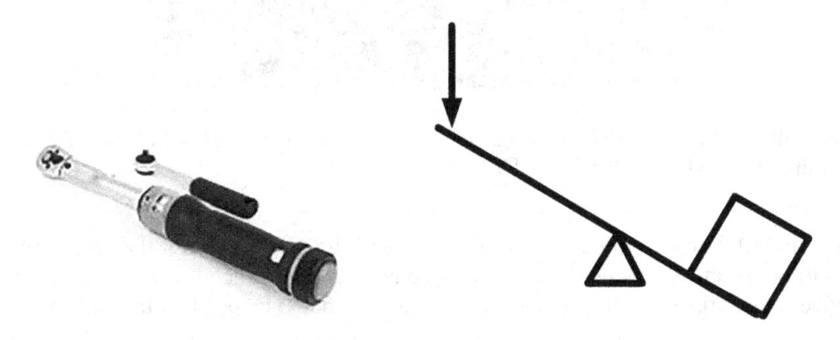

FIGURE 1.10 Wrench and lever work on the same principle.

$$T(\text{newton metres}) = r(\text{metres}) \times F(\text{newtons})$$

Torque makes use of the lever principle. It is easier to turn a longer wrench than a shorter one. That is because less force is needed when it is applied at a larger distance (Figure 1.10).

To avoid confusion, the units of torque are usually given as newton meters (Nm) instead of joules. Power is related to torque by the angular speed of rotation:

$$P = \omega_r T$$

where P is the power in watts, ω_r is the rotational velocity in rad/s, and T is the torque in newton meters (Nm). Note that

$$\omega_r = 2\pi n_r$$

where n_r is in rev/s or

$$\omega_r = \frac{2\pi n_r}{60}$$

where n_r is in rev/min.

Exercise 1.4:

The total energy required to carry a certain load between two railway stations is 60 MJ. What would be the output power in horsepower of the locomotive

FIGURE 1.11 6,400 kW electric locomotive. (With the courtesy of "Skoda Transportation Co.")

(Figure 1.11) engine to carry this load between the two stations if the total time is to be 5 min? Assuming the locomotive engine is 60% efficient and 1 kg of coal gives an energy of 1 MJ, calculate the amount of coal to be used for this transport.

Solution:

$$\text{Required engine output power, } W = \frac{60\,\text{MJ}}{5 \times 60 \times 746} = 268\,\text{HP}$$

$$\text{Amount of coal required} = \frac{60\,\text{MJ}}{0.6 \times 1} = 100\,\text{kg}$$

1.2 ENERGY CONVERSION AND QUALITY OF ENERGY

Much of the electrical and mechanical energy and the heat used in our homes, offices, and factories are generated from fossil fuels such as petroleum, coal, and natural gas which are formed from the remains of dead plants and trees.

That is why they are called fossil fuels. From pollution point of view, coal is the worst of the three. Fossil fuel power stations provide the majority of the electrical energy produced in the world (Figure 1.12).

Since 2000, global electricity demand has grown by 3% a year. Developing economies account for most of this increase (IEA, World Energy Outlook 2018).

China appears to be the leading country in using coal for obtaining energy (Figure 1.13). The United States, Russia, and China have the world's largest coal deposits. Roughly half of the world's oil and natural gas reserves are in the Middle East.

Petroleum is typically a bit cleaner burning, and unlike coal, it is easier to transport as it can be pumped through pipelines. In 2017, world total primary energy supply (TPES) was 13,972 Mtoe (IEA). Oil provides most of the world primary energy (%31.8) (Figure 1.14).

The measure of energy is usually expressed in units of Mtoe (millions of tonnes of oil equivalent). The tonne of oil equivalent (toe) is a unit of energy defined as the amount of energy released by burning one tonne of crude oil.

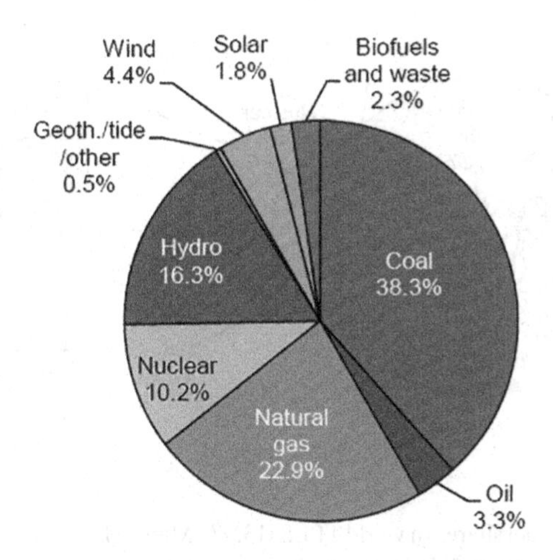

FIGURE 1.12 World gross electricity generation (25,721 TWh) by source, 2017. (IEA; *Electricity Information Overview*, 2019 edition; Page-4, Figure: 4-a.)

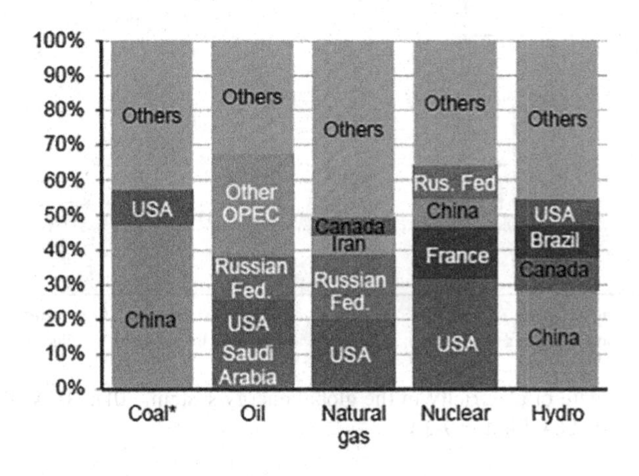

FIGURE 1.13 The largest energy producers by fuel in 2017. (IEA; *World Energy Balances* 2019 edition; Page-5, Figure: 5.)

Electricity accounts for 19% of the total final energy consumption today compared to just over 15% in 2000 (Figure 1.15).

Between 1974 and 2017, world gross electricity production increased from 6,298 to 25,721 TWh, an average annual growth rate of 3.3% (Figure 1.16). Since 1974, global electricity production has grown continuously year on year, except for between 2008 and 2009, when the global financial crisis caused an appreciable decline in production (IEA 2018).

Electricity is mostly consumed in industry and in residences (Figure 1.17). It is a vital source for the development of human civilisation. Therefore, it is natural to expect increased demand in the future. Presently, it is one of the main sources of air pollution in the world.

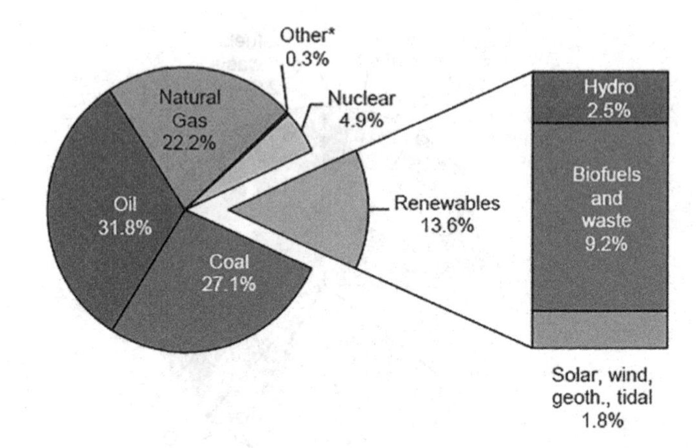

FIGURE 1.14 2017 Fuel shares in world TPES (13,972 Mtoe). (IEA; *Renewables Information Overview* 2019 edition; Page-3, Figure: 1.)

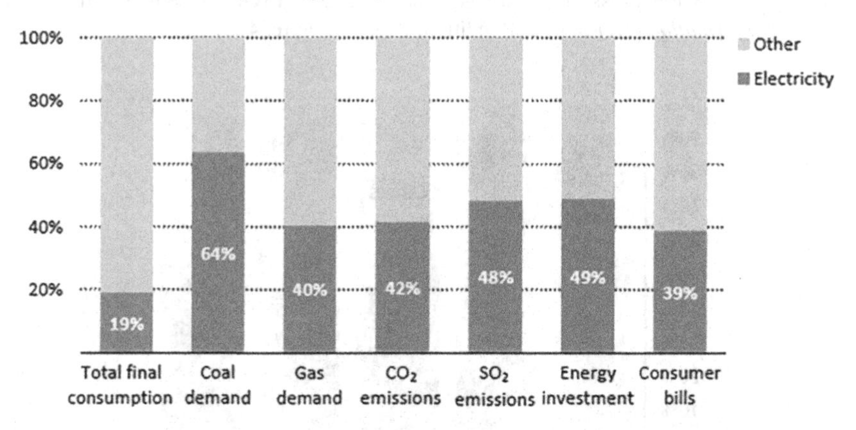

FIGURE 1.15 Share of electricity in the global energy system, 2017. (IEA; *World Energy Outlook* 2018; Page-284, Figure: 7-2.)

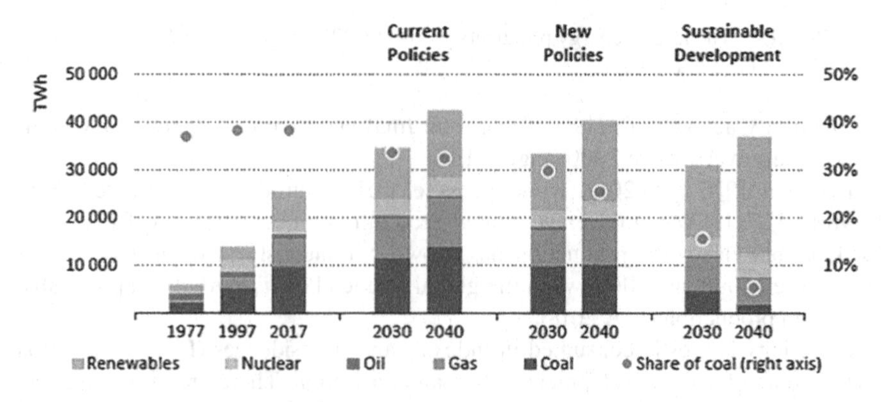

FIGURE 1.16 Global electricity generation by source and scenario. (IEA; *World Energy Outlook* 2018; Page-228, Figure: 5-7.)

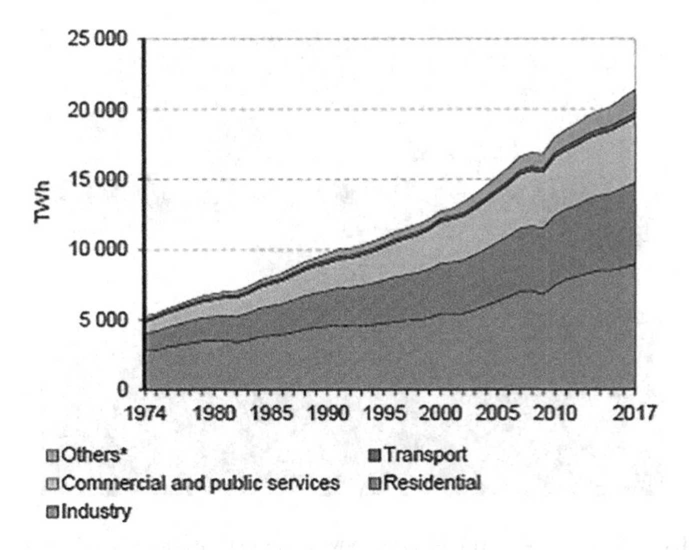

FIGURE 1.17 World electricity final consumption by sector. (IEA; *Electricity Information Overview* 2019; Page-6, Figure: 7.)

In 2017, China was the leader in electricity consumption in the world (IEA 2019) (Figure 1.18), followed by the United States and India. The top ten countries accounted for 68% of the total consumption (21,372 TWh) according to the report of IEA 2019.

Part of our electrical energy comes from hydroelectric stations where gravitational potential energy is converted to electrical energy (Figure 1.19). Hydroelectric stations are efficient about 85%. They are mostly of the reservoir type where the

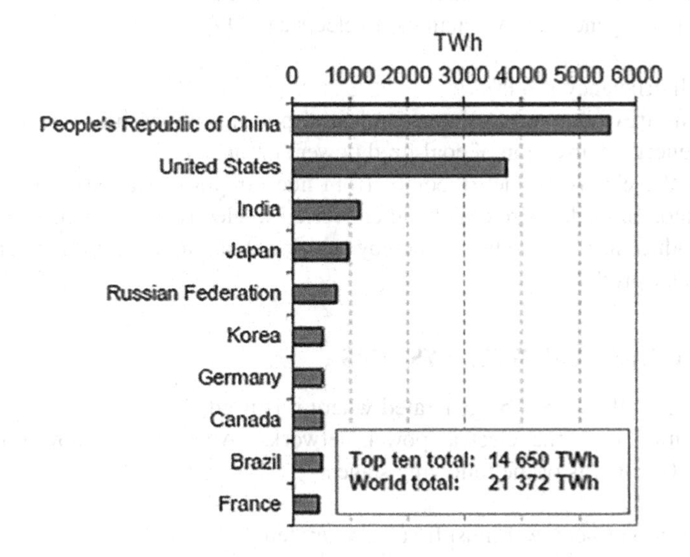

FIGURE 1.18 Top ten electricity-consuming countries, 2017. (IEA; *Electricity Information Overview* 2019; Page-8, Figure: 12.)

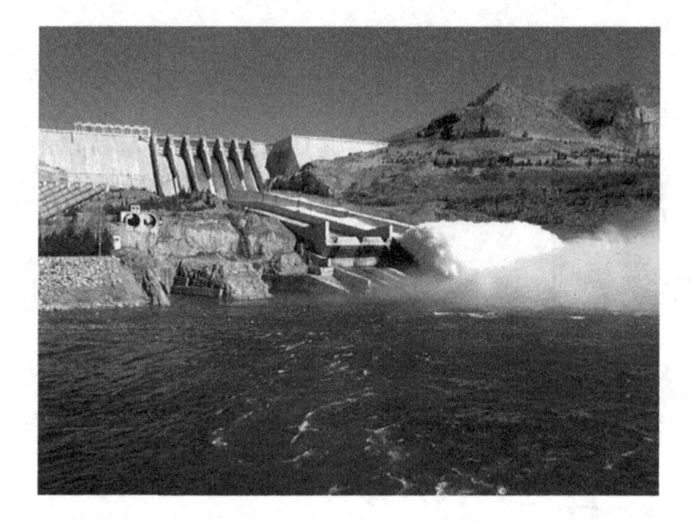

FIGURE 1.19 2,400 MW hydroelectric power station, 169 m height. (Permission of Ajans23.)

water is collected in an artificial lake and released to obtain electrical energy. They can also be run-of-river type and depend on the flow of the river rather than the deposition of water.

The typical efficiency of energy conversion in a coal-fired power station is

- Burning coal to raise steam: Chemical to thermal 88%
- Turning steam turbine: Thermal to mechanical 50%
- Electrical generator: Mechanical to electrical 99%.

The overall efficiency is 44%.

These figures obtained by latest coal-fired power stations illustrate the inefficiency of energy conversion of coal-fired power stations.

Part of the electrical energy comes from nuclear power stations where energy stored in atomic nuclei is released and converted to electricity. A nuclear reactor is used to produce heat to boil water, thereby creating steam to drive turbine generators for electricity production.

1.3 ELECTRICAL POWER SYSTEMS

Electricity usually cannot be generated where it is used. Thus, electric power must be transmitted over the electric power networks. An electrical power network (Figure 1.20) consists of the following stages:

- generation (the power must first be generated)
- transmission (it must then be transmitted to where it will be used)
- distribution (and finally, it must be distributed to the users).

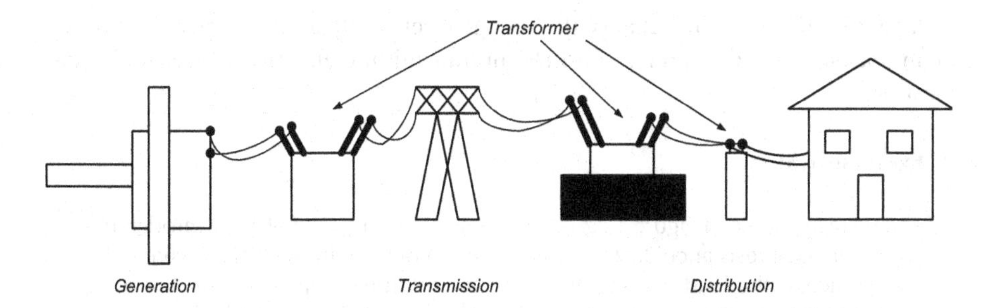

FIGURE 1.20 Electricity transmission.

Consider the simple direct current (DC) system shown in Figure 1.21 which conveys electrical power to a load R_L at a distant location. If R_{line} is the resistance of the transmission line,

- the received voltage (V) at the load resistor will be lower than the generator voltage (E)
- some power will be lost in the lines.

If the power line is very long, the voltage drop and the power dissipated in the line can be large.

If through a DC transformer action we increase the voltage E to αE and decrease the current i by the same ratio, we would be keeping the generated power constant. Through this operation, we would be

- decreasing the voltage drop in the transmission line $\left(v_{\text{line}} = i_{\text{line}} . R_{\text{line}}\right)$ by the same factor α, whereas power dissipated in the same lines $\left(i_{\text{line}}^2 R_{\text{line}}\right)$ would decrease by the square of this ratio α.

As can be seen, it is advantageous to convey electrical power at high voltage levels. This way we can decrease the voltage drop and power loss in the lines. If the voltage E is alternating current (AC), we can raise or lower it by using transformers. If the

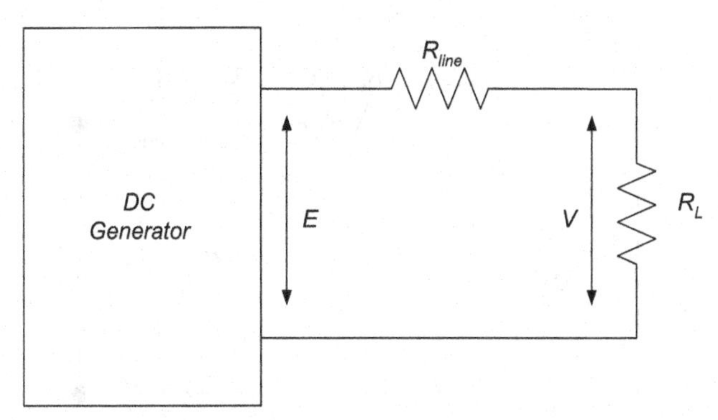

FIGURE 1.21 Electricity transmission analogy.

voltage E is DC, we cannot change the voltage level E with a transformer. That is the main reason why AC is preferred to DC in transmitting electrical power to distant locations.

Exercise 1.5:

The voltage source in Figure 1.22 is supplying 101 V to a load of 1 kΩ through the copper lines of resistance 0.1 Ω/km. If the length of the wire is 100 km, what is the voltage across the load and the power dissipated in the copper wires? Assuming a DC transformer action is possible and source voltage is increased by 100 times whereas current is reduced by the same ratio, what would be the voltage and power percentage regulations on the load?

Solution:

a. Current taken from the source is

$$I = \frac{101}{1010} = 0.1 \text{ A}$$

$$\text{Voltage drop on the wires, } V_{\text{line}} = 0.1 \times 10 = 1 \text{ V}$$

$$\text{Voltage on the load, } V_L = 101 - 1 = 100 \text{ V}$$

$$\text{Voltage regulation} = \frac{100 \text{ V}}{101 \text{ V}} = 0.990$$

$$\text{Power dissipated in the copper lines, } P_{\text{line}} = 1 \text{ V} \times 0.1 \text{ A} = 0.1 \text{ W}$$

$$\text{Power taken by the load, } P_{\text{load}} = 100 \text{ V} \times 0.1A = 10 \text{ W}$$

$$\text{Power regulation} = \frac{10 \text{ W}}{101 \text{ V} \times 0.1 \text{ A}} = \%99$$

b. When the source voltage E increases by 100 and current decreases by 100 (which is known as transformer action),

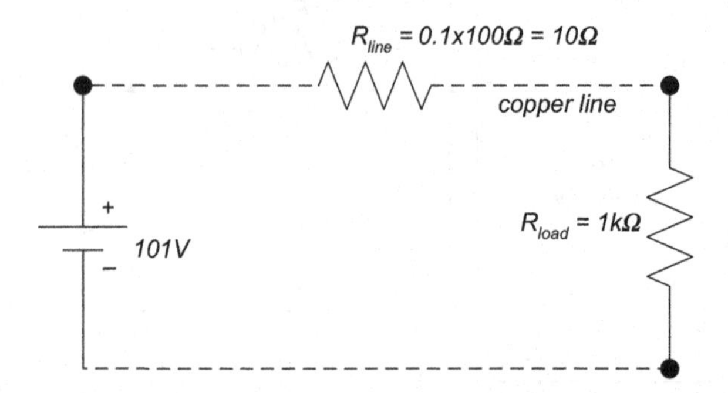

FIGURE 1.22 Effect of copper lines on transmission of electrical power.

$$\text{New current taken from the source, } I = 0.001 \text{ A}$$

$$\text{Voltage drop on the wires, } V_{\text{line}} = 0.001 \times 10 = 0.01 \text{ V}$$

$$\text{Voltage on the load, } V_{\text{load}} = 10{,}100 - 0.01 = 10{,}099.99 \text{ V}$$

$$\text{Voltage regulation} = \frac{10{,}099.99 \text{ V}}{10{,}100 \text{ V}} = 0.9999$$

$$\text{Power dissipated in the copper lines, } P_{\text{line}} = 0.01 \text{ V} \times 0.001 \text{ A} = 0.00001 \text{ W}$$

$$\text{Power taken by the load, } P_L = 10.1 \text{ W} - 0.00001 \text{ W} = 10.099 \text{ W}$$

$$\text{Power regulation} = \frac{10.099 \text{ W}}{10.1 \text{ W}} = \%99{,}99$$

As can be seen from the above example, through this transformer action more power is transferred to the load. This way the transfer of electrical energy through long copper wires with minimum voltage drops and power losses can be achieved.

Figure 1.23 shows the stage-by-stage equivalent circuit of the transmission system in Figure 1.20.

The resulting equivalent circuit viewed from the source is shown in Figure 1.24. Load impedance appears with its true value in the final equivalent circuit, but the impedance of the transmission line is reduced by the factor n^2. This will reduce line losses by the same factor. Voltage regulation on the load will also improve.

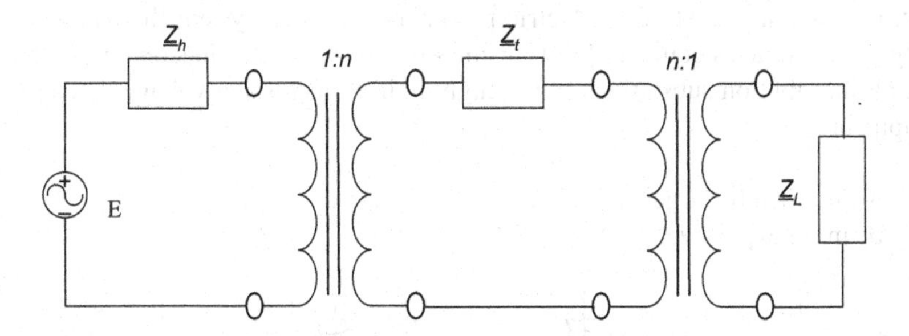

FIGURE 1.23 Stage-by-stage equivalent circuit of Figure 1.19.

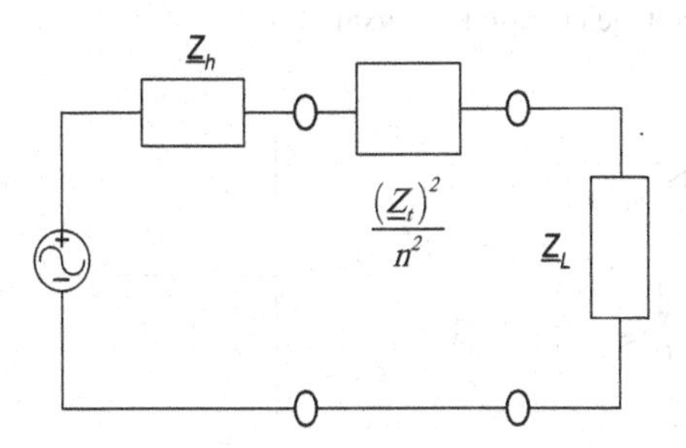

FIGURE 1.24 The final equivalent circuit of Figure 1.20.

For less energy loss and better voltage regulation, voltage

- must be stepped up from the generator to the transmission lines (transmission voltages are above 110 kV)
- then stepped down to the distribution system (distribution voltages are generally below 33 kV)
- with a final step down to 220 V very close to the point of actual use.

Figure 1.25 shows the relatively low voltage local distribution of AC electrical power in residential areas.

Due to construction techniques and material limitations, generators typically can only operate at 15–25 kV. In 1896, the U.S. company Westinghouse demonstrated the feasibility of long-distance transmission of AC power by sending power from hydroelectric generators at Niagara Falls to Buffalo, NY, over 20 miles away.

Alternating current (Figure 1.26) quickly won out over direct current for several reasons:

- ability to transmit over longer distances with lower losses by using transformers.
- AC generators are cheaper to build and maintain than DC generators.
- AC induction motor provides a very cheap, reliable, and rugged motor for the industry.

Subtransmission is part of an electric power transmission system that runs at relatively lower voltages compared to transmission voltages. It is uneconomical to connect all distribution substations to the high main transmission voltage, because the equipment

- would then be larger
- be more expensive.

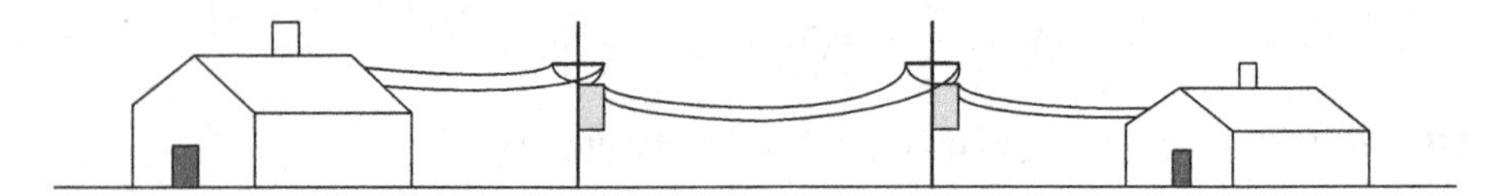

FIGURE 1.25 Low voltage electricity transmission.

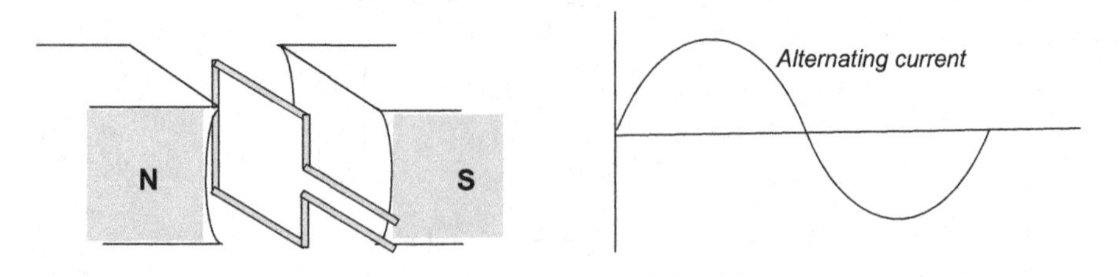

FIGURE 1.26 Electricity generation (AC power).

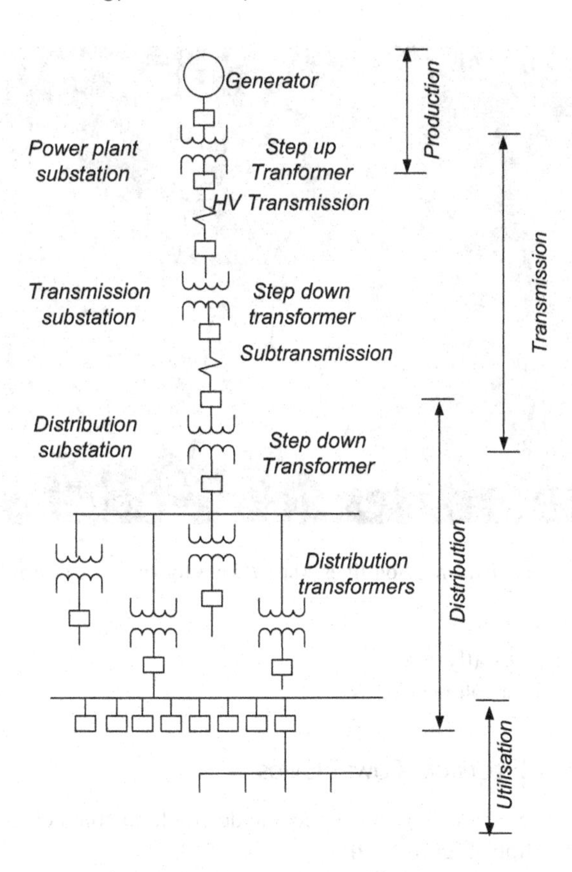

FIGURE 1.27 Electricity transmission and subtransmission.

By using subtransmission that runs at lower voltage levels than the main transmission, only transmission substations connect with this high voltage. It is stepped down and sent to smaller distribution substations in towns and neighbourhoods (Figure 1.27).

1.3.1 Voltage Levels in Electrical Power Systems

- Transmission voltages between 245 and 765kV are called extra high voltages (EHVs) (Figure 1.28).
- Transmission voltages between 115 and 230kV are called high voltages (HVs).
- Voltages between 2.4 and 69kV are called medium voltages (MVs).

There is no fixed cutoff between subtransmission and transmission voltages, or subtransmission and distribution voltages. The voltage ranges may overlap somewhat.

- transmission = usually HV or EHV
- subtransmission = usually HV or MV

FIGURE 1.28 765 kV Transmission lines. (http://www.tonyboon.co.uk/imgs/pages/power-lines.htm.)

- distribution = usually MV
- consumer = low voltage (220 V)

1.3.2 National Electrical Power Grids

As the electric power systems grew, it became desirable to connect to other utilities for improved operation (Figure 1.29).

By connecting all utilities together, it is possible to have all of the generator frequencies synchronised so that peaks occur at the same time. This way, despite the complexity, the network operates in synchronism as a single system.

With some minor exceptions, electricity cannot be stored, so the power generated must equal the power consumed. If the demand for power exceeds the supply, generation plants and transmission equipment can shut down. To reduce the risk of such failures, electric transmission networks are interconnected into

- regional
- national
- continental

wide networks, thereby providing multiple alternative routes for power to flow should failures occur. People tend to use more power at nights and Monday to Friday work-week. In order to meet the load requirements, generation must be greater than the maximum expected demand. Depending on the factors like weather, economy, and others, there will be some minimum load that the utilities will always be providing. This is referred to as the base load, as shown in Figure 1.30.

To meet the base load, the utilities would normally use their most efficient generators that are available since they will be operating continuously. Demand will rise

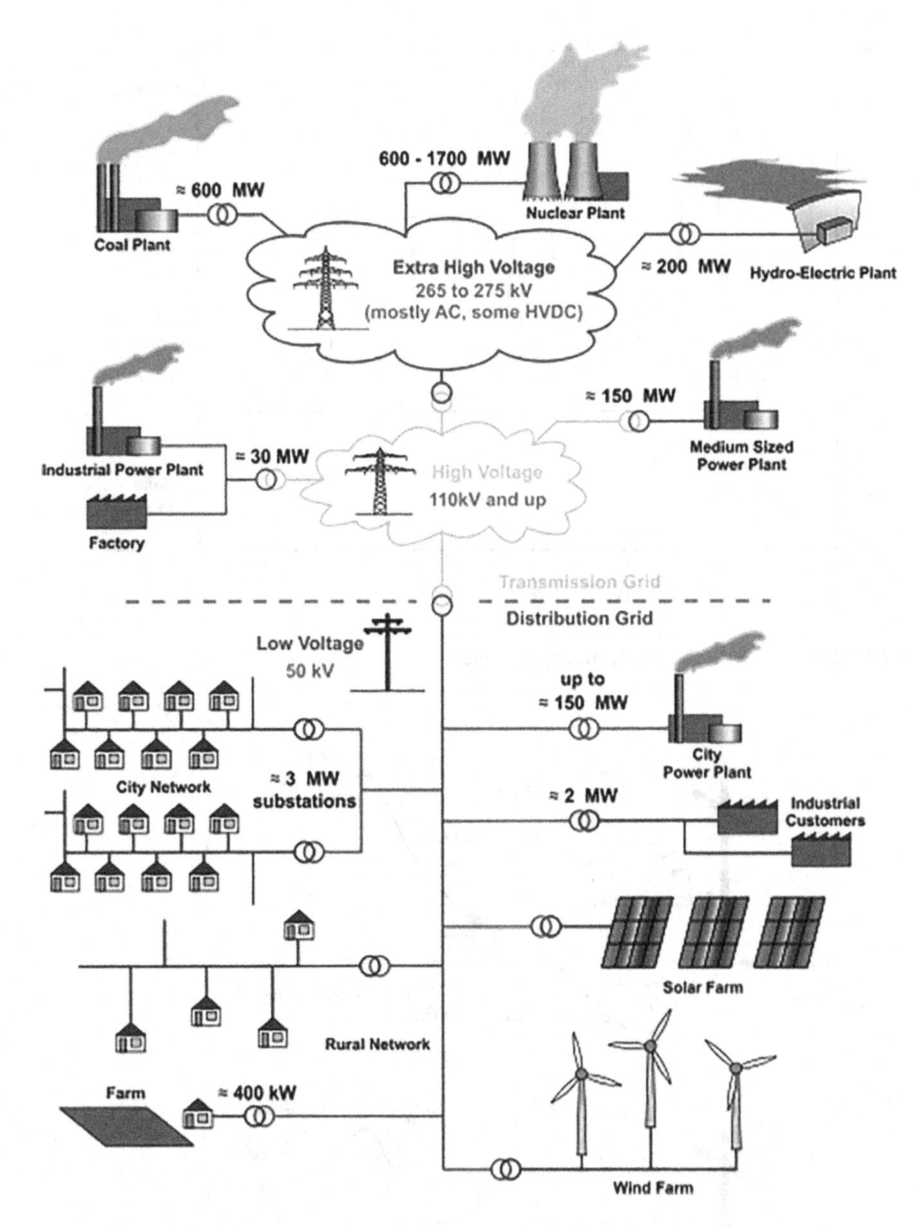

FIGURE 1.29 Electricity grid schematics. ("https://www.wikizeroo.org/")

during the days and evenings, so the utilities must be ready to respond to these surges with enough generators spinning to meet the raised load.

It can take several hours to bring a large unit on line from a dead stop. That is the reason why they are kept spinning. The difference between these units in operation and the base load is named as the spinning reserve. As a solution for this spinning power, we can use pumped storage plants. These are hydroelectric plants having large generators which can also act as pumps (Figure 1.31).

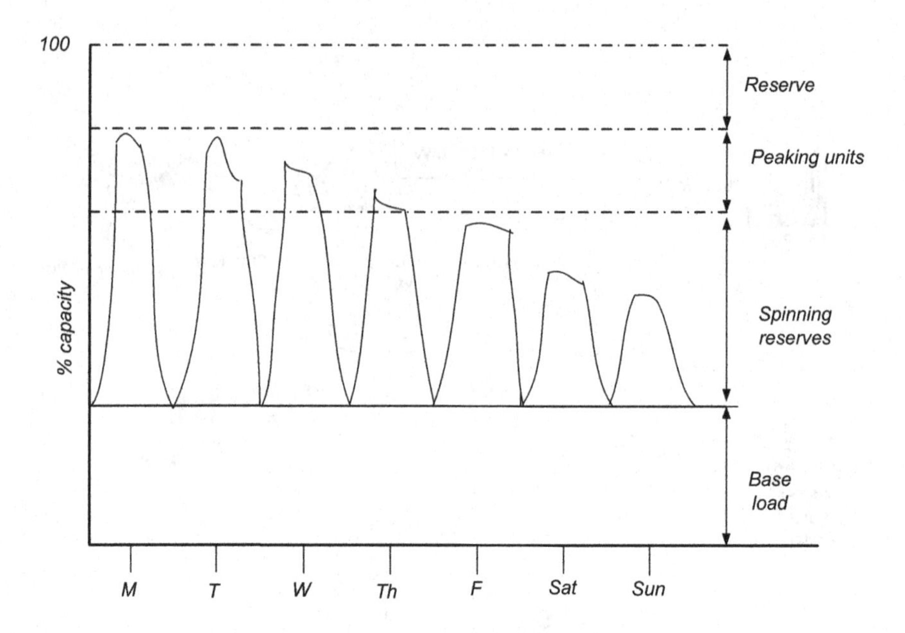

FIGURE 1.30 Variation of electricity demand.

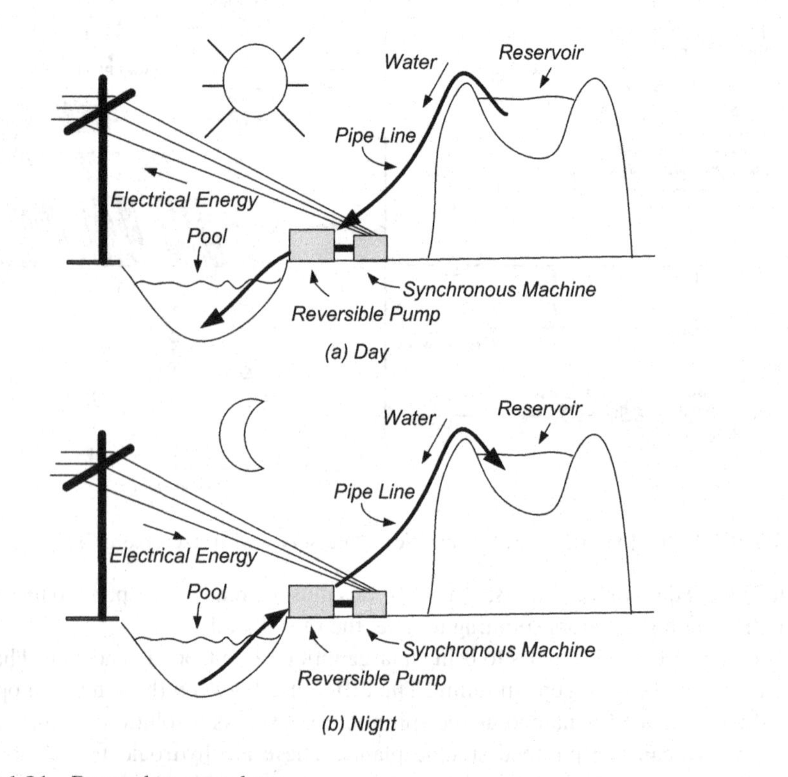

FIGURE 1.31 Pumped storage plants.

When generating, they are driven by water and can be "switched on" in short time (about 30 s) to meet a sudden increase in the demand. Otherwise, they are used to pump water to refill a storage reservoir for future purpose. At present, pumped storage schemes are currently the only means of bulk electricity "storage".

Lastly, sometimes, high demands occur and everything has to be put on line. These occasional high demands are met by units called peaking units. Peaking units are capable of rapid starts. With the help of interconnections between utilities, the sale of energy between companies when they are experiencing a heavy demand or have a unit out of service is made possible.

By interconnecting the utilities, the most efficient available plants could be used to supply the varying loads during the day. This way it is possible to produce electricity at the lowest possible cost.

1.3.3 Electrical Energy Conversion from Coal

In this way of obtaining electrical power, coal stored in a large bunker over the boiler is thrown into the boiler bed (Figure 1.32). The ash and unburned coal fall to a moving bed. An intake fan blows air into the boiler. A fan at the top of the boiler pulls the exhaust out. This gas is filtered (bag house) and then exhausted up the stack.

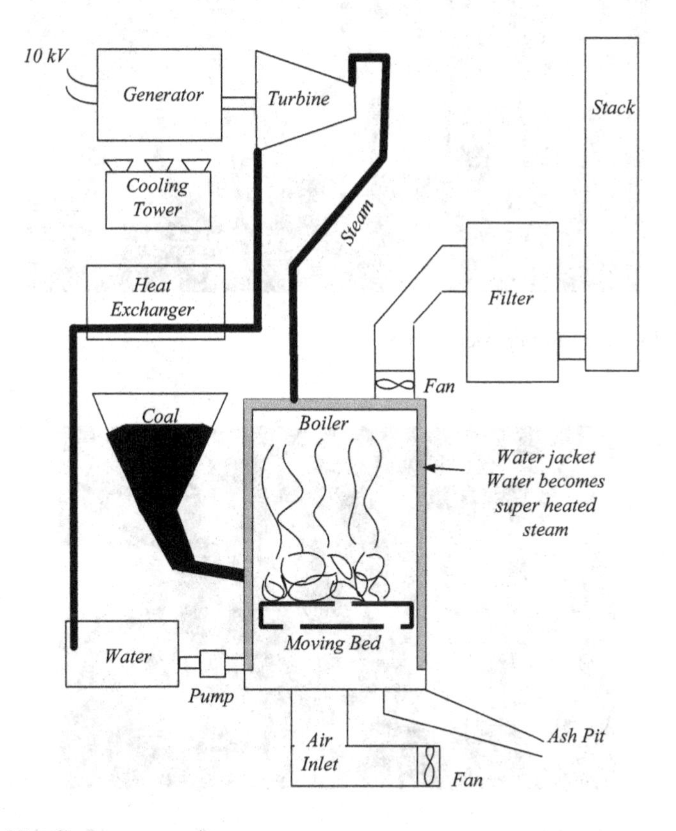

FIGURE 1.32 Coal power station.

- The walls of the boiler are lined with steel tubes into which water is pumped to become superheated steam. If the water used is not absolutely pure, the high temperatures in the boiler can cause tubes to plug up.
- The steam out of the boiler is routed to the turbine where it gives up its energy to drive the generator (Figure 1.34).
- The steam that comes out of the turbine is fully condensed for reuse in the boiler

Waste products, i.e., ash, oxides of sulphur SO_x and nitrogen NO_x, and carbon dioxide, cause major pollution problems (Figure 1.33).

FIGURE 1.33 Coal-fired power station.

FIGURE 1.34 25 MW 850 psi generator and steam turbine. (Courtesy of Teco-Westinghouse.)

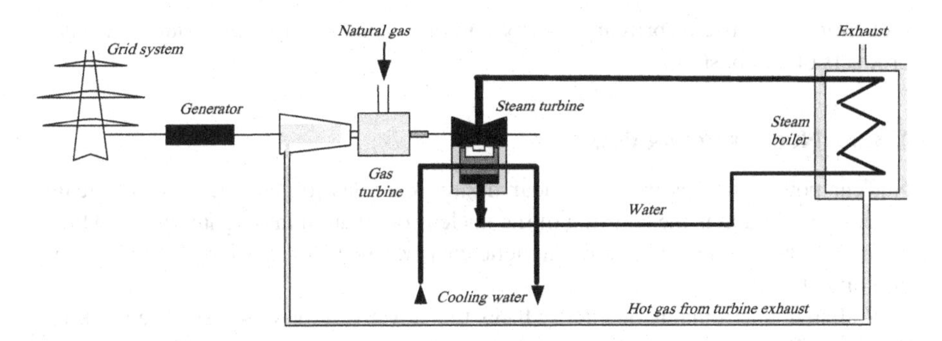

FIGURE 1.35 Combined cycle gas turbine plant (CCGT).

1.3.4 Combined Cycle Gas Turbine (CCGT) Plant

The basic operation of the gas turbine is similar to that of the steam power plant except that air is used instead of water. When a gas turbine is used to drive a generator, the gas and compressed air are fed into the combustion chamber (Figure 1.35). Combustion products drive the turbine. Thermal efficiency is about 30%.

Exhaust gases are at over 500°C. Therefore, they are used to raise steam to drive conventional steam turbine/generator plant. In such CCGT plants, two turbines drive one generator with improved thermal efficiencies (50% or better).

1.3.5 Combined Heat and Power Plant (CHP)

CHP systems (Figure 1.36) make use of the heat produced during the electricity generation process. In this way, the overall efficiencies above 70% can be achieved.

CHP systems supply heat and electrical power directly at the point of use. Therefore, transmission losses are minimum. The high efficiency of CHP means

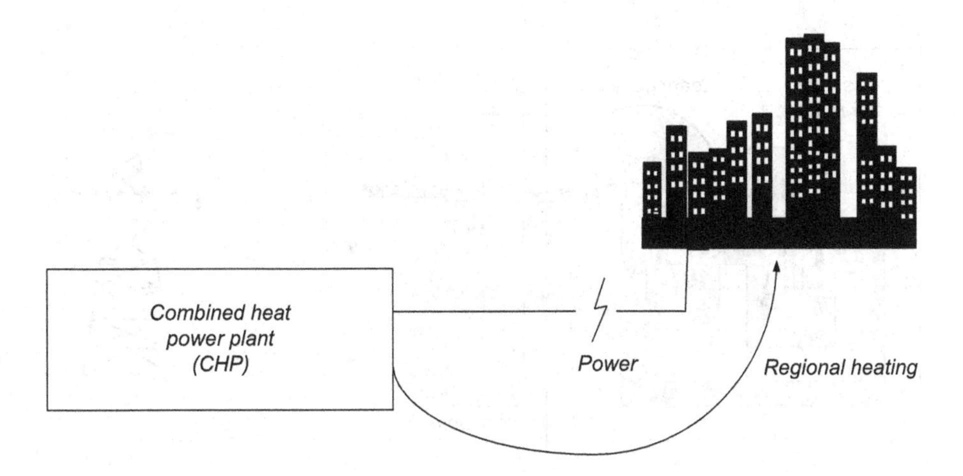

FIGURE 1.36 Combined heat and power plant (CHP).

reduction in the use of primary energy and emissions of carbon dioxide and other products of combustion.

1.3.6 Nuclear Power Plants

Nuclear power stations utilise nuclear fission reactions to generate heat for steam generation. Fission is the splitting of the nucleus of an atom into lighter nuclei which releases large amounts of electromagnetic energy. The process is largely free of carbon dioxide.

Radiated electromagnetic energy heats the power reactor vessel and the working fluid, usually water, which conducts heat away to a steam generator (Figure 1.37). Fossil fuel price increases, global warming from fossil fuel use, and the new safe ways of obtaining nuclear energy may renew the demand for nuclear power.

Primary water is heated to over 300°C and remains a liquid at such a high pressure. It is pumped through the reactor to the steam generator, in which the secondary water boils. The primary coolant is radioactive. Therefore, mixing the primary and secondary fluids is not allowed.

The United States produces the most nuclear energy, about 32% of the world's total nuclear electricity (as of 2017) (Figure 1.38). France produces the highest percentage of its electrical energy from nuclear reactors: 73% as of 2016 (Figure 1.38). In the European Union, Germany provides 13.2% of the electricity. Some countries, such as Austria and Ireland, have no nuclear power stations.

There are now over 449 nuclear power reactors (Figure 1.39) in operation in the world, providing about 10.2% of the world's electricity (as of 2017) (Figure 1.12). Also 52 nuclear plants are under construction (Figure 1.40).

The energy in nuclear fuel is millions of times greater than that contained in a similar mass of a fossil fuel such as oil. However, the fission products are low to medium radioactive and remain so for many years, causing a nuclear waste problem. A 1,000 MW nuclear reactor would produce 800 tonnes of nuclear output per year.

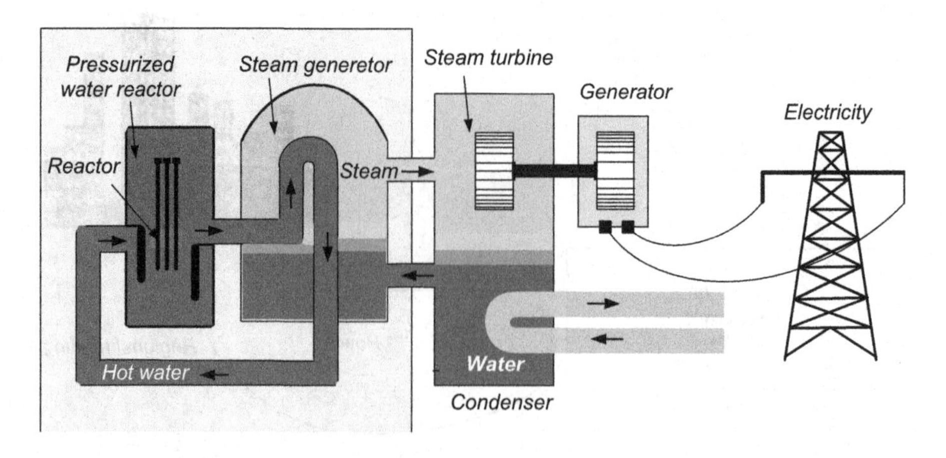

FIGURE 1.37 Nuclear power plant.

Producers	TWh	% of world total
United States	840	32.2
France	403	15.5
People's Rep. of China	213	8.2
Russian Federation	197	7.6
Korea	162	6.2
Canada	101	3.9
Germany	85	3.3
Ukraine	81	3.1
United Kingdom	72	2.8
Sweden	63	2.4
Rest of the world	389	14.8
World	2 606	100.0

Country (top ten producers)	% of nuclear in total domestic electricity generation
France	73.1
Ukraine	49.7
Sweden	40.5
Korea	29.0
United Kingdom	21.3
United States	19.5
Russian Federation	18.1
Canada	15.2
Germany	13.2
People's Rep. of China	3.5
Rest of the world[1]	7.3
World	10.4

FIGURE 1.38 World nuclear electricity production 2016. (IEA; *Key World Energy Statistics* 2018.)

These would have to be deposited until their activities slow down or buried (in land or sea) after mixing with cement. There remains intense political debate over nuclear power.

1.3.7 Electricity from Renewable Energy

The term "renewable energy" describes the energy that reoccurs naturally in the environment, such as

- solar radiation
- wind
- tides
- waves.

The origin of these sources is natural and caused by either the sun or the gravitational effects of the moon and the sun. Such sources are sustainable energy sources since they are not depleted by continued use. However, they are difficult to be converted into more useful forms of energy, particularly electrical energy.

67.9% of the world's renewable energy comes from biofuels and waste (Figure 1.41). The second largest is the hydro by 18.5%.

Country	Number of Reactors	Total Net Electrical Capacity MW
UNITED STATES OF AMERICA	96	97565
FRANCE	58	63130
CHINA	48	45518
JAPAN	37	35947
RUSSIA	36	28355
KOREA, REPUBLIC OF	25	23784
CANADA	19	13554
UKRAINE	15	13107
GERMANY	7	9515
UNITED KINGDOM	15	8923
SWEDEN	8	8613
SPAIN	7	7121
INDIA	22	6255
BELGIUM	7	5918
CZECH REPUBLIC	6	3932
SWITZERLAND	5	3333
FINLAND	4	2784
BULGARIA	2	1966
HUNGARY	4	1902
BRAZIL	2	1884
SOUTH AFRICA	2	1860
SLOVAKIA	4	1814
ARGENTINA	3	1633
MEXICO	2	1552
PAKISTAN	5	1318
ROMANIA	2	1300
IRAN, ISLAMIC REPUBLIC OF	1	915
SLOVENIA	1	688
NETHERLANDS	1	482
ARMENIA	1	375
Total	**449**	**398887**

FIGURE 1.39 IAEA, world number of reactors; power reactor information system (PRIS), September 2019.

In 2017, the world's Total Primary Energy Supply (TPES was 13,972 Mtoe, of which 13.5% or 1,894 Mtoe was produced from renewable energy sources (Figure 1.14). However, this number is continually increasing.

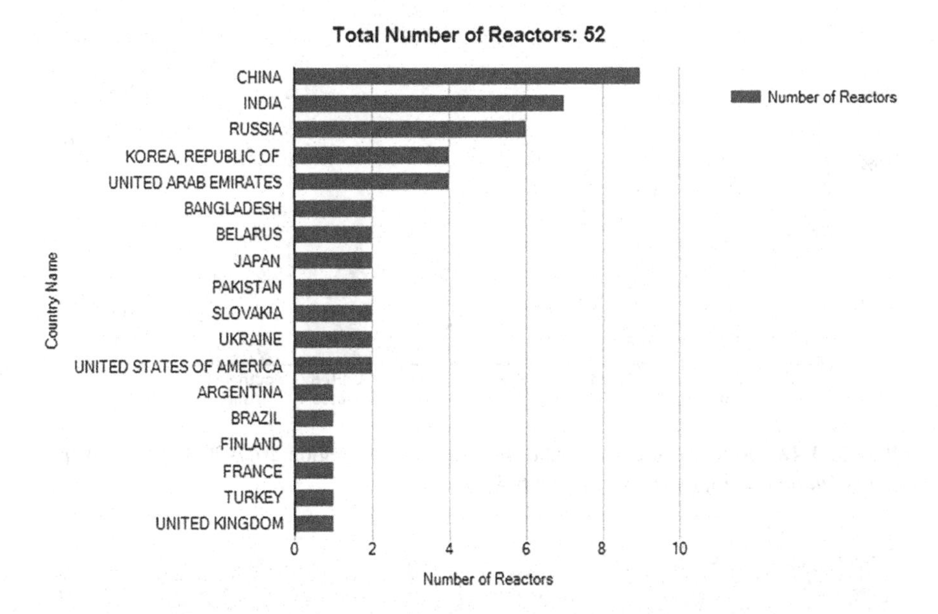

FIGURE 1.40 IAEA, world number of reactors under construction. (PRIS, September 2019.)

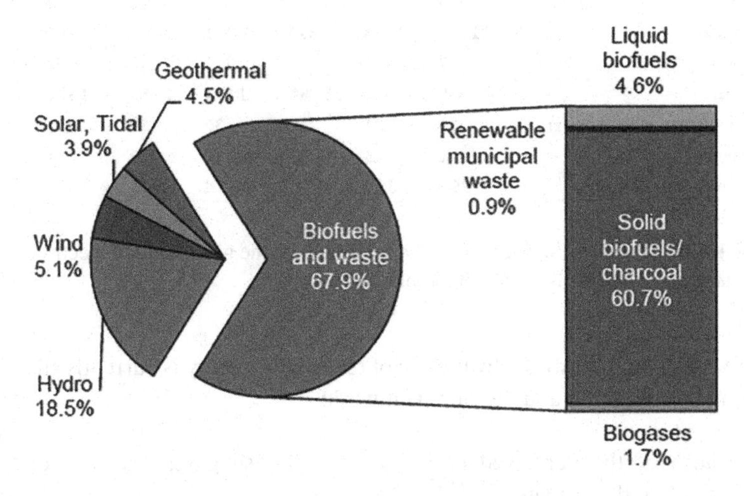

FIGURE 1.41 2017 Product shares in world renewable energy supply (1,894 Mtoe). (IEA; *Overview of Renewables*; Page-3, Figure: 2.)

According to IEA 2017 reports (Figure 1.42), 80% of Brasil's electricity is produced from renewable resources. In this respect Brasil has the higest percentage. Second is the European Union. Solar and wind are the most promising renewable energy resources for renewable electricity production of future (Figure 1.43).

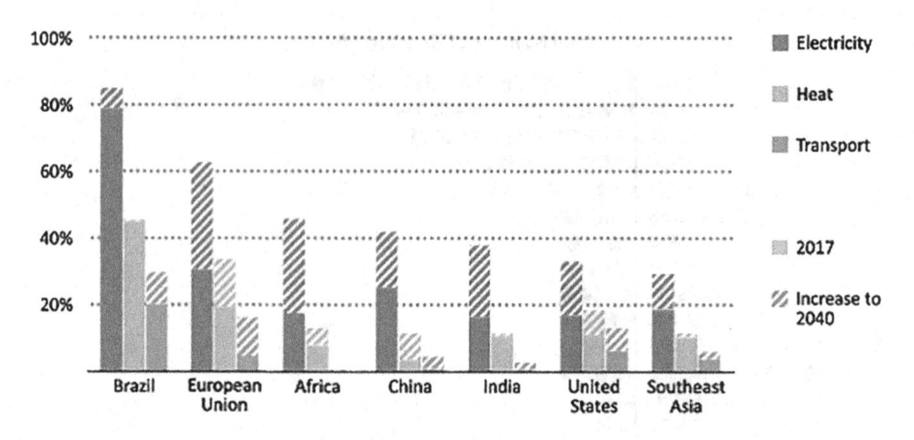

FIGURE 1.42 Renewable energy share by category and region 2017–2040. (IEA; *World Energy Outlook* 2018; Page-253, Figure: 6-4.)

		New Policies		Current Policies		Sustainable Development	
	2017	2025	2040	2025	2040	2025	2040
Electricity generation (TWh)	6 351	9 645	16 753	9 316	14 261	10 917	24 585
Bioenergy	623	890	1 427	873	1 228	1 039	1 968
Hydro	4 109	4 821	6 179	4 801	5 973	5 012	6 990
Wind	1 085	2 304	4 690	2 151	3 679	2 707	7 730
Geothermal	87	129	343	125	277	162	555
Solar PV	435	1 463	3 839	1 334	2 956	1 940	6 409
Concentrating solar power	11	34	222	30	119	54	855
Marine	1	3	52	2	29	4	78
Share of total generation	*25%*	*32%*	*41%*	*30%*	*33%*	*38%*	*66%*

FIGURE 1.43 World renewable energy consumption for the generation of electricity. (IEA; *World Energy Outlook* 2018; Page-250, Table: 6-3.)

There are several reasons why the use of renewable energy is currently the subject of such intense research and development work:

1. because of the perceived effects of greenhouse gases (such as carbon dioxide) on the climate
2. to reduce acidic pollution
3. diminishing reserves of fossil fuels.

1.3.7.1 Hydroelectricity

Hydroelectricity supplies about 16% of the world's electricity (Figure 1.44). Most hydroelectric power comes from the potential energy of water stored in a dam. The potential energy stored is proportional to the

Producers	TWh	% of world total
People's Rep. of China	1 193	28.6
Canada	387	9.3
Brazil	381	9.1
United States	292	7.0
Russian Federation	187	4.5
Norway	144	3.5
India	138	3.3
Japan	85	2.0
Venezuela	68	1.6
Turkey	67	1.6
Rest of the world	1 228	29.5
World	4 170	100.0

Country (top ten producers)	% of hydro in total domestic electricity generation
Norway	96.2
Brazil	65.8
Venezuela	60.1
Canada	58.0
Turkey	24.5
People's Rep. of China	19.2
Russian Federation	17.1
India	9.3
Japan	8.0
United States	6.8
Rest of the world[2]	14.9
World	16.6

FIGURE 1.44 World hydroelectricity production 2016. (IEA; *Key World Energy Statistics* 2018; Page-9.)

- volume of the water stored
- the head (the difference in height between the reservoir surface and the outflow to the turbine).

The water is released through the penstock, driving the water turbine and generator (Figure 1.45). Less common types of hydro schemes use the kinetic energy of water from running rivers. The hydrogenerator has an energy conversion efficiency of about 85%.

1.3.7.2 Electricity from Wind Power

In the late eighteenth century, wind was used in windmills to obtain electrical energy. Today, China, the United States, and Germany are the world leaders in using wind for electrical energy (Figure 1.46). Wind is an inexhaustible resource and is currently the cheapest form of renewable energy. It is expected that at current growth rates, up to 20% of the world's electricity will be produced by wind generators within 30 years.

In addition, there is a growing interest in grid-connected wind generation in homes, farms, and businesses, thereby facilitating exchange of power with the grid.

The most recent wind turbines receive their rotational action through the three blades (Figure 1.47). Small turbines are set to point into the wind. Large turbines generally use a wind sensor in connection with a computer-controlled servomotor to face into the wind.

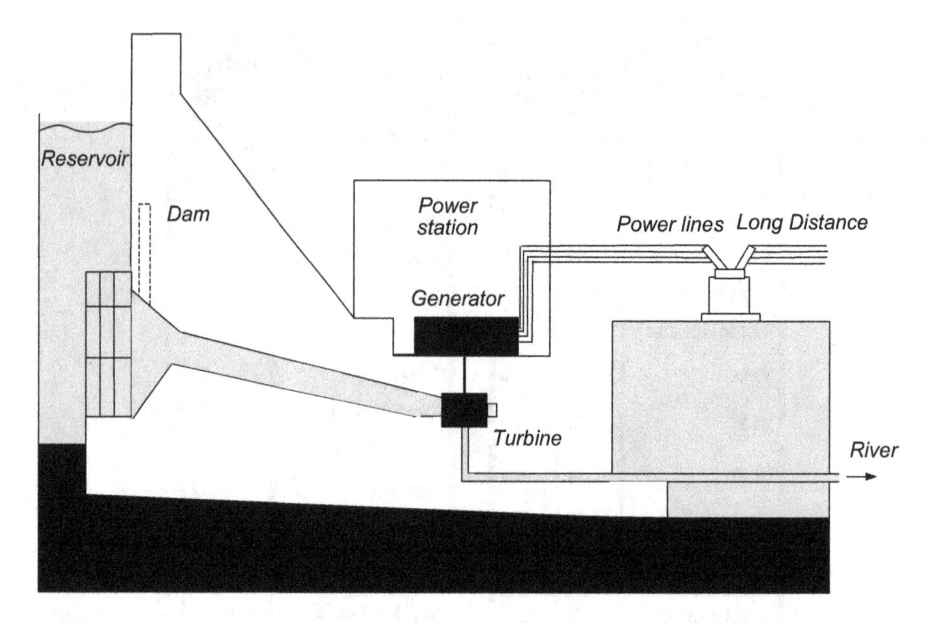

FIGURE 1.45 Hydroelectric power station.

Producers	TWh	% of world total
People's Rep. of China	237	24.8
United States	229	24.0
Germany	79	8.2
Spain	49	5.1
India	45	4.7
United Kingdom	37	3.9
Brazil	33	3.5
Canada	31	3.2
France	21	2.2
Italy	18	1.8
Rest of the world	178	18.6
World	**958**	**100.0**

Country (top ten producers)	% of wind in total domestic electricity generation
Spain	17.8
Germany	12.1
United Kingdom	11.0
Italy	6.1
Brazil	5.8
United States	5.3
Canada	4.6
France	3.9
People's Rep. of China	3.8
India	3.0
Rest of the world[1]	2.2
World	**3.8**

FIGURE 1.46 World wind electricity production 2016. (IEA; *Key World Energy Statistics* 2018; Page-10.)

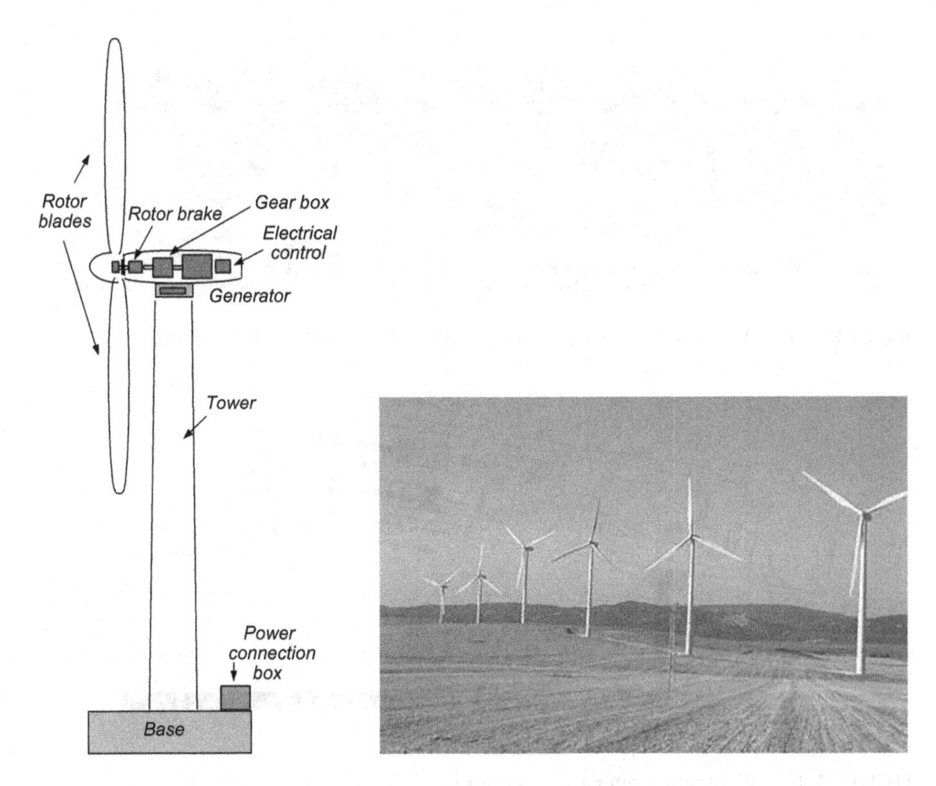

FIGURE 1.47 Wind electricity generators.

A speed-increasing gearbox is used after the rotor blades to drive the generator. Most machines use induction generators which generate electricity at a relatively low voltage (typically 690 V) and high current. A step-up transformer is used to deliver this voltage to the local distribution network.

1.3.7.3 Electricity from Wave Power

In this kind of electricity generation, energy is captured from the ocean surface waves. Wave power generation is still at an experimental stage. The wave energy devices being developed and tested today are highly diverse. Attenuators (Figures 1.48) are multisegment floating structures parallel to the waves and ride the waves like a ship.

Segments are connected to hydraulic pumps (Figures 1.49) to generate power as the waves move across. Power is fed via a common subsea cable to the shore.

A point absorber is a floating buoy inside a fixed cylinder (Figure 1.50). The relative up and down bobbing motion caused by passing waves is used to drive electromechanical or hydraulic energy converters to generate power.

In overtopping devices (Figure 1.51), the incoming waves fill the reservoir like a dam. Water collected is then released to turn hydro turbines to generate electrical power.

In another method, waves outside a chamber constructed in the sea cause water level to oscillate in the chamber up and down which in return acts like a huge piston on the air above the water, pushing it back and forth (Figure 1.52).

FIGURE 1.48 Wave power attenuators. (Courtesy of Pelamis Wave Power Ltd.)

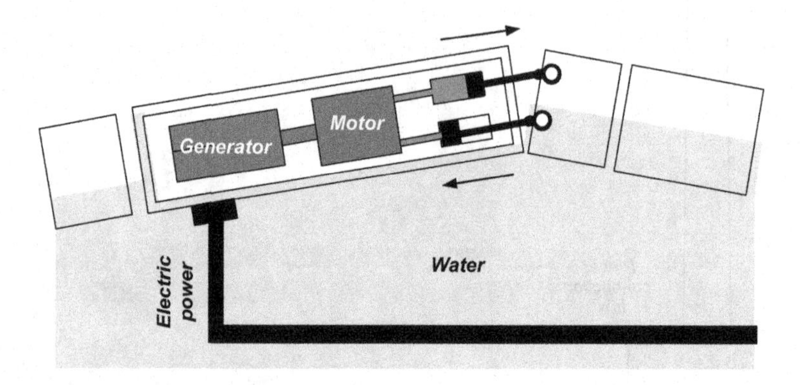

FIGURE 1.49 Wave power attenuator structure.

FIGURE 1.50 A wave power buoy. (Courtesy of Ocean Power Technologies, Inc.)

FIGURE 1.51 Overtopping wave energy. (Courtesy of Boem.)

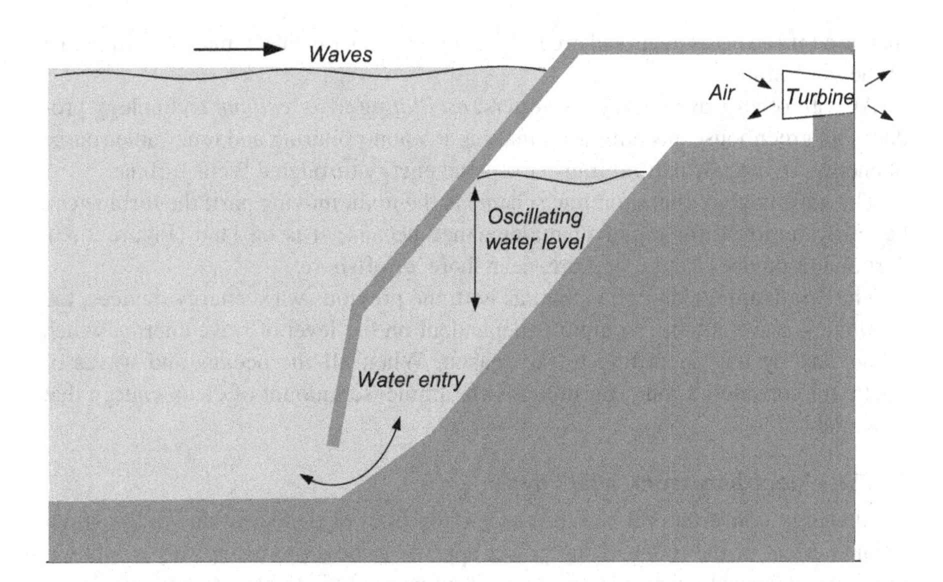

FIGURE 1.52 Oscillating water column.

Water enters through the subsurface opening into the chamber, trapping air connected to the turbine to generate power. As a result, the air inside the enclosure is periodically compressed and decompressed due to this cyclic motion of the water level inside the chamber.

As the direction of air flow changes periodically, a special type of turbine that can rotate always in the same direction irrespective of the direction of air flow (Wells turbine) must be used. The speed of the air flow through the Wells turbine can be

FIGURE 1.53 Wave power generation. (https://www.boem.gov/Ocean-Wave-Energy/.)

increased if the cross-sectional area of the turbines duct is made much less than that of the sea column.

As with other wave energy converters, *oscillating wave column* technology produces no greenhouse gas emissions making it a non-polluting and renewable source of energy, created by natural transfer of wind energy through a Wells turbine.

The advantage of this shoreline scheme is the main moving part; the turbine can be easily removed for repair or maintenance because it is on land (Figure 1.53). Terminator devices can be onshore, near shore, or offshore.

The disadvantage though is that, as with the previous wave energy devices, the oscillating wave column's output is dependent on the level of wave energy, which varies day by day according to the season. When all the oceans and waves of Earth are considered, one can imagine the immense amount of clean energy that is available.

1.3.7.4 Electricity from Tidal Power

Tidal energy converters rely on the twice-daily tides of the ocean due to the gravitational effect of the moon. Tidal power appears to be a promising way of obtaining clean electrical energy. However, environmental effects have to be considered. This subject is still being explored. The first power plant of this sort was constructed in 1966 on the outlet of the river Rance in the northwest of France. It produces 240 MWh electrical energy per year. Tidal power seems to be more predictable and reliable than wind and solar forms. Energy can be extracted from the tidal flow in two principal ways:

a. Tidal barrages

   A difference in water levels (potential energy) across a barrage built over an estuary can be achieved by the twice-daily tides of the ocean. When the water

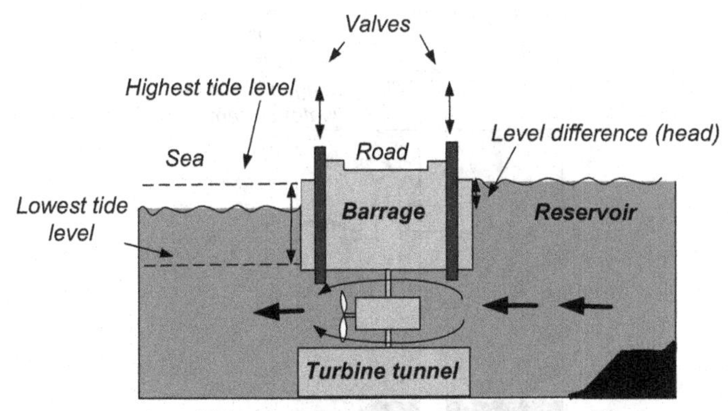

FIGURE 1.54 Water flowing from riverside to sea (low tide).

level is higher on one side of the barrage than the other, water is allowed to flow through the turbines to generate electricity. Turbines located in the inflows/outflows of the barrage are driven by the water flowing in and out of the barrage.

During high tide when the water is level and high on both sides of the barrage, the gates are closed and water is stored in the "river side" of the tidal barrage (Figure 1.54). As the tide falls, the water level on the "sea side" of the dam drops. When it is low enough, the valves are opened for water to flow from the higher side—through turbines—to the lower side. The movement of the water rotates the turbines to generate electricity.

During low tide when the water is level and low on both sides of the barrage, the gates are closed again and the tide raises the water on the "sea side" of the tidal barrage.

When the valves are opened, water flows "upstream" (Figure 1.55). We can use the same turbines in both ways, but in this case, the efficiency would be low and the investment would be high.

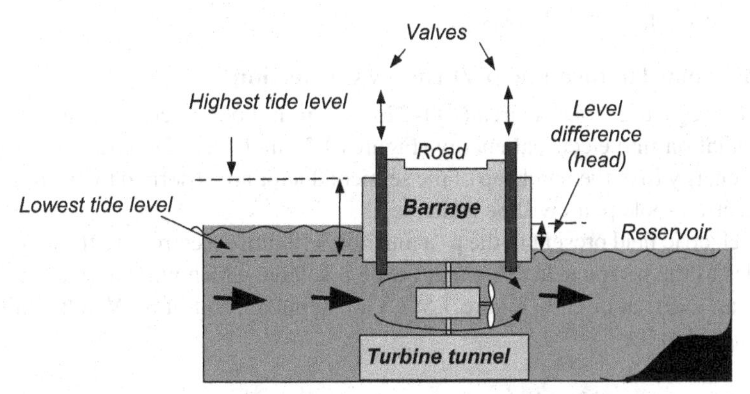

FIGURE 1.55 Water flowing from sea side to river (high tide).

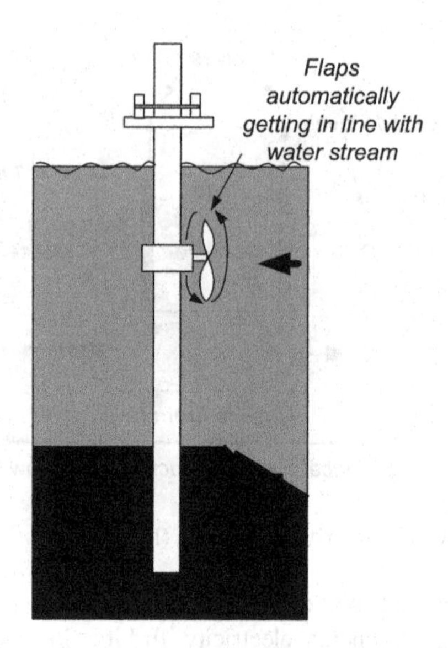

FIGURE 1.56 Tidal streams.

Thus, barrages can effectively generate power four times during the day—during the two fall and the two rising tides. In tidal barrages, generation during the rising tide is generally less efficient than that during the falling tide because of the river flowing against the sea.

b. Tidal streams

The kinetic energy in tidal streams can be extracted directly by the use of submerged turbines which are very similar in operation to wind turbines (Figure 1.56). Tidal stream installations reduce some of the environmental effects of large tidal barrages. This method is not as harmful to natural life as the tidal barrages. Also since they are under the sea, these types of installations are not much affected by bad weather conditions compared to open weather installations.

1.3.7.5 Solar Photovoltaic (PV) Energy Conversion

PV technology uses a large area (100–225 cm^2) p–n diode to convert energy in the sun's radiation into electrical energy (Figures 1.57 and 1.60). When photons with a greater energy than the band gap of the semiconductor are absorbed by the p–n junction, electron–hole pairs will be generated.

The electric field present in the p–n junction will move electrons to the n-type and the holes to the p-type side. This will create a voltage which will cause a current to flow in an external circuit (Figure 1.58). The output current of a PV cell is directly proportional to the

- intensity of incident light
- cell area
- temperature.

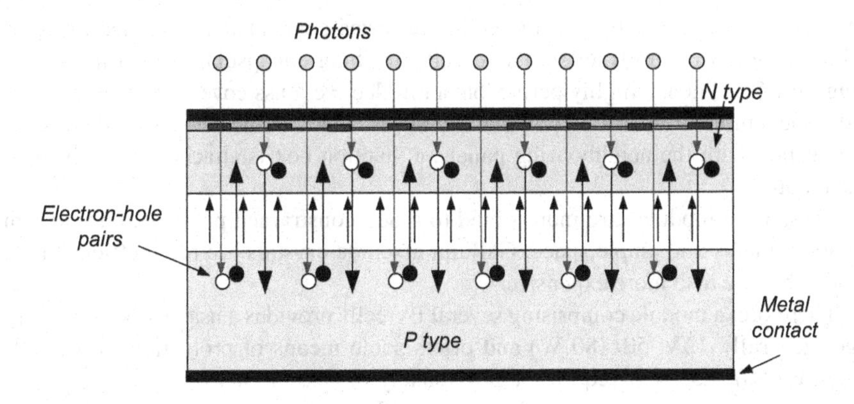

FIGURE 1.57 Large area p–n diode.

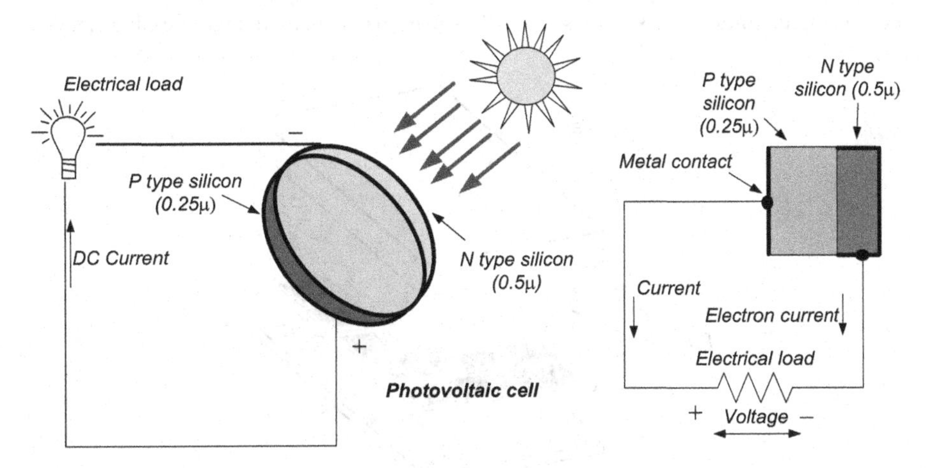

FIGURE 1.58 PV generation of DC voltage.

The main types of PV cells are as follows:

1. Crystalline silicon (most of the PV market)
   There are two types of crystalline silicon cells. Monocrystalline silicon cells have the advantage of higher efficiency (typically 14–15%). Complicated manufacturing process is their disadvantage. Due to the high and complicated technology required for their manufacturing process, monocrystalline silicon cells are expensive. Polycrystalline silicon cells are cheaper but are slightly less efficient (typically 13–15%).
2. Thin films
   There are several types like amorphous silicon. They are cheap but have lower efficiency. PV cells are the building blocks of PV systems (Figures 1.59 and 1.60). They are fragile and susceptible to corrosion, and their leads are delicate. Also, the operating voltage of a single PV cell of diameter 6 cm is about 0.5 V which is too small for many applications.

In crystalline cell PV panels, the cells are series-connected to produce an operating voltage around 30–60 V. These cells are then encapsulated with a polymer (Figure 1.59). Then, a highly permeable antireflective glass cover is put on the front side, and a moisture-resistant polymer (tedlar) is coated at the back. To provide easy connection with the neighbouring panels, a junction box is attached at the back of the module.

Crystalline modules are more suited to space-constrained projects and thin film to hot climates and ample space. Gallium arsenide crystals are more efficient than silicon but are also more expensive.

Therefore, a module comprising several PV cells provides a usable operating voltage (generally 12 V; 50–180 W) and offers some means of protection to the cells (Figure 1.60); 1.5–2 kW requires approximately $15-25\ m^2$ space.

An array is an interconnected system of PV modules that function as a single electricity–producing unit (Figure 1.60). Many different types of PV modules exist. For example, amorphous silicon solar cells are often encapsulated into flexible arrays.

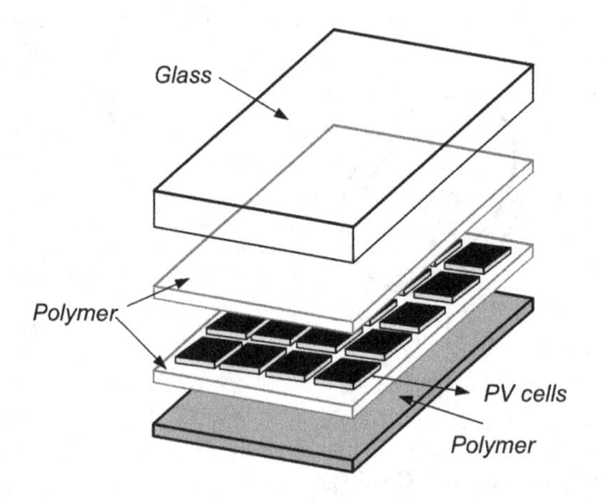

FIGURE 1.59 PV panel.

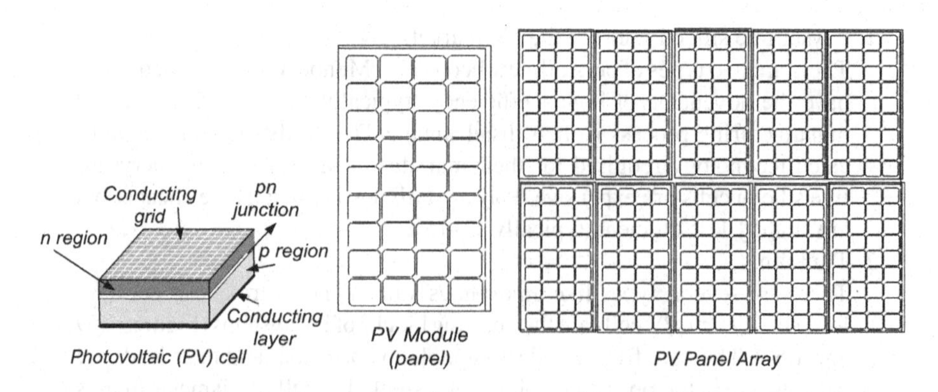

FIGURE 1.60 PV cell, PV module, and PV array.

PV cells have a single operating point (Figure 1.61) where the output power is a maximum. At array or module level, maximum power point trackers (MPPTs) optimise the operating voltage of a PV system to maximise the power.

MPPTs can be purchased separately or specified as an option with battery charge controllers or inverters. The cost and complexity of adding a peak-power tracker should be balanced against the expected power gain and reliability.

At array level (Figure 1.62), a single tracker controls the current through all modules in the array and the function is commonly integrated into the array's inverter. This way maintenance and cost is minimised, and inverter efficiency is high, but not all modules will operate at their maximum power point.

MPP controllers can also be applied to individual modules in an array, so that every module operates at its own maximum power point. These can be placed in microinverters for each module. However, multiple small controllers inherently cost more than a single centralised one. MPP of individual modules is most suited to PV systems where the value of produced electricity is high. Advantages of PV are as follows:

- PV produces no greenhouse gases.
- it can be installed in a wider variety of places (on rooftops).
- additional modules can be added incrementally.
- a utility-scale power plant can be constructed in months.
- it requires very little maintenance.
- it produces electricity during the afternoon when demand is highest.
- it is often sited close to electricity users, which reduces transmission costs.

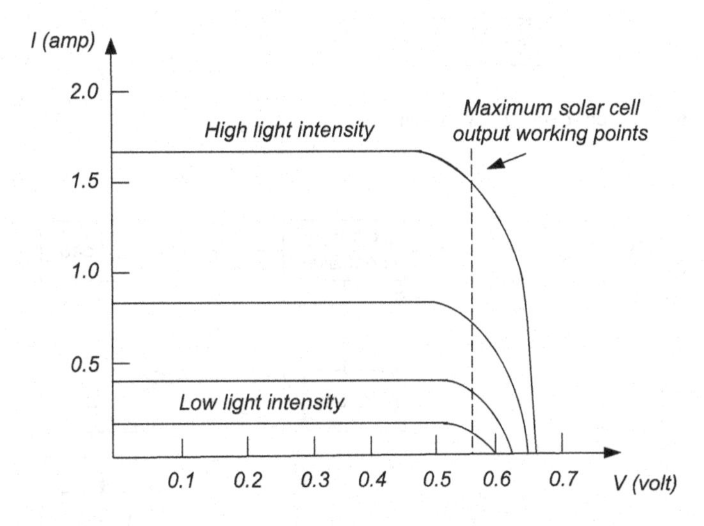

FIGURE 1.61 Graph showing maximum output power points of PV cells. (CRC Press, *Solar Energy Sciences and Engineering Applications,* Napoleon Enteria and Aliakbar Akbarzadeh.)

FIGURE 1.62 Array of PV panels. (Courtesy of Ener Power.)

Types of PV Systems

In general, there are four types of PV systems in use:

1. Direct-coupled PV systems (Figure 1.63)
   PV array is connected directly to the load. The load can operate only under solar radiation. As a typical application of this type of system, water is pumped up to a pond on a hill storing water for future use.
2. Stand-alone PV systems (Figure 1.64)
   These are used in rural areas that are not easily accessible and have no connection to power grids. Inverter in the system converts the DC to AC. Charge controller prevents the batteries from overcharging or over-discharging.

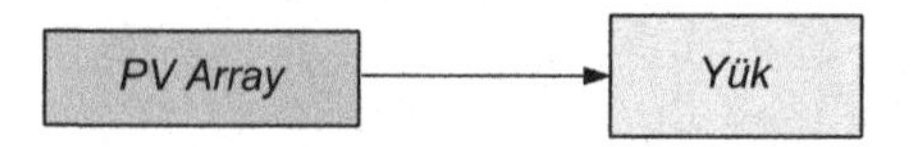

FIGURE 1.63 Direct-coupled PV system.

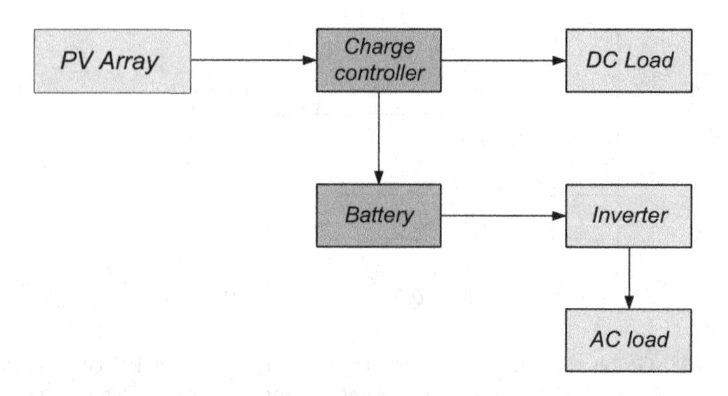

FIGURE 1.64 Stand-alone PV systems.

3. Grid-connected system (Figure 1.65)

   In this type of system, the grid acts as an energy storage system, and there is no need for battery storage. The PV system is connected to the local electricity network. During the day, electricity generated by the PV system can be used immediately or be sold to one of the electricity supply companies. In the evening, power is bought back from the network.

4. Hybrid-connected systems (Figure 1.66)

   In the hybrid-connected system, there is a back-up generator usually of a diesel type or utility grid type besides the PV generator. Both DC and AC loads can be fed simultaneously.

   The major problem limiting the widespread use of photovoltaics (PVs) is the high cost of manufacturing the sheets of semiconductor materials needed for power systems. However, with falling costs and government incentives, PV has become an economic source of electricity in many places (Figure 1.68) and the great majority of PV systems are now grid-connected.

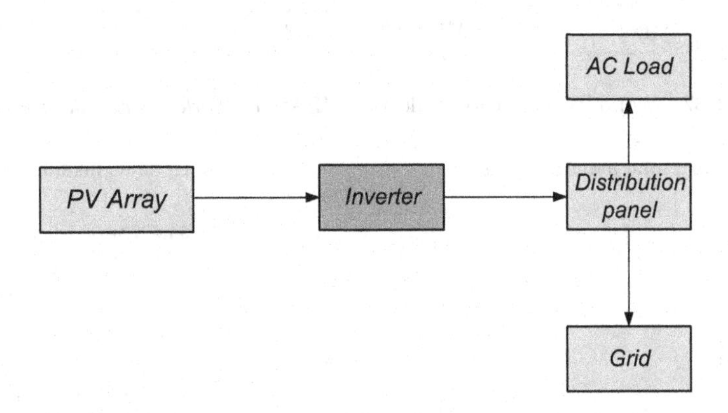

FIGURE 1.65 Grid-connected system.

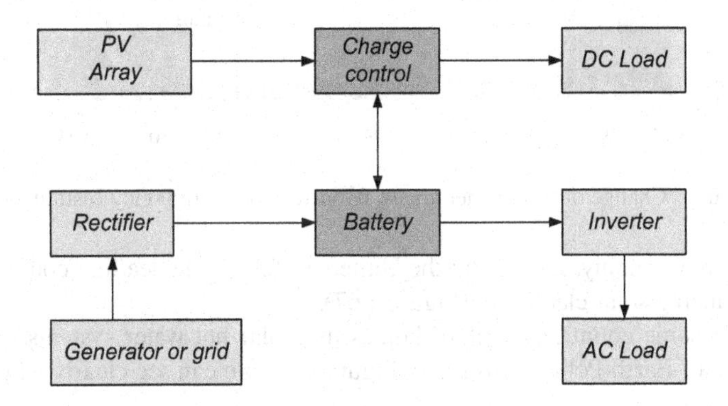

FIGURE 1.66 Hybrid connected systems.

Producers	TWh	% of world total
People's Rep. of China	75	22.9
Japan	51	15.5
United States	47	14.2
Germany	38	11.6
Italy	22	6.7
India	14	4.3
United Kingdom	10	3.2
France	8	2.5
Spain	8	2.5
Australia	6	1.9
Rest of the world	49	14.7
World	328	100.0

Country (top ten producers)	% of solar PV in total domestic electricity generation
Italy	7.6
Germany	5.9
Japan	4.8
United Kingdom	3.1
Spain	2.9
Australia	2.4
France	1.5
People's Rep. of China	1.2
United States	1.1
India	1.0
Rest of the world[1]	0.6
World	1.3

FIGURE 1.67 World PV electricity production. (IEA; *Key World Energy Statistics* 2018.)

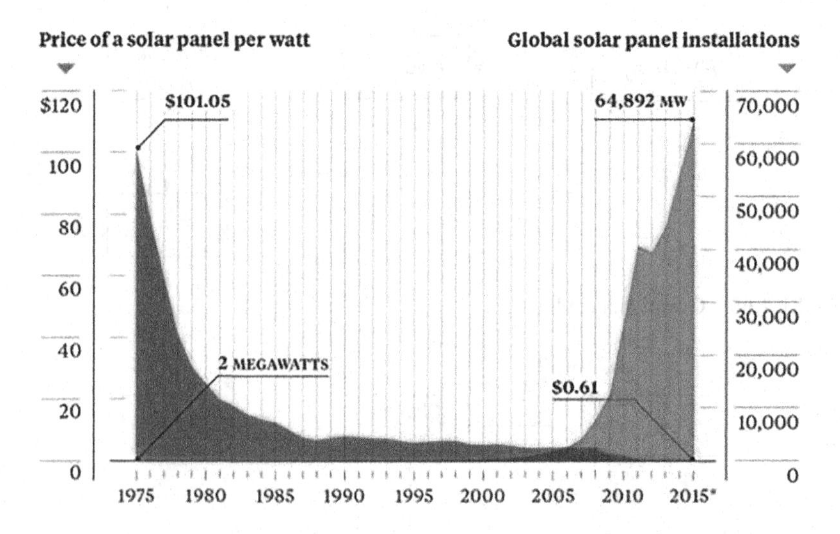

FIGURE 1.68 Change of solar panel prices. (Courtesy of Earth Policy Institute.)

China, Germany, Japan, and the United States are the leading countries in producing solar electricity (Figure 1.67).

In some countries, 90% of homes use solar hot water systems.[1] In the chart of Earth Policy Institute in Figure 1.68, we can see clearly why solar

[1] Del Chiaro, Bernadette; Telleen-Lawton, Timothy. "Solar Water Heating".

energy is the energy of tomorrow. There has been a tremendous drop in the price of a solar panel per watt.

1.4 SUSTAINABLE ENERGY

Of the four greenhouse gases (GHGs) (carbon dioxide, methane, nitrous oxide, and sulphur hexafluoride), carbon emission from energy use is a major cause of climate change because of their accumulation in the atmosphere. This has caused a number of constraints on the amount and type of energy we can all use. The use of coal, oil, and gas appears to be major causes of the release of carbon dioxide gas (Figure 1.69).

There are of course other reasons to prefer "clean" energy, but climate change comes on top of such concerns as a universal and very challenging problem. In this respect, nuclear is still a questionable sustainable energy source. As long as the security concerns about nuclear power are not reduced and the storage of radioactive nuclear waste is not resolved, nuclear energy will remain a questionable sustainable energy for future.

There is a large potential for increasing the share of solar as well as of wind in total energy consumption over the coming years. Biomass and geothermal energy can also contribute to a more sustainable energy future. For companies and governments, moving towards clean energy (no carbon emission) gradually will develop an advantage over latecomers.

Energy availability and use has been a driver of economic development worldwide for centuries. Until recently, it was the availability of energy, such as the limited supply of fossil fuels or of water required for hydropower, that was considered a serious long-term constraint on the use of energy.

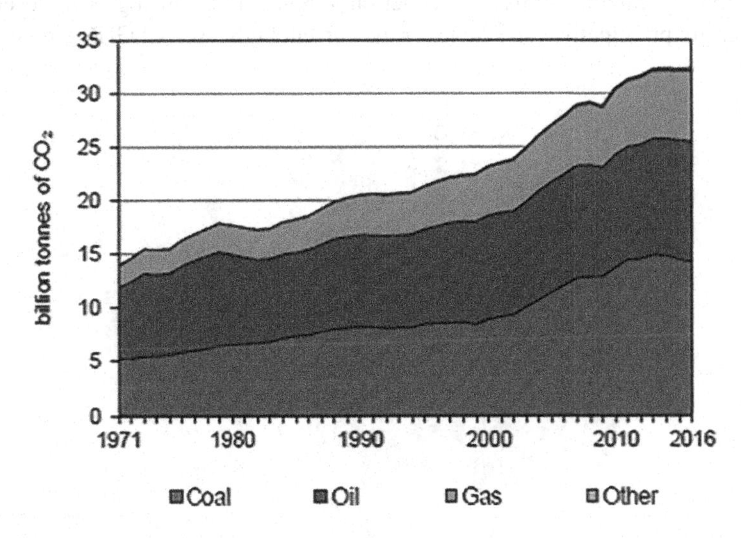

FIGURE 1.69 CO_2 emission by fuel [IEA; *CO_2 Emission from Fuel Combustion* (*Highlights* 2018 edition).]

Over the past few decades, the realisation that the carbon emissions from energy use and other human activity were a serious cause of climate change because of their accumulation in the atmosphere has led us to understand that the necessity to protect the planet's climate constitutes another fundamental constraint on the amount and type of energy we can all use. As far as CO_2 emission for electricity production is concerned, China leads the world followed by the United States and India (Figure 1.70).

There exist optimistic as well as more pessimistic scenarios. But the fact that the emissions of carbon and other heat-trapping gases lead to climate change is now beyond reasonable doubt. Sooner or later all countries will have to participate in this revolution. The young, in particular, will demand policies and actions that will allow their children and grandchildren to live in a world where climate continues to allow a decent life and further economic development.

Given this agreed need for an energy revolution, the countries that succeed in planning ahead, in moving towards clean energy, in learning which technologies to use as well as in using the best regulatory and price policies, will develop an advantage over latecomers. They will be able to avoid sudden and disruptive changes by acting early and will develop a competitive advantage in the know-how relating to the energy revolution, perhaps contributing themselves to inventions that can lead to good economic and financial returns.

There is a huge potential to improve energy efficiency and increase the contribution of solar (PV and concentrated solar power or CSP) as well as of wind in total energy consumption over the coming decades. Efficient use of these resources will require smart grids and appropriate tariffs. Biomass and geothermal energy can also contribute to a more sustainable energy future.

The Kyoto Protocol is an international agreement linked to the United Nations Framework Convention on Climate Change, which commits its parties by setting internationally binding emission reduction targets. Recognising that developed countries are principally responsible for the current high levels of GHG emissions in

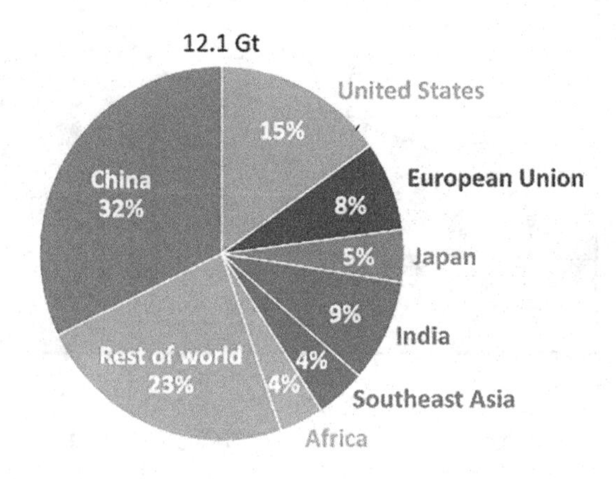

FIGURE 1.70 CO_2 emissions for electricity production by region. (2016. IEA; *World Energy Outlook* 2018; Page-320, Figure: 7-33.)

the atmosphere as a result of more than 150 years of industrial activity, the protocol places a heavier burden on developed nations.

The Kyoto Protocol was adopted in Kyoto, Japan, on 11 December 1997 and entered into force on 16 February 2005. The detailed rules for the implementation of the protocol were adopted at COP 7 in Marrakesh, Morocco, in 2001, and are referred to as the "Marrakesh Accords". Its first commitment period started in 2008 and ended in 2012.

Although the Kyoto Protocol represented a landmark diplomatic accomplishment, its success was far from assured. Reports issued in the first two years after the treaty took effect indicated that most participants would fail to meet their emission targets. Even if the targets were met, however, the ultimate benefit to the environment would not be significant, according to some critics, since China, the world's leading emitter of greenhouse gases, and the United States, the world's second-largest emitter, were not bound by the protocol (China because of its status as a developing country and the United States because it had not ratified the protocol). Other critics claimed that the emission reductions called for in the protocol were too modest to make a detectable difference in global temperatures in the subsequent several decades, even if fully achieved with U.S. participation. Meanwhile, some developing countries argued that improving adaptation to climate variability and change was just as important.

Paris Agreement is a pact sponsored by the United Nations to bring all countries together in the fight against climate change. Participating nations made a historic pact on 12 December 2015, in Paris, France,

- to adopt green energy sources
- cut down on greenhouse gas emissions
- limit the rise of global temperatures.

Under the agreement, every country will have an individual plan to fight against its GHG emissions. Of the 196 negotiating countries that signed the agreement, 181 parties have ratified it. In June 2017, U.S. President Donald Trump announced his intention to withdraw his country from the agreement.

PROBLEMS OF CHAPTER 1

Problem 1.1: An electrical domestic waste plant of efficiency 35% is to be installed in a certain region to supply 10 GWh of energy annually to the community in the region. If the domestic waste used releases 10 GJ of energy per tonne of waste and the yearly waste output of each house is 1.2 tonnes,

a. calculate the required number of houses.
b. for a plant load factor of 70%, what would be the required generator kW rating?
c. what fraction of the houses found in (a) can be supplied from this power plant if the annual demand per household is 5,000 kWh?

Problem 1.2: If 1 kg of coal gives an energy of 15 MJ, and a power plant has a power output of 80 MW at an efficiency of 50%, calculate the amount of coal in tonnes that has to be burnt in one hour?

Problem 1.3: One kilogram of the coal used in a certain coal-fired power station of 55% efficiency gives 3,500 kcals of energy. If the plant is using 650 tonnes/h of coal and the yearly total energy output of the plant is 5,500 GWh (1 kWh = 860 kcals),

a. what is the power output of the plant?
b. what is the yearly working hours of the plant?

Problem 1.4: In a hydroelectric power plant, water from a lake is made to fall down a height of 30 m at a rate of 1,000 tonnes per hour to produce 60 kW of electric power. What is the efficiency of this plant? How many hours in a year does this plant have to work in order to supply a yearly requirement of 100 MWh of energy?

Problem 1.5: In an hydroelectric power plant of 120 m of water head, the flow rate of the water is 20 m^3/s. If the forced pipe efficiency is 90%, turbine efficiency is 89% and the generator efficiency is 95%,

a. what is the power output of the plant?
b. how many days in a year does this plant have to work in order to supply yearly 100 GWh of energy to certain a region?

Answers to odd-numbered questions:
1.1: 8,571; 1,631 kW; 23%
1.3: 1,454,942 kW; 3,780 h
1.5: 17.9 MW; 233 days

2 Magnetic Circuits

2.1 INTRODUCTION TO MAGNETIC CIRCUITS

It is important to analyse and understand magnetic field quantities for a good understanding of devices like transformers and electrical machinery. In most electrical machines, except permanent magnet machines, the magnetic field (or flux) is produced by passing an electric current through coils wound on ferromagnetic materials. Therefore, a good understanding of magnetic materials and circuits is necessary to carry on to the subject of electromechanical energy conversion devices.

2.2 MAGNETIC FIELDS AND CURRENT-CARRYING CONDUCTORS

We all have noticed some strange effects that occur around permanent magnets. In the vicinity of permanent magnets, there are forces that act on magnets, irons, and current-carrying conductors. It is interesting to note that similar effects are also observed in the neighbourhood of a current-carrying conductor (Figure 2.1).

In addition to these, if the current flowing in the conductor is changing with time, a voltage is induced in the nearby circuits. The neighbourhood of the conductor (or the permanent magnet) is considered as possessing a force field. It is a region of energy storage able to produce forces and do work. This field is called the magnetic field (Φ) (Figure 2.1).

The magnetic field can be mapped by drawing lines, which indicate the direction of the force on the north magnetic pole of a permanent magnet. The density of the

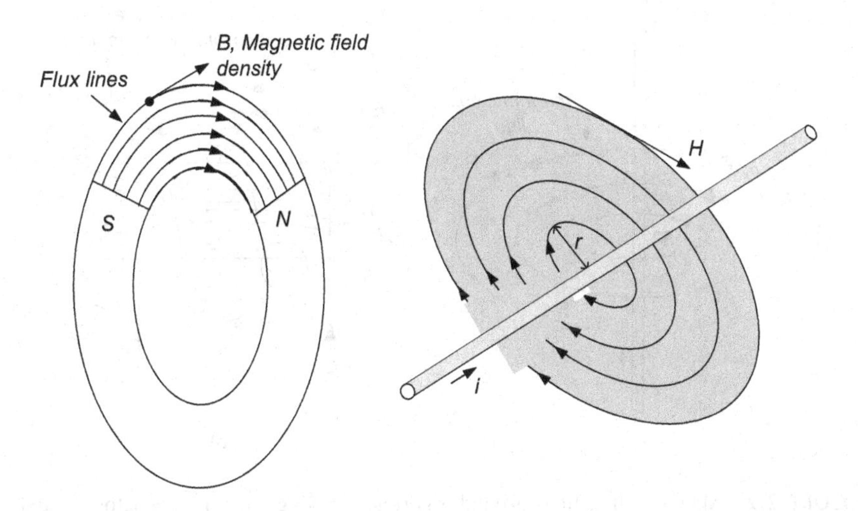

FIGURE 2.1 Magnetic field around permanent magnets and current-carrying conductors.

lines at a particular point is made proportional to the force exerted on the magnetic pole. In this respect, the lines resemble somewhat the contour lines on a topographical map. Lines are densest where the force is greatest. These lines are called magnetic field (Φ) lines or magnetic flux lines (Figure 2.1). In this book, we shall be using the following descriptions in the same meaning:

$$\text{Magnetic field} = \text{Magnetic flux} = (\Phi)(\text{Measured in Webers})$$

When the magnetic flux (Φ) (or magnetic field) is distributed over an area A, the flux density at any point in that area is given as

$$B = \frac{\Phi}{A}\,\text{Wb/m}^2\,(\text{T})$$

In Figure 2.2a,

1. the flux density decreases outward from the conductor.
2. for any point on the circles, the flux density (B) is constant.

If a current-carrying wire is formed into a multiturn coil wound around an iron core of ferromagnetic material (Figure 2.2b), the magnetic field is greatly intensified. The direction of the lines can be found with the help of a right-hand rule, which can be stated differently for each case:

1. "If the conductor in Figure 2.2a is grasped by the right hand (Figure 2.3a), with the thumb pointing in the current direction, the flux direction will be in the direction of the fingers wrapping around the conductor".

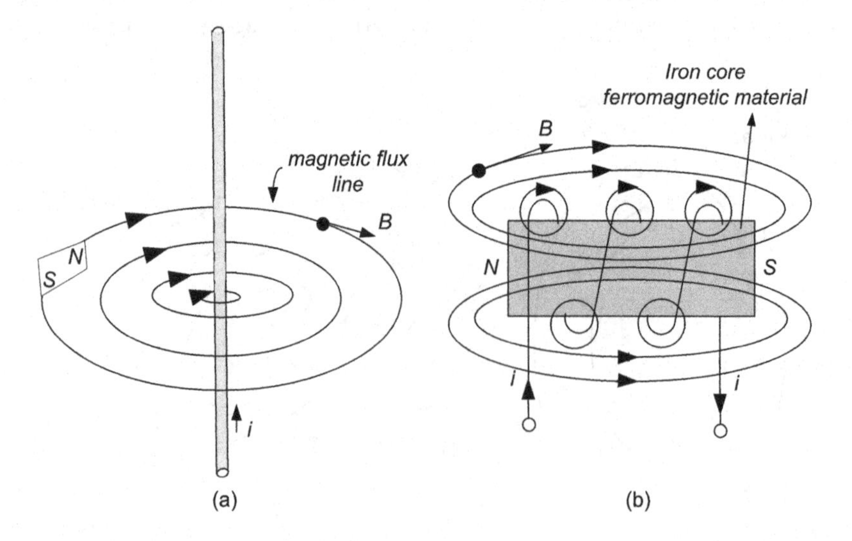

FIGURE 2.2 Magnetic flux lines around a current-carrying straight conductor (a) and a coil (b).

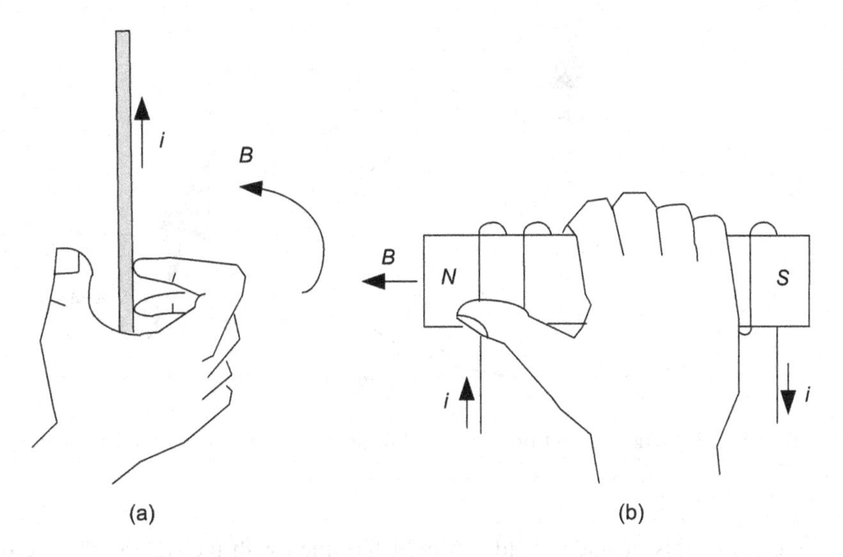

FIGURE 2.3 Right-hand rules for current-carrying conductor (a) and coil (b).

2. Similarly, if the coil in Figure 2.2b is grasped in the right hand with the fingers pointing in the direction of the current (Figure 2.3b), the thumb will point in the direction of the magnetic field. Notice that outside the core, field lines are from north (N) to south (S) pole, and inside the core, they are from south to north pole.

2.3 MAGNETIC FIELDS OF ELECTRONS AND MAGNETIC MOMENT

In both cases of permanent magnet and current-carrying conductor, the basic source of the magnetic field is the same. The cause of the magnetic field happens to be the current flowing or the "electrical charges in motion". Electrons move around the nucleus like the Earth circles around the sun (Figure 2.4). This electron motion is a small electric current, and anywhere there is a current, there is a magnetic field (moment).

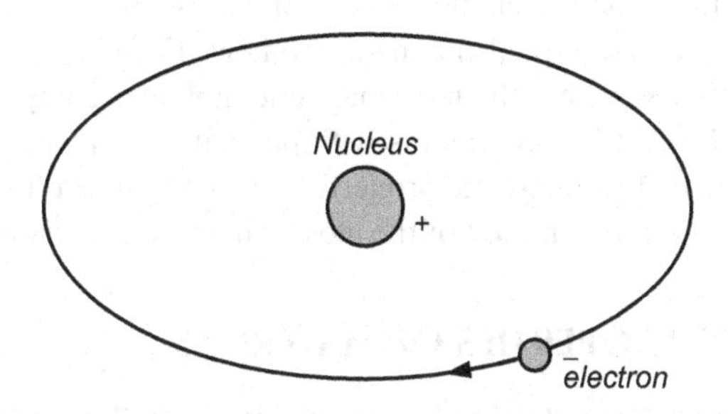

FIGURE 2.4 The structure of an atom.

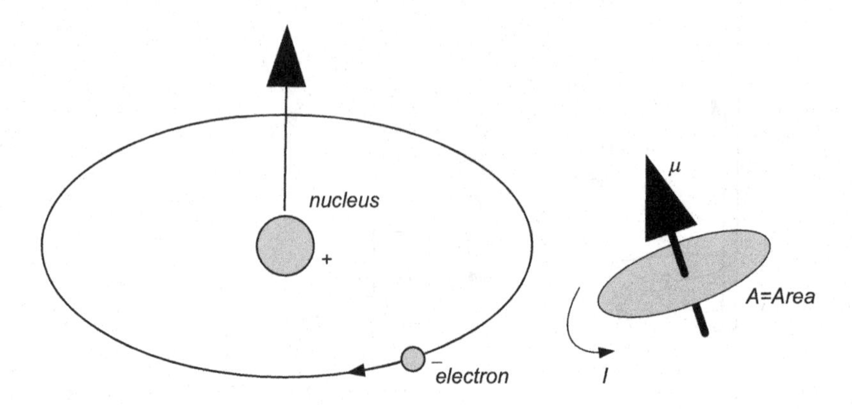

FIGURE 2.5 Magnetic moment of electrons flowing in their orbits around the nucleus of an atom.

Direction of this magnetic field can be determined with the right-hand rule of Figure 2.3b. The magnetic moment of a ring of current can be considered to be a vector quantity with direction perpendicular to the current loop (area) (Figure 2.5) in the direction indicated by the right-hand rule applied just like to a wound coil of n turns. Anything that is magnetic, like a bar magnet or a loop of electric current, has a magnetic moment.

The magnetic moment of a ring of current is given by

$$\mu = \text{Magnetic moment} = A.I$$

where A is the area of the current loop, and I is the current flowing in the loop. When this current loop is placed in a magnetic field B, the torque on it is given by Lorentz's force law as

$$T = \mu.B.\sin\theta$$

where θ is the angle between $\mathbf{B}$ and $\boldsymbol{\mu}$. According to Lorentz's force law, this torque tends to line up the magnetic moment with the magnetic field B.

Electrons also spin around their axes like the Earth. If the electron charge is considered to be distributed over the surface of the electron, the spinning creates another current and an additional magnetic moment. For the case of a completely filled electron shell or subshell, the total magnetic moments completely cancel each other out (Figure 2.6). Thus, only atoms with partially filled electron shells have a net magnetic moment. The magnetic properties of materials are in large part determined by the nature and magnitude of the atomic magnetic moments.

2.4 MAGNETIC PROPERTIES OF MATERIALS

There are two atomic-level contributors to magnetic properties of elements—orbital angular momentum and spin.

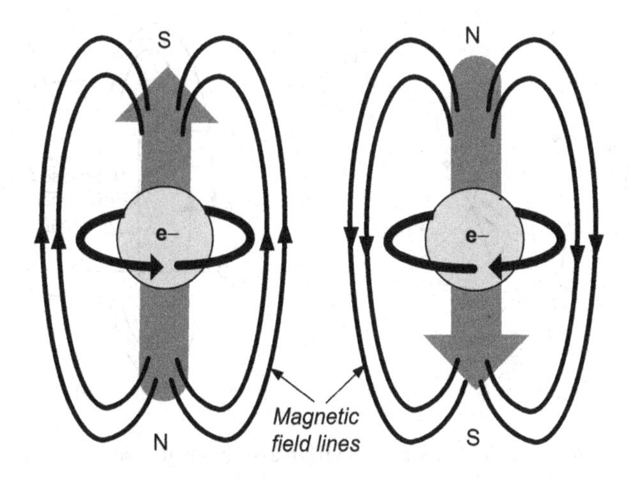

FIGURE 2.6 Two electron magnetic moments cancelling each other.

- in about two-thirds of the elements, the orbital and spin moments cancel, so the atom has no net magnetic moment.
- in all but five of the remaining elements, the neighbouring atoms cancel each other's magnetic moment.
- there are two naturally found elements that have a net magnetic moment. These two exhibit permanent magnetic properties and are named as ferrimagnetic materials (magnetite and loadstone).
- the other three are iron, nickel, and cobalt (including their alloys and oxides) which have net magnetic moment when placed in a magnetic field. These materials are called ferromagnetic.

Ferromagnetic materials can be magnetised and are also strongly attracted to a magnet. The behaviour of ferromagnetic materials such as iron, cobalt, and nickel (or their alloys and oxides) can be explained with the help of magnetic domains. A magnetic domain is a very small region in which all the magnetic dipoles are aligned as in Figure 2.7a.

The direction of alignment of the magnetic dipoles varies randomly from one domain to the next. Owing to these random alignments, a virgin ferromagnetic material is in a non-magnetised state (Figure 2.7a). When the ferromagnetic material is placed in an external magnetic field (Figure 2.7b), all the dipoles tend to align along that magnetic field. One way to place the magnetic material in a magnetic field is to wind a current-carrying wire around it as indicated in Figure 2.7b. Those materials that are feebly repelled when placed near the north pole of a magnet (or solenoid) are called diamagnetic. Material does not retain the magnetic properties when the external field is removed. Bismuth, silver, and copper are diamagnetic materials. Substances that are pulled towards the centre of the magnet with a feeble force are called paramagnetic. Paramagnetic materials include magnesium, molybdenum, lithium, and tantalum.

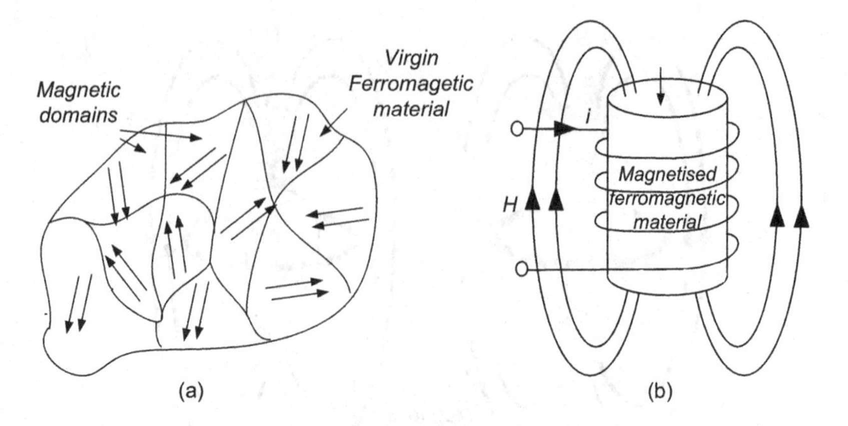

FIGURE 2.7 Ferromagnetic materials: (a) virgin and (b) magnetised.

2.4.1 Types of Ferromagnetic Materials

Ferromagnetic materials can be divided into two types:

- magnetically "soft" materials like annealed iron (Figure 2.8b), which can be magnetised but when the magnetising source is removed lose their magnetisation.

  Soft magnetic materials are mainly used in magnetic cores of motors, generators, transformers, and inductors. They have high permeability, low magnetic losses, and a low coercivity.
- Magnetically "hard" materials which can be magnetised and do stay magnetised. Hard ferromagnetic materials when placed in an external magnetic field do get magnetised, and when the magnetising source is removed, they

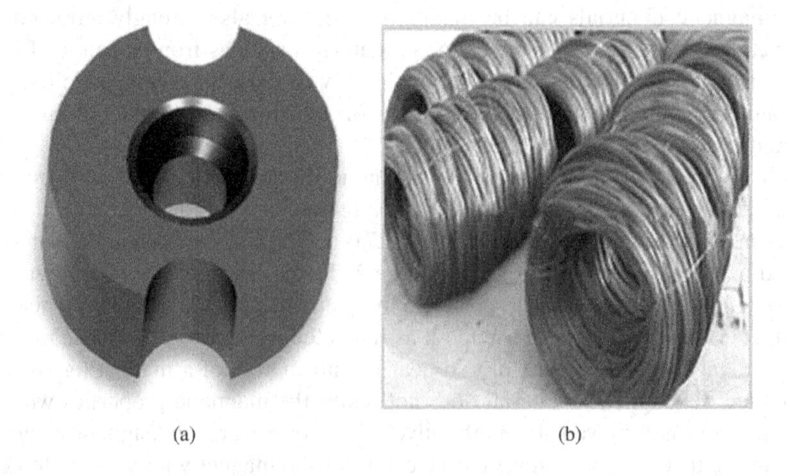

FIGURE 2.8 Soft ferromagnetic materials: metal powder (a) (http://www.innovativesintered.com) and annealed iron (b) http://www.ecvv.com.

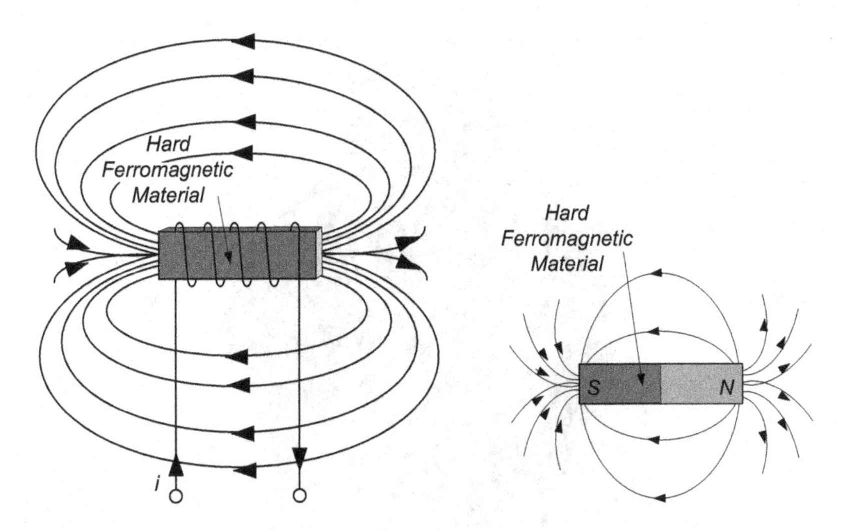

FIGURE 2.9 The making of permanent magnets from a hard ferromagnetic material.

don't lose their magnetisation. This is the way permanent magnets are made (Figure 2.9). A certain amount of energy is stored in the magnet, and it becomes permanently magnetised. Hard magnets are also referred to as permanent magnets. They are used in automative industry (starter motors, wiper motors), in telecommunications (loudspeakers, microphones) and in electronics and instrumentation. Alnico 5 is the most important material used to create permanent magnet.

2.4.2 History of Magnetism

Magnetism has originated from lodestones that occur naturally as a permanent magnet which attracts other metals. The word "magnet" in Greek meant "stone from magnesia". Magnesia used to be a part of ancient Greece (Ionia) where lodestones were found. Today, it lies in the west coast of Turkey, and it is called Manisa. Loadstone is an extremely rare form of the mineral magnetite. Both magnetite and lodestone are naturally found iron oxides (iron ores). Magnetite is attracted to magnets and when naturally magnetised becomes a loadstone. So we can say all loadstones are magnetites but not all magnetites are loadstones.

This "magic" property of pulling metals (Figure 2.10) was known to many ancient cultures and a powerful lodestone has always been an expensive item.

The loadstone has an important place in the history of magnetism. It is believed that around first century BC the Chinese has discovered that a freely suspended elongated lodestone (Figure 2.11) would align its long axis in the north–south direction (in the direction of Earth's magnetic field). This was the beginning for the compass of today. Later compasses were made of iron needles, magnetised by striking them with a lodestone. The Chinese also appear to have discovered that this characteristic could be transferred to an iron piece if it is touched by a loadstone.

FIGURE 2.10 Loadstone with attracted iron pins (Wikipedia).

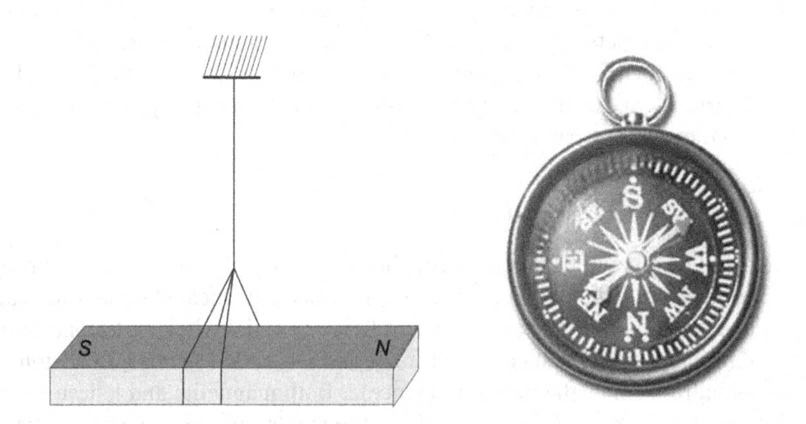

FIGURE 2.11 Compasses of today and old times.

Before the compass, it was difficult to determine where you were heading to if there were no visible landmarks and especially in the dark or fog. With the compass, ships were able to sail to sea without the fear of getting lost. This has greatly contributed to world trade and the discovery of the unknown parts of the world. The first recorded appearance of the use of the compass in Europe is believed to be in 1190.

Ferrimagnetic materials, which include ferrites and the oldest magnetic materials magnetite and lodestone are similar to but weaker than ferromagnetics.

2.5 MAGNETIC CIRCUITS AND PERMEABILITY

Ferromagnetic materials (iron alloys with cobalt, tungsten, nickel, aluminium, and more) normally are much more receptive to magnetic flux than is air or free space. This property is useful in making them carry the magnetic flux from one point to another. This way the majority of the magnetic flux would confine in the ferromagnetic material. This action is like confining water to a pipe when transferring it from one place to another (Figure 2.12). Ferromagnetic materials are used in the making of magnetic circuits just like the copper wires used in electrical circuits. In fact, there is a similarity between electrical current flowing in a wire and the magnetic flux in a ferromagnetic material. We may think of the flux Φ like the current in electric circuits.

Permeability is a measure of this receptiveness of the ferromagnetic materials to having magnetic flux setup in them (Figure 2.13). For free space, the permeability μ_0 is $4\pi \times 10^{-7}$ in the rationalised mks system. Its unit is Henry/meter.

The permeability of ferromagnetic materials is up to thousands of times that of free space. This is why flux tends to concentrate in such materials. This in fact is called the relative permeability of the material. Permeability is often stated in numbers relative to the permeability of the air. The permeability of a diamagnetic material is slightly less than that of free space. Paramagnetic substances exhibit slightly greater permeabilities than that of free space.

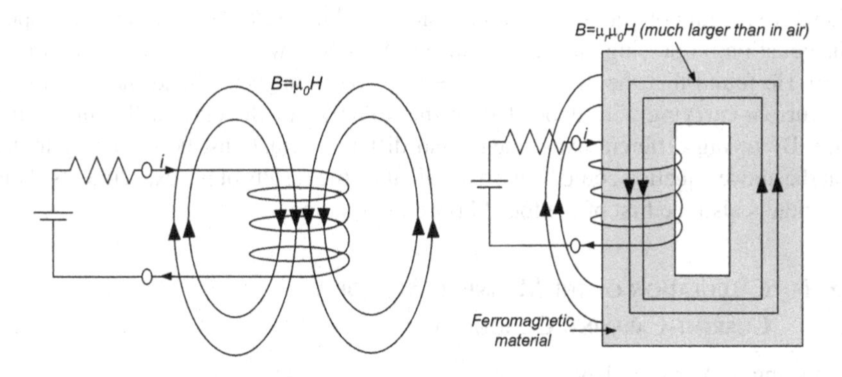

FIGURE 2.12 Confining magnetic field in a ferromagnetic material.

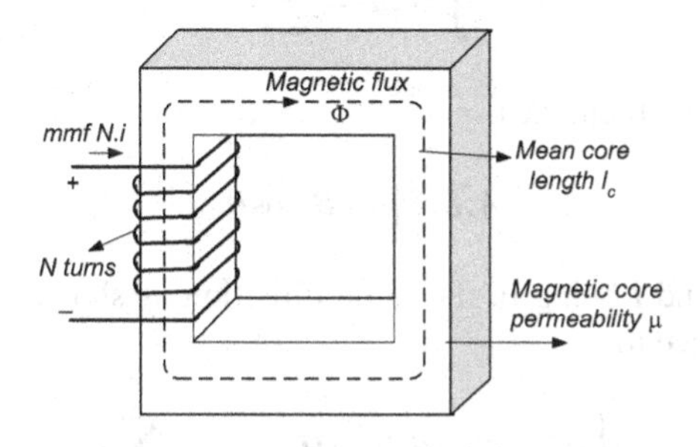

FIGURE 2.13 Permeability is a measure of magnetic receptiveness.

2.6 AMPERES CIRCUITAL LAW

Ampere's circuital law states the relationship between the current and the magnetic field created by it. If magnetic field intensity is $\boldsymbol{H}$ and

$$\boldsymbol{H} = \frac{\boldsymbol{B}}{\mu}$$

Ampere's circuital law states that "the line integral of the magnetic field intensity around a closed loop is equal to the algebraic sum of the currents flowing through the area enclosed by the path". Therefore, we can state "Ampère's circuital law" mathematically as

$$\oint \boldsymbol{H}.\boldsymbol{dl} = \sum i$$

Here, $\sum i$ is known as the magnetomotive force mmf,

$$\sum i = \text{magnetomotive force} = \text{mmf}$$

Therefore the unit of magnetic field intensity H is Ampère/meter. Ampère developed this equation by passing an electric current through a wire (wound around an iron core). He found that the direction of the force on a magnet around the vicinity of the current-carrying wire depended on the direction of the current flowing in the wire. By trying different experiments and different equations, he gradually found that the above equation could be used to calculate the results of his experiments. This equation is also the first of the four Maxwell's equations.

2.6.1 Calculation of the Magnetic Field of a Current-Carrying Conductor

According to Ampère's law,

$$\oint \boldsymbol{H}.\boldsymbol{dl} = \sum i$$

We remember from mathematics,

$$\boldsymbol{A}.\boldsymbol{B} = |\boldsymbol{A}|.|\boldsymbol{B}|\cos\theta,$$

For the case of $\boldsymbol{H}$ and $\boldsymbol{l}$ being in the same direction as shown in Figure 2.14, this equation would reduce to

$$\oint |\boldsymbol{H}_r|.|\boldsymbol{dl}|.\cos\theta = |\boldsymbol{H}_r|.2\pi r = \sum i$$

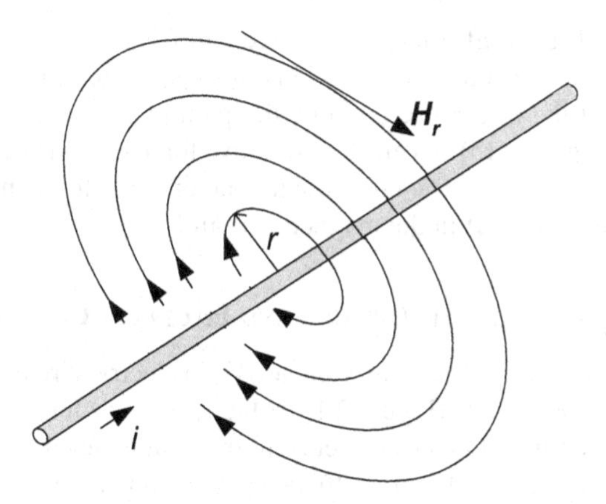

FIGURE 2.14 Magnetic intensity $\boldsymbol{H}$ reduces proportionally with radius r.

Hence,

$$\rightarrow |\boldsymbol{H}_r| = \frac{\sum i}{2\pi r}$$

Note: *The dot product of two vectors* **A** *and* **B** *is defined as* $\mathbf{A.B} = |\mathbf{A}||\mathbf{B}|\cos\theta$ *where* θ *is the angle between vector* **A** *and vector* **B**.

For the magnetic circuit in Figure 2.15, if there are N turns of current flowing in the winding, then

$$\oint \boldsymbol{H}.\boldsymbol{dl} = Ni$$

$$Ni = F = \text{Magnetomotive force responsible for the magnetic field}$$

Here, $\boldsymbol{l}$ is the length of the magnetic circuit, not that of the coil. If there are more than one winding, we should add them algebraically to obtain the total mmf. The mmf is measured in amp-turns and can be described as the magnetic potential difference that forces flux around the magnetic circuit.

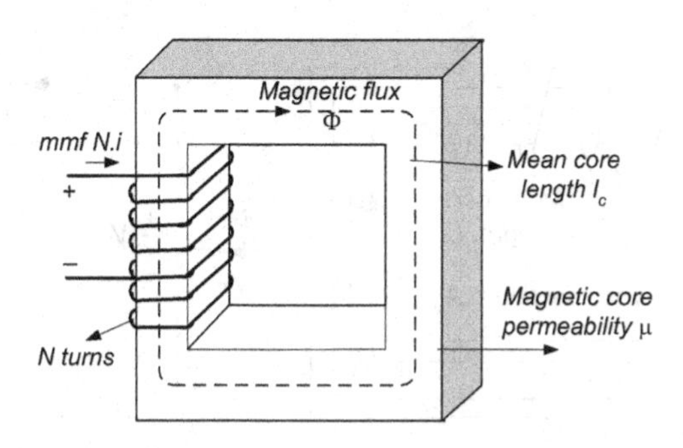

FIGURE 2.15 A magnetic circuit.

We may think of the mmf F like the voltage emf (E) in electric circuits and flux Φ like the current I in electric circuits. This analogy is helpful in trying to solve magnetic circuits. Like in electric circuits, the path of magnetic flux is known as magnetic circuit (Figure 2.15). As an electromotive force (emf) is necessary to cause the flow of current i in an electric circuit, a magnetomotive force (mmf) is required to create the magnetic flux Φ in the magnetic circuit.

2.6.2 Analogy of Magnetic Circuits and Electrical Circuits Reluctance

We may think of the mmf "F" like the voltage "V" in electric circuits and flux Φ like the current i in electric circuits (Figure 2.16). Magnetomotive force (mmf) is actually the flux-producing ability of an electric current in a magnetic circuit.

For the magnetic circuit in Figure 2.16, because H and l are in the same direction ($\cos\theta = 1$) if H_c is the average magnitude of H and l_c is the mean flux path length and A_c is the crcss-sectional area,

$$F = Ni = H_c.l_c$$

$$= \frac{B}{\mu}l_c = \frac{\Phi/A_c}{\mu}l_c = \Phi\left(\frac{l_c}{\mu A_c}\right) = \Phi.\mathcal{R}$$

Here, $\mathcal{R}$ is defined as the reluctance of the magnetic circuit of Figure 2.16. It bears similarities to resistance in the DC circuits. Like the resistance in the electric circuits, reluctance $\mathcal{R}$ is

- directly proportional to length l_c
- inversely proportional to cross-sectional area A_c
- dependent on the material (μ) of the magnetic circuit.

With l and A in m and m^2, respectively,

$$\mathcal{R} = \text{Reluctance} = \frac{l_c}{\mu A_c} = \frac{N.i}{\Phi}\text{At/Wb}$$

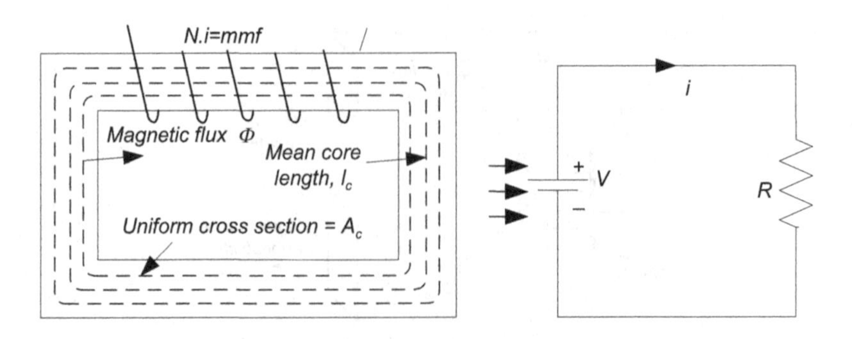

FIGURE 2.16 Analogy between electric and magnetic circuits.

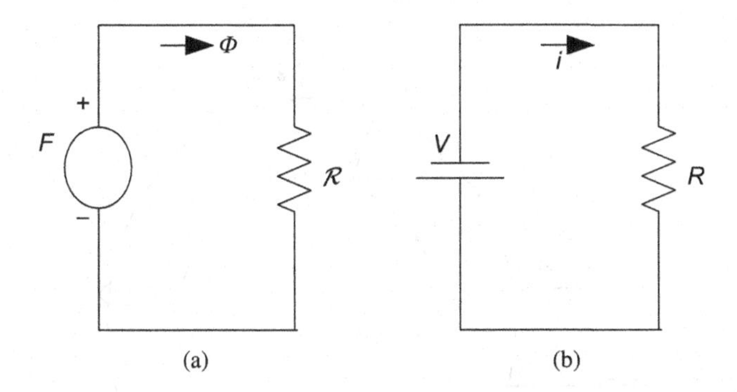

FIGURE 2.17 (a) Magnetic circuit and (b) equivalent electric circuit.

The quantity μ is called the permeability of the material. Permeance is defined as

$$P = \frac{1}{\mathcal{R}}$$

$$F = \text{Magnetomotive force} = \Phi.\mathcal{R}$$

can be regarded as ohm's law for a magnetic circuit (Figure 2.17) where

- F replaces the voltage V.
- Φ replaces the current i.
- $\mathcal{R}$ replaces the resistance R.

The close analogy between the DC-resistive electric circuits and magnetic circuits would be a good advantage when solving magnetic circuits.

2.6.3 Toroidal Ring

Let us consider a toroidal ring of a ferromagnetic material with a coil of wire wound around it (Figures 2.18 and 2.19). When a current flows in the coil, a magnetic flux will be confined mainly in the ring.

The flux lines will be concentric circles (Figure 2.19), and the area of the path will be the same at any perpendicular section. If the width of the ring is small compared to? with its two diameters:

- the length of the flux path will be essentially the same along any circle.
- the flux will be distributed uniformly over the area.

The magneto motive force (mmf) is given by

$$\text{mmf}(F) = N.i \text{ At}$$

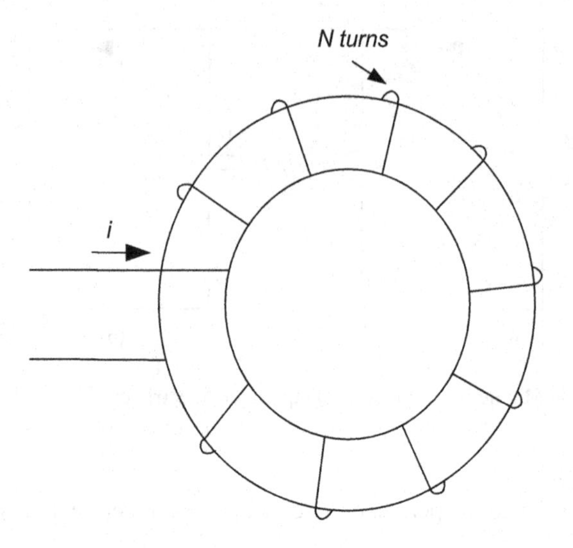

FIGURE 2.18 Toroidal ring of uniform cross section.

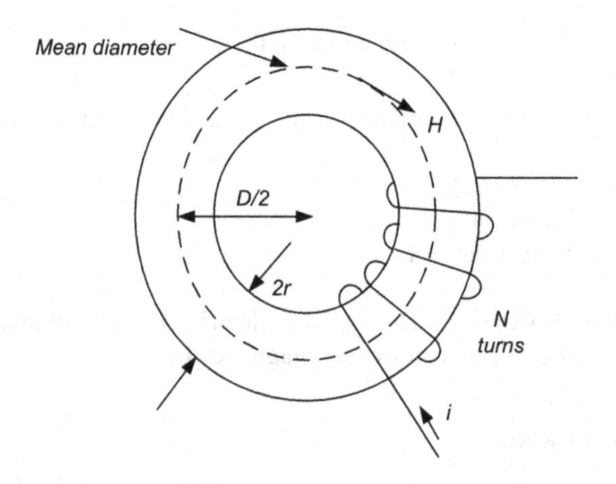

FIGURE 2.19 A toroidal ring with small diameters.

The reluctance is given by

$$\mathcal{R} = \frac{l}{\mu A} = \frac{2\pi \frac{D}{2}}{\mu \pi r^2} = \frac{D}{\mu r^2}$$

Hence,

$$\Phi = \frac{N.i}{\mathcal{R}} = \frac{N.i}{D/\mu r^2} = \frac{N.i.\mu r^2}{D}$$

Exercise 2.1:

Consider a toroidal core with an inside diameter of 0.60 cm and an outside diameter of 0.90 cm. The core cross section is rectangular, and its height is 0.35 cm. It is made of ferrite, having a permeability 100 times that of free space. The core flux density is to be 0.2 T. Determine the core flux and the number of ampere-turns (mmf) which must be wound on the core to produce this flux density.

Solution:

The cross-sectional area of the toroidal core is

$$A = 0.35 \times \frac{0.90 - 0.60}{2} = 0.35 \times 0.15 = 0.0525 \text{ cm}^2 = 525 \times 10^{-8} \text{ m}^2$$

Hence, the magnetic flux is

$$\Phi = B.A = 0.2 \times 525 \times 10^{-8} = 105 \times 10^{-8} \text{ Wb}$$

$$\text{The mean length of the magnetic flux} = \pi \times 0.75 \times 10^{-2} = 2.355 \times 10^{-2} \text{ m}$$

Then,

$$\mathcal{R} = \frac{l}{\mu A} = \frac{2.355 \times 10^{-2}}{4\pi \times 10^{-7} \times 100 \times 525 \times 10^{-8}} = 3571.4 \times 10^4 \text{ At/Wb}$$

$$\text{mmf} = \Phi \times \mathcal{R} = 105 \times 10^{-8} \times 3571.4 \times 10^4 = 37.5 \text{ At}$$

Exercise 2.2:

A coil has 200 turns wound over a ring having a mean circumference of 500 mm and a uniform cross-sectional area of 600 mm². Assume the core has $\mu_r = 1$ and the current flowing through the coil is 3.0 A, calculate

a. the magnetic field strength
b. the flux density
c. the total flux.

Solution:

$$\text{Mean circumference} = 500 \text{ mm} = 0.5 \text{ m}$$

$$H = \frac{3 \times 200}{0.5} = 1200 \text{ At/m}$$

$$\text{Magnetic flux density, } B = \mu_0 H = 4\pi \times 10^{-7} \times 1200 = 1507.2 \ \mu\text{T}$$

$$\text{Cross-sectional area} = 600 \text{ mm}^2 = 600 \times 10^{-6} \text{ m}^2$$

$$\text{Total magnetic flux} = B.A = 1507.2 \times 10^{-6} \times 600 \times 10^{-6} = 0.9 \ \mu\text{Wb}$$

2.7 MAGNETIC CIRCUITS WITH AIR GAPS

Energy conversion devices which include a moving element must have air gaps in their magnetic circuits (Figure 2.20). When air gap length g is much smaller than the dimensions of the adjacent core faces, magnetic flux Φ and magnetic flux density B can be assumed to be the same in the core and air gap. As a result, rules of the magnetic circuit analysis can be used in both the core and the air gap.

If the air gap length becomes large, the flux would tend to "leak out" of the air gap (Figure 2.21) and Φ and B will no longer be the same in the air gap and the core. In such cases, the rules of magnetic circuit analysis will no longer be applicable.

There are empirical methods to account for this fringing effect. Adding the gap length to each of the two dimensions of the air gap cross-sectional area could be one way of these correction methods in short air gaps. In practice, we neglect the effect of fringing fields. If fringing is neglected, we can take

$$A_{\text{gap}} = A_{\text{core}}$$

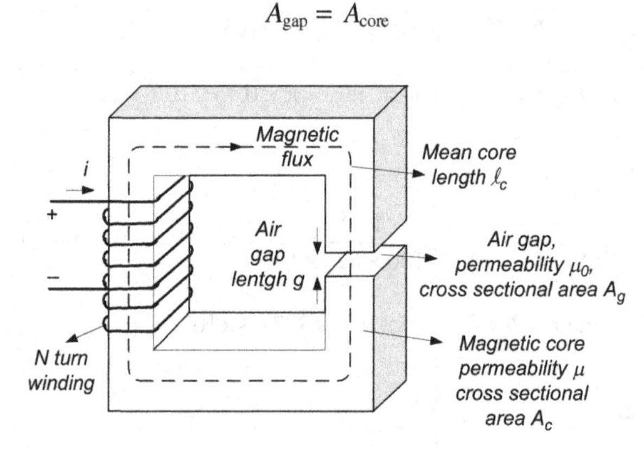

FIGURE 2.20 Magnetic circuit with an air gap.

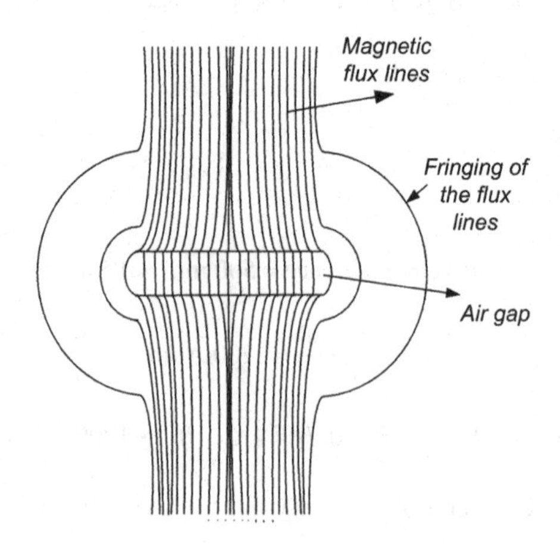

FIGURE 2.21 Fringing of the flux lines in the air gap.

Exercise 2.3:

The magnetic circuit in Figure 2.20 has $A_c = A_g = 10\,\text{cm}^2$, $g = 0.060\,\text{cm}$, $l_c = 40\,\text{cm}$, and $N = 1{,}000$ turns. Assuming core material's relative permeability as 80,000, find reluctances $\mathcal{R}_{\text{core}}$ and $\mathcal{R}_{\text{gap}}$. If the magnetic field density B is 1 T, determine the values of magnetic flux Φ and current i.

Solution:

a. Calculation of the reluctances:

$$\mathcal{R}_{\text{core}} = \frac{l_c}{\mu_r \mu_0 A_c} = \frac{0.4}{80000 \times 4\pi \times 10^{-7} \times 10 \times 10^{-4}} = 3.98 \times 10^3 \text{ At/Wb}$$

$$\mathcal{R}_{\text{gap}} = \frac{g}{\mu_0 A_g} = \frac{6 \times 10^{-4}}{4\pi \times 10^{-7} \times 10 \times 10^{-4}} = 4.78 \times 10^5 \text{ At/Wb}$$

$$\Phi = 1 \times 10 \times 10^{-4} = 1 \text{ mWb}$$

$$i = \frac{F}{N} = \frac{\Phi\left(\mathcal{R}_{\text{core}} + \mathcal{R}_{\text{gap}}\right)}{N}$$

$$= \frac{10 \times 10^{-4}\left(3.98 \times 10^3 + 4.78 \times 10^5\right)}{1000}$$

$$= \frac{10 \times 10^{-4}\left(4.82 \times 10^5\right)}{1000} = 0.482 \text{ A}$$

2.8 CIRCUIT LAWS OF MAGNETIC CIRCUITS

As we have said earlier, there exists an analogy between electrical and magnetic circuits. The equivalent of the two Kirchhoff laws in magnetic circuits can be stated as follows:

1. The net amount of the flux entering and leaving a junction in a magnetic circuit equals to zero (analogous to the Kirchhoff current law or KCL) (Figures 2.22 and 2.23).

$$\Phi - \Phi_1 - \Phi_2 = 0$$

2. Around any closed path in a magnetic circuit, the net mmf to force the flux through the closed circuit is the algebraic sum of the mmfs (analogous to the Kirchhoff voltage law or KVL) (Figure 2.24).

$$F = \text{net mmf} = N_1 i_1 + N_2 i_2$$

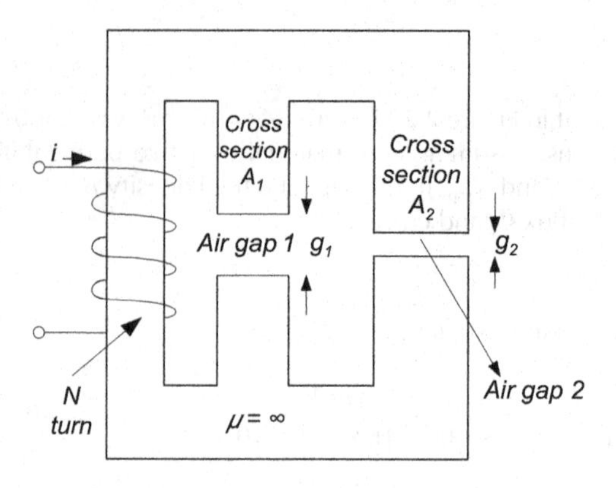

FIGURE 2.22 Magnetic circuit.

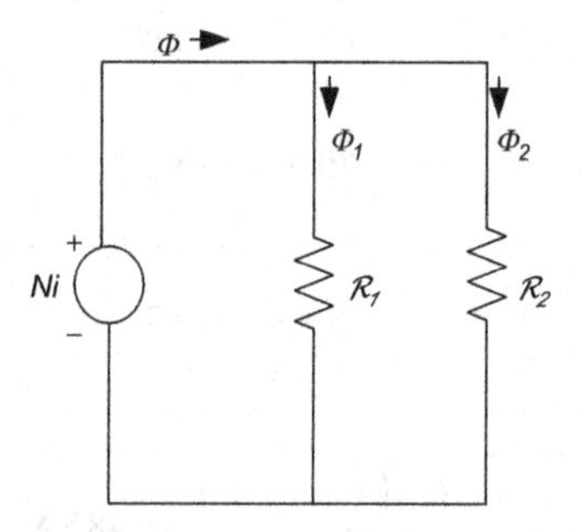

FIGURE 2.23 Electrical analog of Figure 2.22.

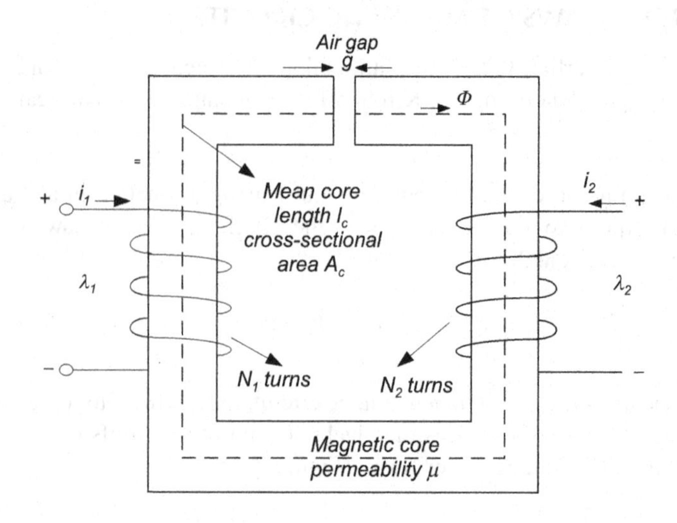

FIGURE 2.24 Algebraic sum of mmfs.

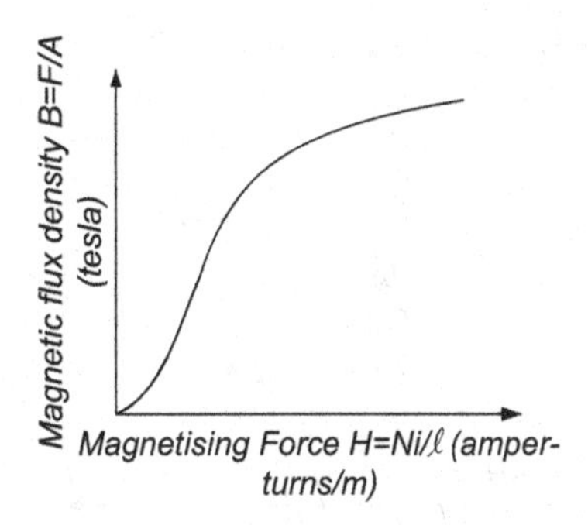

FIGURE 2.25 B-H curve of iron.

However, there are certain differences between DC resistive electrical circuits and magnetic circuits. For example, although resistance is associated with power loss, its equivalent in magnetic circuit reluctance is not. What is more, magnetic fluxes take leakage paths, whereas electric currents on the contrary do not.

Permeability and reluctance of a magnetic circuit made of a ferromagnetic material vary with the flux in that branch (Figure 2.25). For large values of flux, magnetic circuits seem to saturate. In other words for large values of mmf, larger change in mmf is required to produce the same change in flux for the smaller values of flux. Accepting the material permeability, a constant (B changes linearly with H) often gives acceptable results with reasonable accuracy. This is generally an acceptable approach. But for quantitative analysis, graphical methods are often preferred. As an aid to this approach manufacturers often give a curve of flux density B as a function of magnetic field intensity H together with the delivery of the material. It is simply called a B-H curve or magnetisation curve (Figure 2.25).

The B-H curve is independent of the dimensions of the material and is dependent on the material only. Throughout this book, we should keep in mind that the B-H curve for the air is linear (μ is constant).

2.9 CHARACTERISTICS OF MAGNETIC MATERIALS, HYSTERESIS

The B-H curve displays the magnetic properties of a material and shows how the material will respond to an external magnetic field. This information is important when designing a magnetic circuit.

When the mmf on the toroidal ring in Figure 2.19 is varied from $+F_1$ to $-F_1$, the magnetic intensity H in the magnetic circuit varies correspondingly from $+H_1$ to $-H_1$. If this cycle is repeated a number of times, the flux density then varies cyclically from $+B_1$ to $-B_1$. However, this does not happen as a single-valued function of H. B varies with H around the loop $a_1bcdefa_1$ as shown in Figure 2.26.

Decreasing magnetising force H_1 to zero would not decrease the magnetism on the material to zero. If magnetic field intensity H_1 is removed a residual magnetism equal to 0b remains on the material. This residual magnetism is removed only if the magnetising force is reversed and made equal to 0c (the coercive force). This phenomenon is known as hysteresis, and the loop $a_1bcdefa_1$ is called a hysteresis loop.

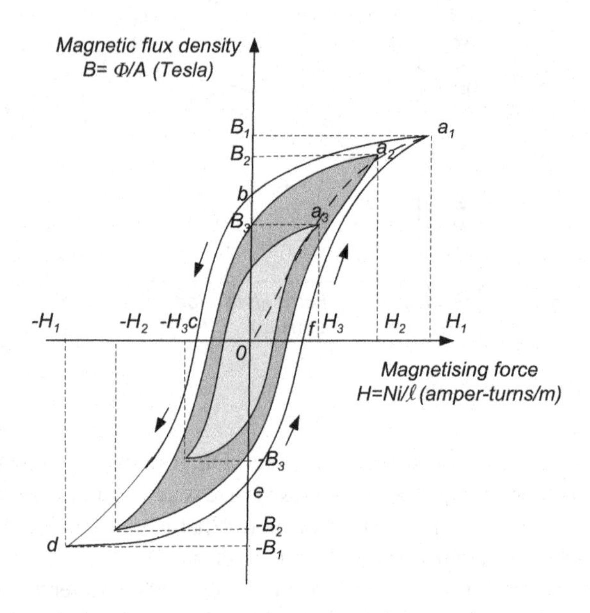

FIGURE 2.26 Hysteresis loop.

The coercivity can be thought of as the magnitude of the mmf required to demagnetise the material. From the hysteresis loop of Figure 2.26, it is clear that there is an element of uncertainty in the results of the analysis of the magnetic circuits. There are two values of *B* against a specific value of *H*, and which value is applicable depends on the previous magnetic history of the circuit. To determine the true value of *B*, we must look whether *H* is ascending or descending. The line $Oa_3a_2a_1$ joining the tips of successive hysteresis loops is usually taken to be the average (normal) magnetisation curve of that particular material. This is known as a DC or normal magnetisation curve.

Normal magnetisation curves for some common materials such as silicon sheet steel, cast iron, and cast steel are shown in Figure 2.27a. We must emphasise the effects of mechanical handling and heat treatment which greatly changes the magnetic properties of the magnetic materials. Therefore, different lots of a magnetic material may have varying magnetic properties. High precision must not be expected in magnetic circuit calculations. Alloys of iron with other metals are used when special magnetic properties are desired. For example, for a permanent magnet the material should have a relatively high remanent flux density. Alnico may fit well for this case. When very high permeability is desired, nickel–iron alloys known as Permalloy or Hipernik may be used. Perminvar which is a nickel–cobalt–iron alloy displays constant permeability over a large range of magnetisation.

Square loop magnetic materials (Figure 2.27b) are another group of magnetic materials, which are important for electronic control circuits and computers. These materials have a nearly rectangular hysteresis loop (see Figure 2.27b). The core is almost a bistable magnetic element because it can be in one of two states. Square-loop materials are used in switching circuits, as storage elements in computers and in special types of transformers in electronic circuits.

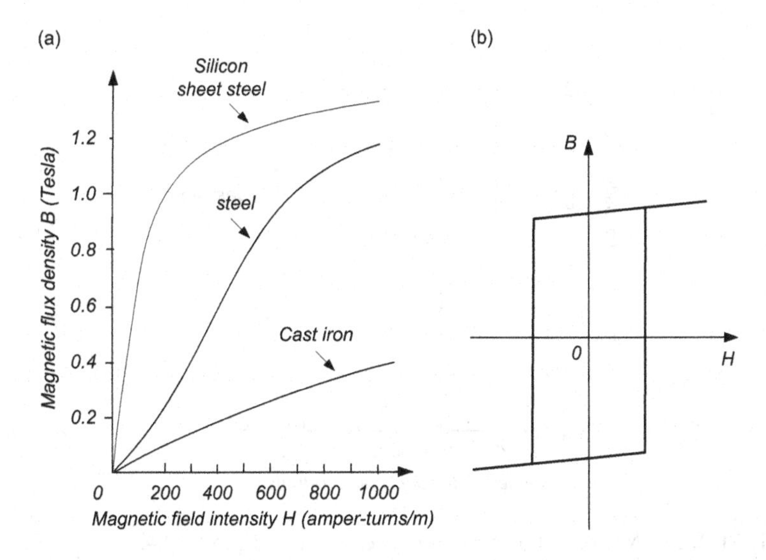

FIGURE 2.27 B-H curves of certain materials. (a) BH Curves of certain materials (b) Square loop magnetic material properties.

Exercise 2.4:

The core in the relay circuit of Figure 2.28 is cast steel (see Figure 2.29 for the magnetisation curve), and the mean core path is $l_c = 36\,\text{cm}$. A current i is made to flow in the coil of 500 turns to create a magnetic field of 1 T which is just enough to keep the air-gap lengths at 1 mm each.

a. find the current i that would have to flow in the coil.
b. calculate the permeability and relative permeability of the core.
c. if there is no air gap in the core, find the current in the coil for the same flux density (1 T) in the core (fringing can be neglected).

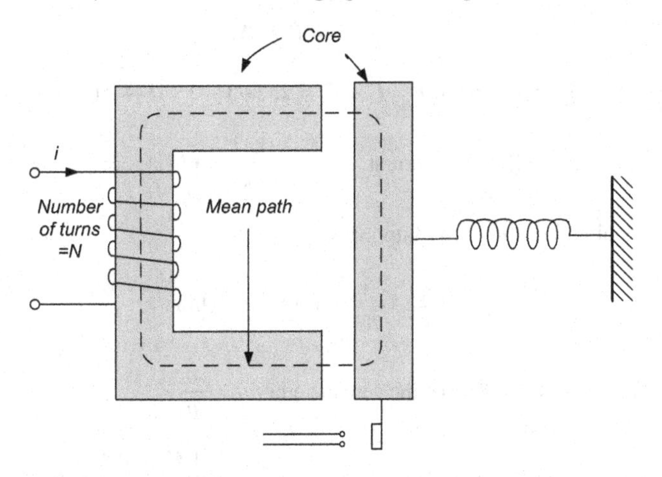

FIGURE 2.28

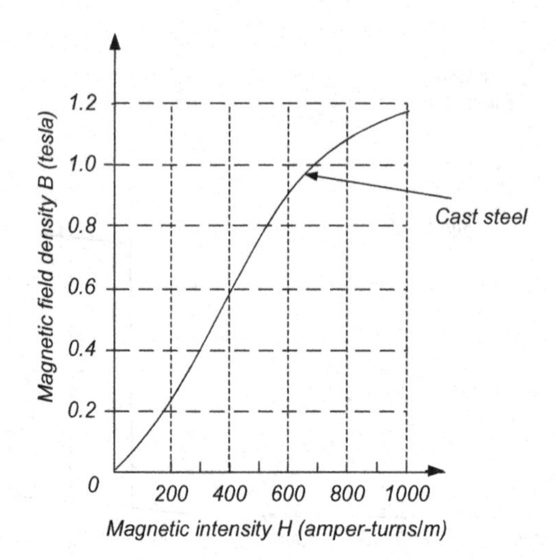

FIGURE 2.29 Magnetisation curve for the cast steel in Exercise 2.4.

Solution:

a. Magnetic field densities in the air gap and the core should be the same. From the B-H curve of the cast iron (Figure 2.29),

$$B_c = 1\,\mathrm{T} \rightarrow H_c = 700\ \mathrm{At/m}$$

$$\text{mmf in the core, } \rightarrow F_c = H_c l_c = 700 \times 0.36 = 252\ \mathrm{At}$$

$$\text{mmf in the air gap, } \rightarrow F_g = H_g\left(2l_g\right) = \frac{B_g}{\mu_0} \times 2l_g$$

$$= \frac{1}{4\pi \times 10^{-7}} \times 2 \times 10^{-3}$$

$$= 1592\ \mathrm{At}$$

$$\text{Necessary mmf, } F = F_c + F_g = 252 + 1592 = 1844\ \mathrm{At}$$

$$\text{Necessary current, } i = \frac{F}{n} = \frac{1844}{500} = 3.69\ \mathrm{A}$$

b. Permeability of the core material is

$$\mu_c = \frac{B_c}{H_c} = \frac{1}{700} = 1{,}42 \times 10^{-3}\ \mathrm{H/m}$$

$$\text{Relative permeability of the core material, } \mu_r = \frac{\mu_c}{\mu_0}$$

$$= \frac{1.42 \times 10^{-3}}{4\pi \times 10^{-7}} = 1130$$

c.

$$i = \frac{252}{500} = 0.5 \text{ A}$$

Exercise 2.5:

In the previous exercise, if the current i is 0.6 A for an air gap length of 1 mm, calculate the magnetic field density in the air gap.

Solution:

The B-H characteristic of air gap is linear, but the B-H characteristic of the core is not linear. Therefore, the mmf created by the 0.6 A current in the coil of 500 turns will be

$$N.i = H_g l_g + H_c l_c = \frac{B_g}{\mu_0} l_g + H_c l_c$$

$$B_c = B_g = -\mu_0 \frac{l_c}{l_g} H_c + Ni \frac{\mu_0}{l_g}$$

This represents a straight line called the "load line" (see Figure 2.30). If we plot this equation on the B-H plot, the slope would be

$$\text{slope, } g = -\mu_0 \frac{l_c}{l_g} = -4\pi \times 10^{-7} \times \frac{360}{1} = -4.52 \times 10^{-4}$$

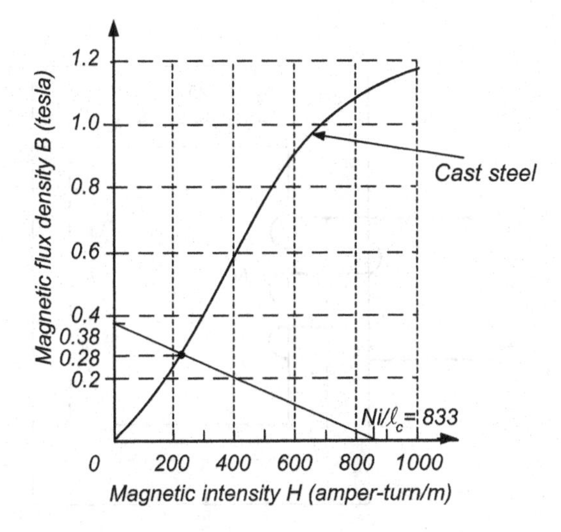

FIGURE 2.30 B-H Curve and load line for Exercise 2.5.

Intersection on the *B* axis would be

$$B = Ni\frac{\mu_0}{l_g} = \frac{500 \times 0.6 \times 4\pi \times 10^{-7}}{10^{-3}} = 0.38 \text{ T}$$

Intersection on the *H* axis would be

$$H = \frac{Ni}{l_c} = \frac{500 \times 0.6}{0.36} = 833 \text{ At/m}$$

The load line would intersect the B-H curve at $B = 0.28$ T.

2.10 MAGNETICALLY INDUCED VOLTAGES; SELF-INDUCTANCE

Previously, we saw that when a current passes through a straight wire, a magnetic field appears around it which is proportional to the amount of current that passes through the wire (Figure 2.31). If we reverse the operation and move the piece of wire in a magnetic field, we will notice a voltage induced in the wire. We can create the same effect by altering the magnetic field surrounding the piece of wire. In fact, it is this simple discovery of Faraday that led to the electrical power generation for the utilities of today. Transformers, motors, and generators all depend on electromagnetic induction of voltage for their operation.

The effect is described by Faraday's law as "any change in the magnetic field around a coil of wire will induce a voltage (emf) in the coil". For an N-turn coil, this can be expressed mathematically as

$$e = n\frac{d\Phi}{dt} = \frac{d\lambda}{dt}$$

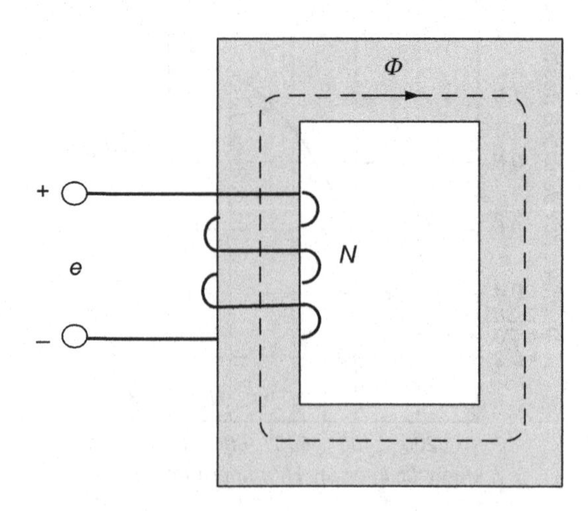

FIGURE 2.31 Magnetic field produced by current flowing in the coil.

Here, $\lambda = n\Phi$ is the magnetic flux linkage, Φ is the magnetic flux linking the coil, and t represents the time. In the above equation, they both are expressed in webers (Wb), and time t is in seconds. The direction of the induced voltage is described by Lenz's law. According to Lenz's law, the direction of the induced voltage is such as to produce a current in the short-circuited coil opposing the flux change.

In the previous chapter, when discussing magnetic circuits with DC excitation, no consideration was given to this induced voltage as it is then zero. However, it plays an important role in circuits with AC excitation and in all types of electrical machines.

The induced voltage (e) appears in any circuit which may be linked by the changing flux, including the circuit giving rise to the flux (Figure 2.32).

Thus, current in a circuit produces a magnetic field linking the circuit. Any change in current with time causes corresponding changes of the magnetic flux and induces a voltage in the circuit.

$$e = N\frac{d\Phi}{dt} = N\frac{d\Phi}{di}\frac{di}{dt} = L\frac{di}{dt}$$

Here,

$$L = N\frac{d\Phi}{di}$$

where L is the self-inductance (or simply the inductance) of the electrical circuit. If Φ is in webers (Wb), then self-inductance L is stated in Henries (H). In a ferromagnetic circuit if no significant part of the path of the flux Φ is in saturated iron, then i is directly proportional to Φ:

$$L = \frac{N\Phi}{i} = \frac{NBA}{i} = \frac{N\mu HA}{i}$$

$$= \frac{N\mu HA}{Hl/N} = \frac{N^2}{l/\mu A}$$

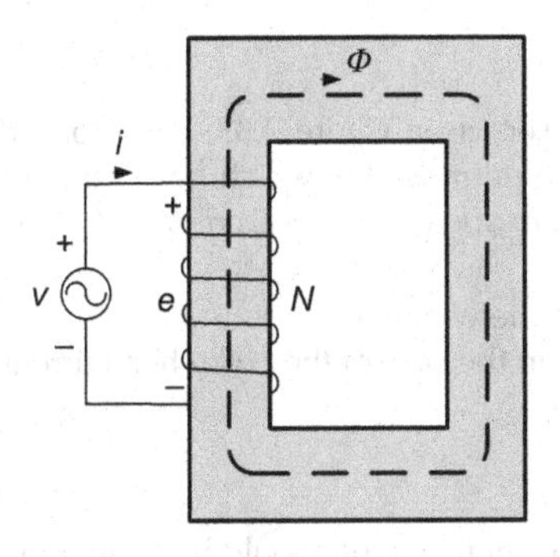

FIGURE 2.32 Induced voltage e due to changing flux.

Hence for a static magnetic circuit with *N* turns, composed of magnetic material of constant magnetic permeability,

$$L = \frac{N^2}{l/\mu A} = \frac{N^2}{\mathcal{R}}$$

Here, $\mathcal{R}$ is the total reluctance of the magnetic circuit including any air gaps. Remember that for air gaps, the magnetic permeability is also a constant. The units of reluctance can be shown as H^{-1} meaning the inverse of Henries. For a static magnetic circuit of constant magnetic permeability, the inductance *L* is fixed and the induced voltage equation becomes

$$e = L\frac{di}{dt}$$

In electromechanical energy conversion devices, magnetic circuit is not static and therefore inductances usually change with time. Therefore, it would be better to write the above equation as follows:

$$e = L\frac{di}{dt} + i\frac{dL}{dt}$$

In practise, in static magnetic circuits with air gaps,

- area *A* is generally accepted to be the same in the core and gap.
- nonlinear effects of the magnetic core material can be ignored (μ_c and R_c constant).
- *B* (magnetic field density) is the same in the core and gap.
- Air gap dominates the circuit ($\mu_0 = \mu_{gap} =$ constant).
- *H* (magnetic field intensity) is different in the core and gap (μ is different).
- Φ = Flux = *B.A* is generally accepted to be the same in the core and gap.

Exercise 2.6:

In the magnetic circuit shown in Figure 2.33, a coil of 500 turns is wound on a magnetic core of infinite permeability which has two parallel air gaps of lengths $x_{g1} = 1$ mm and $x_{g2} = 2$ mm and areas $S_1 = 2\,\text{cm}^2$ and $S_2 = 4\,\text{cm}^2$, respectively. Find

a. the inductance of the winding
b. the flux density *B* in the core of the coil when current $i = 5$ A flows in the coil.

Solution:

The equivalent of the magnetic circuit would be as shown in Figure 2.34

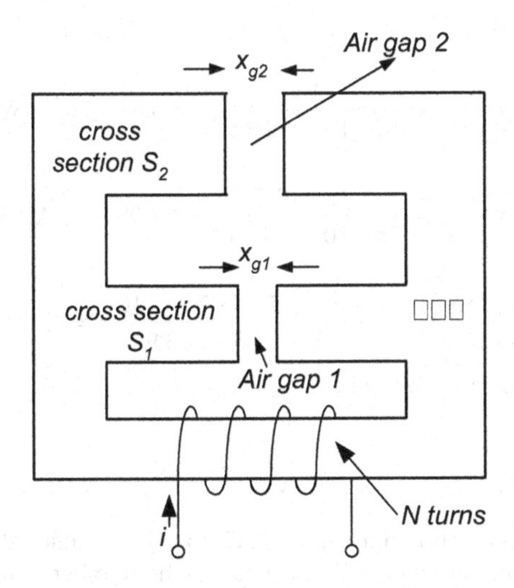

FIGURE 2.33

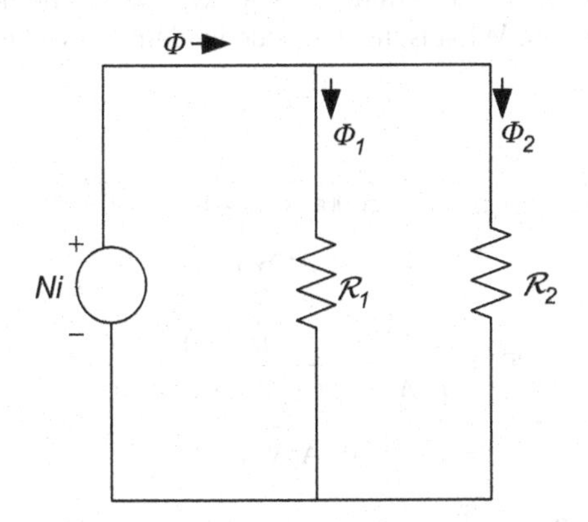

FIGURE 2.34

a.

$$L = \frac{\lambda}{i} = \frac{N\Phi}{i} \quad ; \quad \Phi = \frac{Ni}{\dfrac{\mathcal{R}_1\mathcal{R}_2}{\mathcal{R}_1 + \mathcal{R}_2}}$$

$$L = \frac{N}{i} \times \Phi = \frac{N}{i} \times \frac{Ni}{\dfrac{\mathcal{R}_1\mathcal{R}_2}{\mathcal{R}_1 + \mathcal{R}_2}} = \frac{N^2(\mathcal{R}_1 + \mathcal{R}_2)}{\mathcal{R}_1\mathcal{R}_2} = \mu_0 N^2 \left[\frac{S_1}{x_{g1}} + \frac{S_2}{x_{g2}}\right]$$

$$L = 4\pi \times 10^{-7} \times 500^2 \left[\frac{2 \times 10^{-4}}{10^{-3}} + \frac{4 \times 10^{-4}}{2 \times 10^{-3}}\right] = \pi \times 10^{-1} \times 0.4 = 126 \text{ mH}$$

b.

$$\mathcal{R}_1 = \frac{x_{g1}}{\mu_0 S_1} = \frac{10^{-3}}{4\pi \times 10^{-7} \times 2 \times 10^{-4}} = 3.98 \times 10^6 \text{ At/Wb}$$

$$\mathcal{R}_2 = \frac{x_{g2}}{\mu_0 S_2} = \frac{2 \times 10^{-3}}{4\pi \times 10^{-7} \times 4 \times 10^{-4}} = 3.98 \times 10^6 \text{ At/Wb}$$

$$\Phi = \frac{500 \times 5}{\dfrac{(3.98 \times 10^6)(3.98 \times 10^6)}{3.98 \times 10^6 + 3.98 \times 10^6}} = \frac{5 \times 5 \times 7.96 \times 10^{-4}}{3.98 \times 3.98} = 12.6 \times 10^{-4} \text{ Wb}$$

Exercise 2.7:

In the magnetic circuit shown in Figure 2.35, the cross-sectional area of the core is 10 cm². The mean core length l_c is 40 cm and the number of turns in the coil is $N = 400$. Assuming no fringing effects in the air gap, calculate the inductance of the winding for the relative permeability of the core material of $\mu_r = 75{,}000$ if the air gap length $l_g = 0.6$ mm. If current i is 5 A, what would be the magnetic field density B in the air gap? What is the magnetic field intensity H in the core?

Solution:

$$\mathcal{R}_c = \frac{l_c}{\mu_r \mu_0 A_c} = \frac{0.4}{75{,}000 \times 4\pi \times 10^{-7} \times 10 \times 10^{-4}}$$

$$= 4{,}246 \text{ At/Wb}$$

$$\mathcal{R}_g = \frac{l_g}{\mu_0 A_g} = \frac{0.6 \times 10^{-3}}{4\pi \times 10^{-7} \times 10 \times 10^{-4}}$$

$$= 4.77 \times 10^5 \text{ At/Wb}$$

$$\mathcal{R}_{\text{total}} = 4.77 \times 10^5 + 4244 = 4.81 \times 10^5 \text{ At/Wb}$$

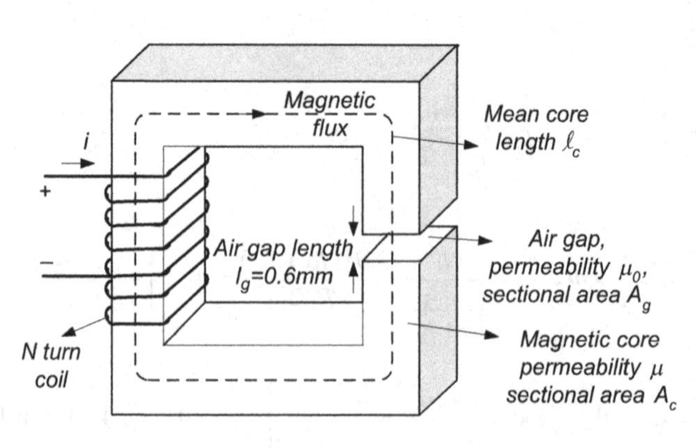

FIGURE 2.35

Hence, the inductance L would be

$$L = \frac{N^2}{\mathcal{R}_{\text{total}}} = \frac{400^2}{4.81 \times 10^5} = 0.33 \text{ Henry}$$

$$\Phi = \frac{Ni}{\mathcal{R}} = \frac{400 \times 5}{4.81 \times 10^5} = 4.16 \text{ mWb}$$

$$B_g = B_c = \frac{\Phi}{A_g}$$

$$= \frac{4.16 \times 10^{-3}}{10 \times 10^{-4}} = 4.16 \text{ T}$$

$$H_c = \frac{B_c}{\mu_r \mu_0}$$

$$= \frac{4.16}{7500 \times 4\pi \times 10^{-7}} = 441.6 \text{ At/m}$$

We see in Exercise 2.7 that the effect of the core on the total value of the reluctance is negligible. Therefore to avoid unnecessary complications, we can assume the permeability of the core as infinity ($\mu_c \to \infty$) when calculating the inductance of similar magnetic circuits.

This way we assume that inductance is only dependent on the dimensions of the air gap that is present in the core. A coil wound around magnetic forms is called inductor in which i is directly proportional to Φ.

$$L = \frac{N\Phi}{i}$$

L is named as the self-inductance of the inductor and is independent of current and dependent only on the geometry of the circuit element (Figure 2.36a) and the permeability of the magnetic medium.

2.11 MUTUAL INDUCTANCE

Figure 2.37 shows a magnetic circuit with an air gap and two windings. We saw previously that the total mmf acting on the magnetic circuit would be given by the addition of the net ampere-turns of both windings. The total mmf is therefore

$$\text{mmf} = N_1 i_1 + N_2 i_2$$

and with the reluctance of the core neglected and assuming no fringing of the magnetic field in the air gap so that $A_c = A_g$ the core flux is

$$\Phi = (N_1 i_1 + N_2 i_2)\frac{\mu_0 A_c}{g}$$

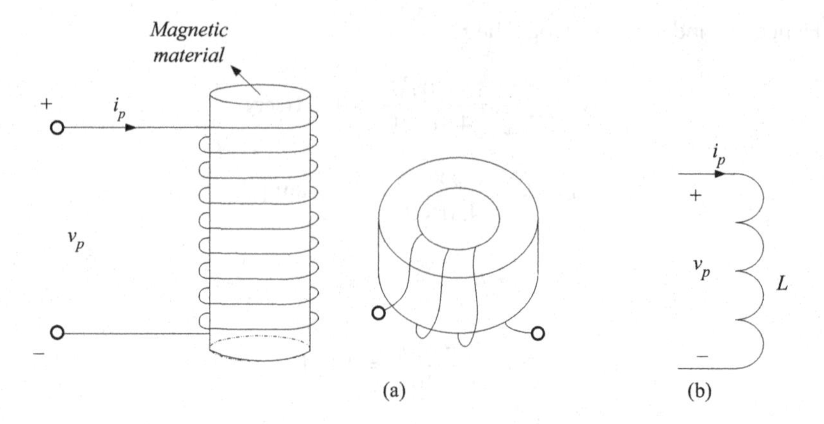

FIGURE 2.36 (a) Various inductors and (b) symbol of inductor.

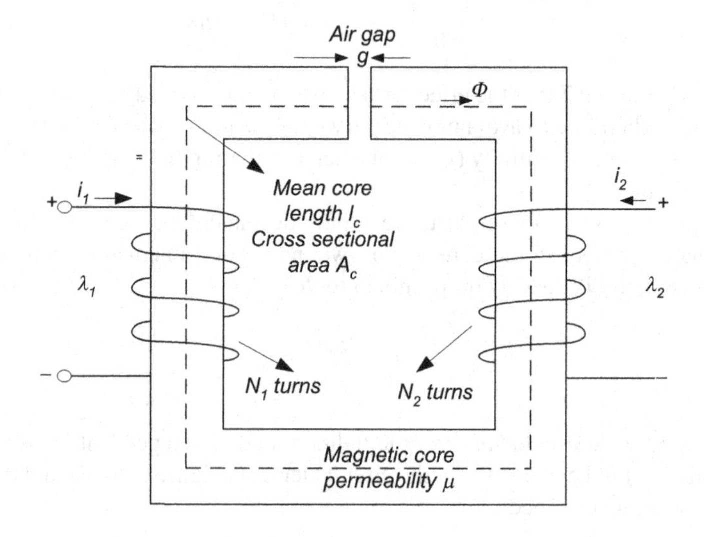

FIGURE 2.37 Magnetic circuit with two windings.

The total magnetic flux linkage in coil 1

$$\lambda_1 = N_1\Phi = N_1\left(N_1 i_1 + N_2 i_2\right)\frac{\mu_0 A_c}{g}$$

$$= N_1^{\,2}\left[\frac{\mu_0 A_c}{g}\right] i_1 + N_1 N_2\left[\frac{\mu_0 A_c}{g}\right]$$

We can write this as

$$\lambda_1 = L_{11} i_1 + L_{12} i_2$$

where

$$L_{11} = N_1^2\left[\frac{\mu_0 A_c}{g}\right]$$

is the self-inductance of coil 1, and $L_{11}i_1$ is the flux linkage of coil 1 due to its own current i_1. The mutual inductance between coils 1 and 2 is

$$L_{12} = \text{Mutual inductance} = N_1N_2\left[\frac{\mu_0 A_c}{g}\right]$$

and $L_{12}i_2$ is the flux linkage of coil 1 due to current i_2 in the other coil. Similarly, the flux linkage of coil 2 is

$$\lambda_2 = N_2\Phi = N_2(N_1i_1 + N_2i_2)\frac{\mu_0 A_c}{g}$$

$$= N_2^2\left[\frac{\mu_0 A_c}{g}\right]i_2 + N_1N_2\left[\frac{\mu_0 A_c}{g}\right]i_1$$

We can write this as

$$\lambda_2 = L_{22}i_2 + L_{21}i_1$$

where,

$$L_{21} = L_{12} = \text{Mutual inductance} = M = N_1N_2\left[\frac{\mu_0 A_c}{g}\right] = \frac{N_1N_2}{\mathcal{R}}$$

and $L_{21}i_1$ is the flux linkage of coil 2 due to current i_1 in the other coil. Also

$$L_{22} = N_2^2\left[\frac{\mu_0 A_c}{g}\right]$$

is the self-inductance of coil 2, and $L_{22}i_2$ is the flux linkage of coil 2 due to its own current i_2. If all the flux produced by one coil is linked with the other coil, mutual inductance can also be expressed as

$$M = \frac{N_1N_2}{\mathcal{R}} = \sqrt{L_{11}L_{22}}$$

All of the flux produced by one coil may not fully link with the other coil. In other words, there may be some flux leakage between the windings. Therefore, below equation is a better definition of mutual inductance where k is known as the coupling coefficient of the two coils.

$$M = k\sqrt{L_{11}L_{22}}$$

If i and Φ change proportionally with each other, L and M are constants. In short, we can express the flux linking each coil as

$$\lambda_1 = L_1 i_1 \pm M i_2$$

$$\lambda_2 = L_2 i_2 \pm M i_1$$

According to Faraday's law, the induced voltages in each coil would then be

$$e_1 = \frac{d\lambda_1}{dt} = L_1 \frac{di_1}{dt} \pm M \frac{di_2}{dt}$$

$$e_2 = \frac{d\lambda_2}{dt} = L_2 \frac{di_2}{dt} \pm M \frac{di_1}{dt}$$

2.12 DOT CONVENTION

According to the dot convention, "currents entering the dotted terminals produce aiding magnetic flux" (Figure 2.38). This is in accordance with right-hand rule. Thus, if both currents enter (or if both leave) the dotted terminals, the mutual flux linkages add to the self-flux linkages. In contrast, if one current enters a dotted terminal and the other leaves, the mutual flux linkages have a minus sign.

Exercise 2.8:

Two coils are wound on a toroidal core of reluctance 2×10^7 At/Wb as shown in Figure 2.39. Find the self-inductances and mutual inductance of the coils. Assume that coefficient of coupling $k = 1$.

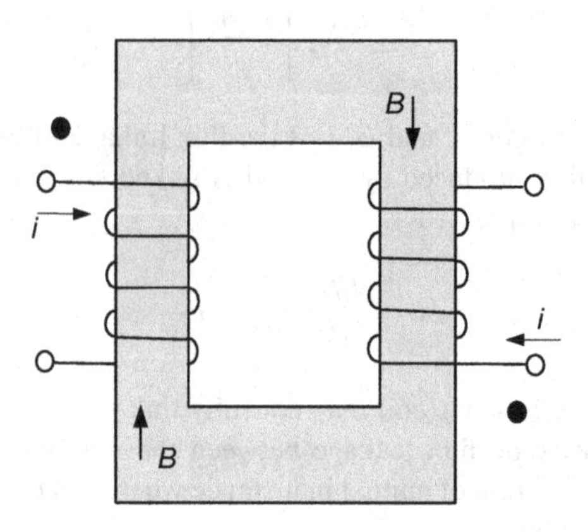

FIGURE 2.38 Dot convention.

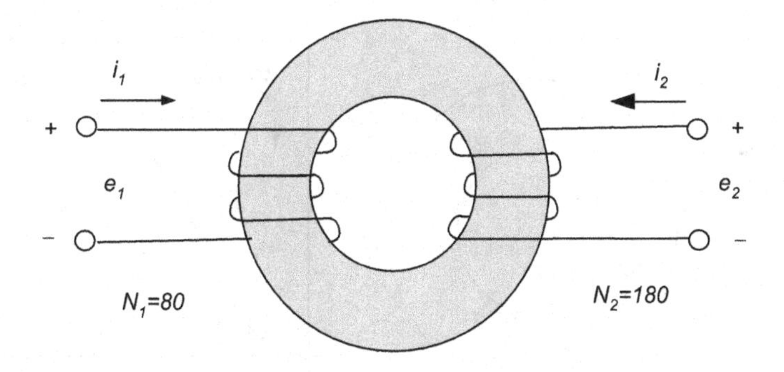

FIGURE 2.39

Solution:

$$L_1 = \frac{N_1^2}{\mathcal{R}} = \frac{80^2}{2\times10^7} = 0.32\text{ mH}$$

$$L_2 = \frac{N_2^2}{\mathcal{R}} = \frac{180^2}{2\times10^7} = 1.62\text{ mH}$$

$$M = \sqrt{L_1L_2} = \sqrt{0.32\times1.62} = 0.72\text{ mH}$$

Alternatively,

$$\Phi_1 = \frac{80\times i_1}{2\times10^7} = 4\times10^{-6}i_1$$

$$\lambda_{21} = 180\times\Phi_1 = 180\times4\times10^{-6}i_1$$

$$M = \frac{\lambda_{21}}{i_1} = 180\times4\times10^{-6} = 0.72\text{ mH}$$

2.13 ENERGY STORED IN PER UNIT VOLUME OF THE MAGNETIC FIELD AND HYSTERESIS LOOP

Let us consider the energy flow to the coil which we assume to have no resistive losses as shown in Figure 2.40a. As the current into the coil increases, the flux density increases and this induces a voltage, resulting in energy flow (W) into the coil. Now let us calculate this energy flow at time t by integrating the power flow into the coil between times $t = 0$ and t.

$$W = \int_0^t v.i.dt = \int_0^t N\frac{d\Phi}{dt}.i.dt = \int_0^t Ni.d\Phi$$

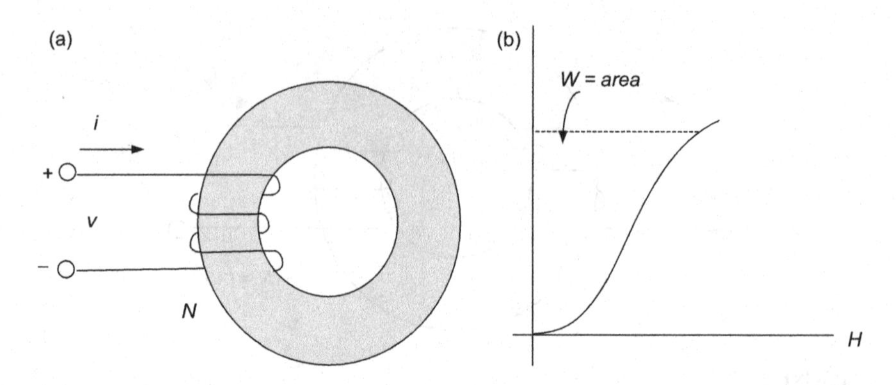

FIGURE 2.40 (a and b) Energy flow into a coil.

Now according to Ampere's circuital law, we can replace Ni (mmf) in this equation by Hl where

$$l = \text{mean core length}$$

and $d\Phi$ by AdB where

$$A = \text{cross-sectional area}$$

Then, the above equation becomes

$$W = \int_0^t HlA.dB$$

Now, since lA gives us the volume of the core, we can write that

$$w = \text{Energy stored per unit volume of the core} = \int_0^t H.dB \tag{2.1}$$

As shown in Figure 2.40b, energy per unit volume delivered to the coil (w) is equal to the area between B-H curve and the B-axis. When H is reduced back to zero, stored energy decreases to zero (energy given back to source). This released energy (cdec) (Figure 2.41) is less than the stored energy (abcdea) in the core.

The difference (abce) is the energy converted to heat in the process of magnetising the core. This is called the hysteresis loss. Therefore, the area of the hysteresis loop is equal to the energy converted to heat during magnetisation of the core's per unit volume per cycle. On the other hand, the released energy (cdec) is the reactive energy which is cyclically supplied and recalled back by the excitation source to the magnetic circuit. Notice that this reactive power is not dissipated but swings back and forth between the magnetic core and the source.

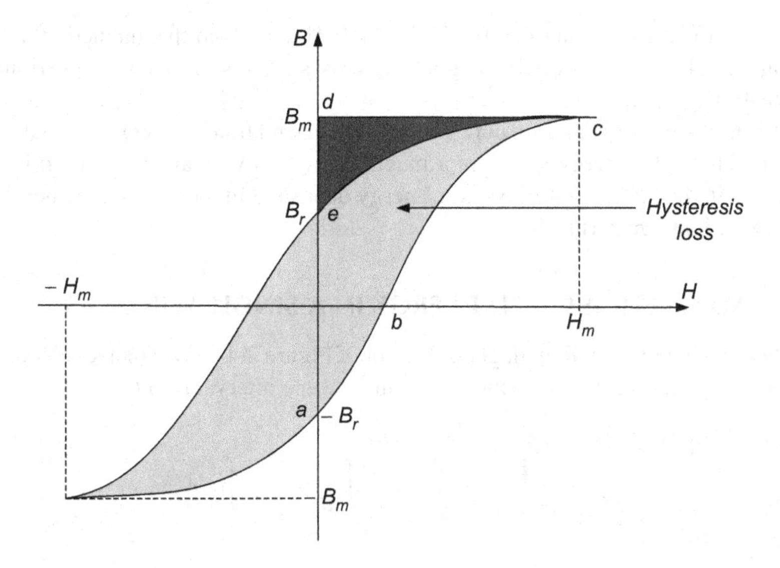

FIGURE 2.41 Hysteresis curve.

As long as the cores are operated below their saturation levels, the linear approximation would be sufficiently accurate. In such cases, the equation between magnetic field B and magnetic intensity H (Figure 2.42) would be

$$B = \mu H$$

Substituting this into equation (2.1) and integrating, we obtain the energy per unit volume of the core W_v:

$$W_v = \frac{W}{Al} = \int_0^t H.dB = \int_0^t \frac{B}{\mu}.dB = \frac{B^2}{2\mu}$$

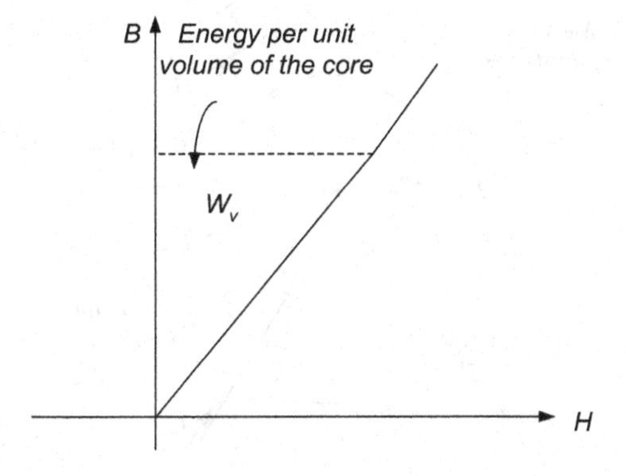

FIGURE 2.42 Linear B-H curve and energy deposited per unit volume.

This in fact is the area enclosed by the linear B-H curve and the magnetic flux axis *B* (Figure 2.42). Notice that the magnetic energy stored is inversely proportional to permeability μ of the core.

The permeability of an iron core is several thousand times greater than that of the air. Therefore, the energy stored in a magnetic system with air gap is mainly concentrated in the air gap of the system. Energy deposited in the core can be neglected because of its large permeability.

2.14 MAGNETIC STORED ENERGY IN A SINGLE WINDING

Consider the single winding magnetic circuit of Figure 2.43. The change in magnetic stored energy ΔW in the magnetic circuit in the time interval t_1 to t_2 is

$$\Delta W = \int_{t_1}^{t_2} p.dt = \int_{t_1}^{t_2} vi.dt = \int_{\lambda_1}^{\lambda_2} \frac{d\lambda}{dt} i.dt = \int_{\lambda_1}^{\lambda_2} i.d\lambda$$

For a single winding system of constant inductance (i.e., a linear system), this equation becomes

$$\Delta W = \int_{\lambda_1}^{\lambda_2} i.d\lambda = \int_{\lambda_1}^{\lambda_2} \frac{\lambda.}{L} d\lambda = \frac{1}{2L}\left(\lambda_2^2 - \lambda_1^2\right)$$

The total magnetic stored energy at any given value of λ can be found from setting λ_1 equal to zero:

$$\Delta W = \int_{\lambda_1}^{\lambda_2} \frac{\lambda.}{L} d\lambda = \frac{1}{2L}\lambda^2 = \frac{1}{2} i^2 L$$

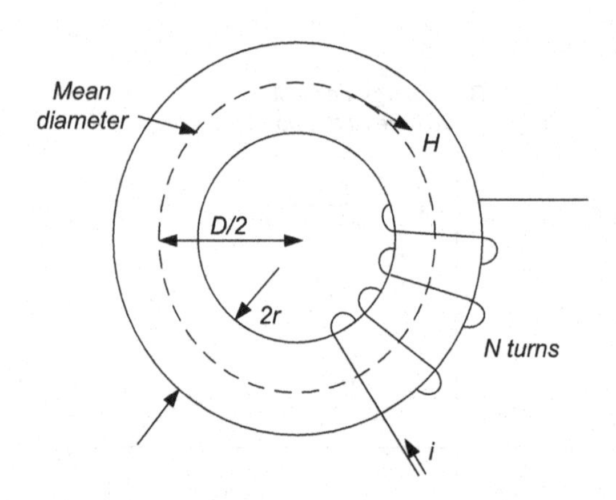

FIGURE 2.43 Single winding magnetic system.

The same result can also be obtained by multiplying the energy stored per unit volume of the magnetic field w_f with the volume of the magnetic material V:

$$W_v = \frac{B^2}{2\mu} \times \text{Volume}(V)$$

$$= \frac{(\mu H)^2}{2\mu} \times A \times l = \frac{\mu H^2}{2} \times A \times l = \frac{\mu}{2} \times \left(\frac{Ni}{l}\right)^2 \times A \times l$$

$$= \frac{i^2}{2} \times \frac{\mu A N^2}{l} = \frac{Li^2}{2}$$

- assume now there is an air gap in the magnetic circuit (Figure 2.44).
- the magnetic core is ideal, ($\mu = \infty$) (no mmf drop).
- the stored magnetic energy contained entirely in the air gap.
- the flux density B is the same in the iron core as in the air gap.
- the λ–i characteristic of the air gap is essentially linear.

With these assumptions, the total magnetic energy of the magnetic circuit is

$$\text{Energy in the air gap} = W_{fg} = \frac{B^2}{2\mu_0} \times \text{Air gap volume}(V)$$

$$= \text{Energy per unit volume of the air gap} \times \text{air gap volume}$$

We can see the plot of this equation in Figure 2.45.

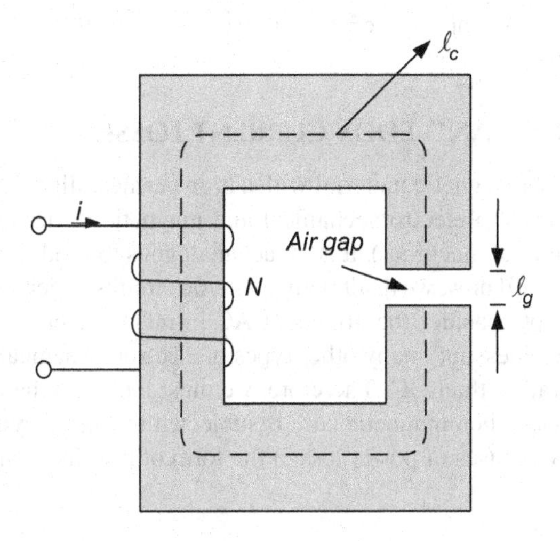

FIGURE 2.44 Magnetic circuit with air gap..

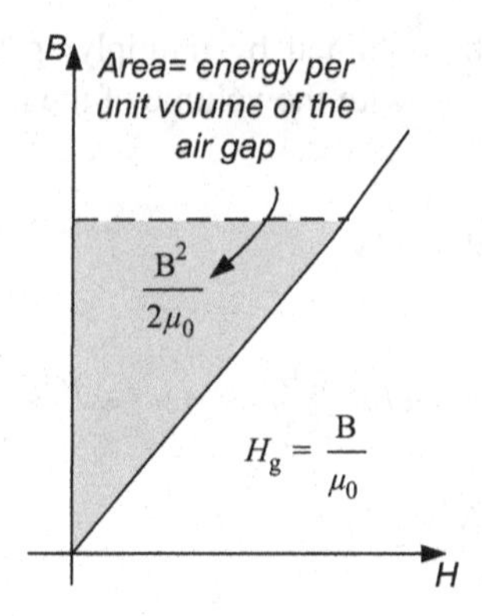

FIGURE 2.45 Energy stored per unit volume of the air gap.

Exercise 2.9:

The air gap inside a square cross section magnetic core has a cross-sectional area of $3\times3\,\text{cm}^2$ and the air gap length is 0.6 cm. If the coil wound on the core has 450 turns and the current driven through it is 2 A, calculate the magnetic field density B_g in the air gap? (Assume the magnetic permeability of the core as infinity.) How much energy is deposited in the air gap?

$$2\times450 = H_g.l_g = \frac{B_g}{\mu_0}\times l_g \quad \to B_g = \frac{900\times4\pi\times10^{-7}}{0.6\times10^{-2}} = 0.19\ \text{T}$$

$$w_v = \frac{B_g^{\;2}}{2\mu_0} = \frac{0.19^2}{2\times4\pi\times10^{-7}} = 14.3\times10^3\ \text{J/m}^3$$

$$W = w_v \times \text{Air gap volume} = 14\times10^3\times\left(3\times3\times0.6\times10^{-6}\right) = 0.076\ \text{J}$$

2.15 HYSTERESIS AND EDDY CURRENT LOSSES

A magnetic core is a magnetic material with a high permeability used to confine and guide magnetic fields in electromechanical and magnetic devices (electromagnets, transformers, electrical machines). It is in fact analogous to conductive wires in electrical circuits. Up until now, we dealt with magnetic circuits which are excited by DC sources. We did not consider the effects of AC in magnetic circuits. Transformers and electrical machines, and many other types of electromechanical devices operate from AC source rather than DC. Therefore, we must look into the effects of AC on the magnetic cores. When magnetic core is subjected to time-varying flux densities (AC), there are two causes of power loss in the form of heat in the iron core:

1. hysteresis loss
2. eddy current loss.

These two losses comprise the total core loss which is generally termed as iron loss. These losses are important because they play an important role in the performance of the electromechanical devices.

2.15.1 Hysteresis Loss

This is associated with the phenomenon of hysteresis. Hysteresis loss is due to heat that occurs in reversing the directions of the magnetic domains when trying to magnetise ferromagnetic materials. All ferromagnetic materials exhibit this phenomenon called hysteresis which is defined as the lagging of magnetisation or flux density (*B*) behind the magnetising force (*H*) (Figure 2.46).

We can also define it as the energy dissipated in a magnetic substance on the reversal of its magnetism. As we saw earlier, the volumetric energy converted to heat due to hysteresis per cycle is equal to the area of the hysteresis loop. The power loss due to hysteresis would then be

$$\begin{aligned} P_{\text{hysteresis}} &= \text{Number of loops traversed per second} \\ &\quad \times \text{Area of the hysteresis loop} \times \text{Volume} \\ &= \text{Frequency} \times \text{Area of the hysteresis loop} \times \text{Volume} \end{aligned}$$

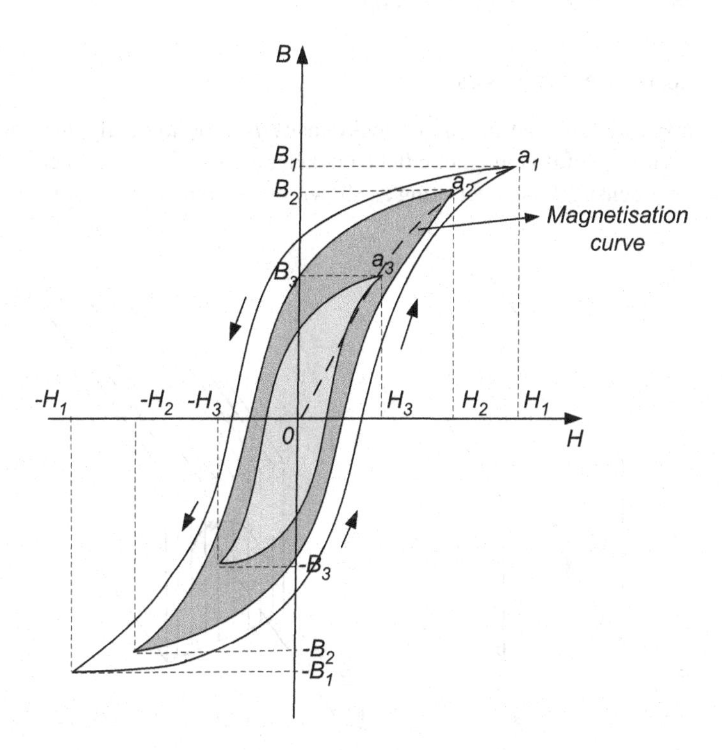

FIGURE 2.46 Hysteresis loss due to changing mmf.

Exercise 2.10:

A 60 Hz AC source is connected to a coil wound on an iron core. If the hysteresis loop of the core material has an area of 70 J/m^3 and the volume of the core is 250 cm^3, calculate the hysteresis power converted to heat.

Solution:

$$\text{Hysteresis loss per cycle} = 250 \times 10^{-6} \times 70 = 0.0175\ \text{J}$$

$$\text{Hysteresis power loss} = 0.0175 \times 60 = 1.05\ \text{W}$$

Hysteresis loss per unit volume of core material may be approximated empirically by the following expression:

$$P_h = f k_h B_m^n \text{ watts}$$

where

- k_h is a characteristic constant of the core.
- f is the frequency.
- B_m is the maximum flux density.
- n is the Steinmetz exponent (1.5 to over 2.0, usually taken as 1.6).
- $k_h B_m{}^n$ is the area of hysteresis loop.

2.15.2 Eddy Current Losses

Suppose in Figure 2.47 the magnetic field density B is periodically changing with time. In the vicinity of this magnetic field, consider a cross section of a core through which the flux density B is penetrating as shown in the same figure. Consider a path

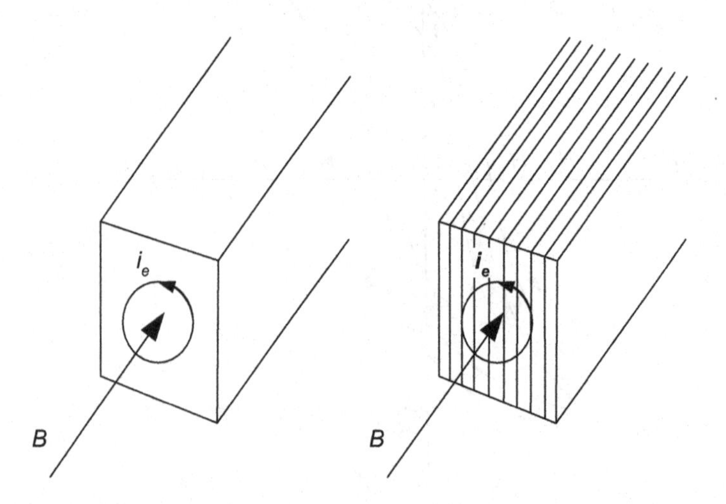

FIGURE 2.47 Eddy currents.

in this cross section. Voltage will be induced in the path because of the time variation of flux. Consequently, a current i_e, known as an eddy current, will flow around the path.

Because core material has resistance, this eddy current will cause a power loss in the core material. Eddy currents circulate in the core material and according to Lens' law oppose changes in flux density in the material. Therefore to counteract this demagnetising effect, the current (and hence the magnetic intensity H) in the exciting winding must increase.

Therefore, the resultant "dynamic" B-H loop under AC operation is somewhat "larger" than hysteresis loop for smaller frequency conditions, and this effect increases as the excitation frequency is increased. It is for this reason that the characteristics of electrical steels vary with frequency and manufacturers' supply the relevant information with their delivery of magnetic steel.

Eddy current loss can be reduced in two ways:

1. A high-resistivity core material may be used. Addition of a few percent of silicon (4%) to iron will increase the resistivity significantly.
2. A laminated core may be used. The thin laminations are insulated from each other. The lamination is made in the plane of the flux. In transformers and electric machines, the magnetic cores that carry time-varying flux are normally laminated. The laminated core structure is shown in Figure 2.47.

The eddy current loss per volume in a magnetic core subjected to a time-varying flux is expressed as follows:

$$P_e = K_e f^2 B_{max}^2 \text{ watts}$$

Here, K_e is a constant whose value depends on the type of material and its lamination thickness. The lamination thickness varies from 0.5 to 5 mm in electrical machines and from 0.01 to 0.5 mm in devices operating at higher frequencies. Hysteresis and eddy current losses, added together, are known as core loss.

$$P_c = P_h + P_e$$

In DC and AC electrical machines that have a magnetic core and a time-varying flux, core loss occurs and the loss appears as heat in the core. Transformers and many machines have constant core losses regardless of load.

2.16 MAGNETIC CIRCUITS WITH AC EXCITATION

In AC electric machines, the voltages and fluxes vary sinusoidally with time. In Figure 2.48, assume that the core flux $\Phi(t)$ varies sinusoidally with time. Thus,

$$\Phi(t) = \Phi_{max} \sin \omega t$$

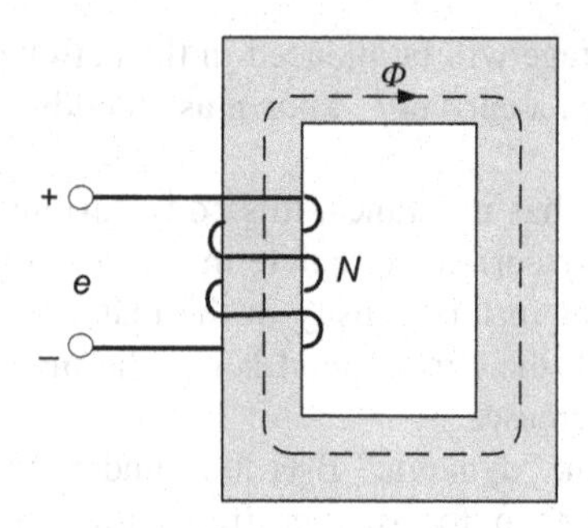

FIGURE 2.48 Magnetic circuit with AC excitation.

Here, Φ_{max} is the amplitude of the core flux, $\omega = 2\pi f$ angular velocity, and f frequency of the core flux. According to Faraday's law, the induced voltage in the coil would be

$$e(t) = N\frac{d\Phi}{dt} = N\Phi_{max}\omega\cos\omega t$$

$$= e_{max}\cos\omega t$$

Note that if the flux changes sinusoidally, the induced voltage changes cosinusoidally (induced voltage leading the flux-excitation current by 90°). The waveforms of e and Φ and their phasor representation would be as shown in Figure 2.49. The root-mean-square (rms) value of the induced voltage is

$$E_{rms} = \frac{E_{max}}{\sqrt{2}} = \frac{N\Phi_{max}\omega}{\sqrt{2}} = 4.44Nf\Phi_{max}$$

The induced voltage and the coil resistance drop oppose the impressed voltage. The resistance drop does not exceed a few percent of the impressed voltage in AC machines, and most of transformers and AC electromagnetic devices. Resistance

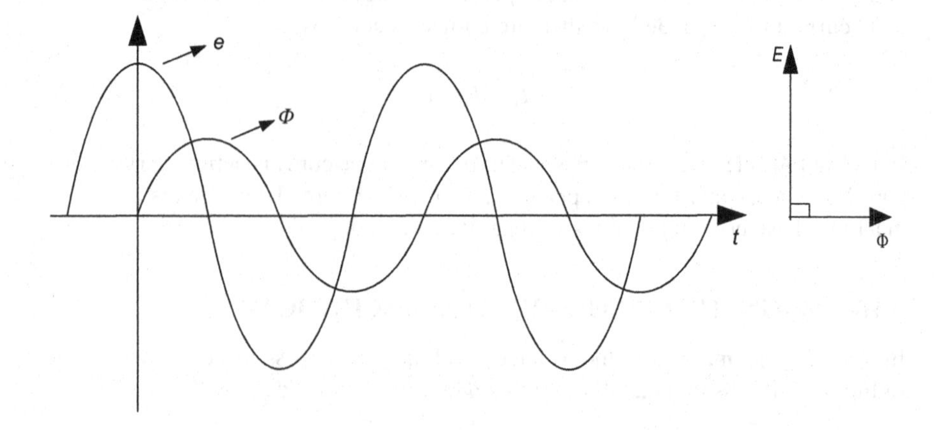

FIGURE 2.49 Phase difference between magnetic field and induced voltage.

drop may be neglected, and the impressed and induced voltages may be considered equal. The flux Φ_m is then determined by the impressed voltage in accordance with

$$E_{rms} = 4.44 N f \Phi_{max}$$

2.17 PERMANENT MAGNET

A permanent magnet is capable of maintaining a magnetic field without any excitation mmf provided to it. They are made from ferromagnetic materials such as iron, nickel, and cobalt and are created when the material is placed inside of a magnetic field (Figure 2.50).

When the magnetic field is removed, the object remains magnetised. Permanent magnets have their own permanent magnetic field which does not turn on and off like electromagnets do.

They are characterised by the following properties (Figure 2.51):

- a large B-H loop
- high value of remanent (or residual) magnetic flux B_r
- high coercive force H_c so that it cannot be easily demagnetised by stray magnetic fields.

Permanent magnets are often referred to as "hard iron" and other magnetic materials that lose their magnetism after removing the magnetising mmf as "soft iron".

2.17.1 Approximate Design of Permanent Magnets

Let us assume that in Figure 2.52a, the hard iron piece be magnetised to the residual flux density (B_r) denoted by point r in Figure 2.53.

If the mmf is removed, the permanent magnet would stay magnetised at the residual flux density (B_r) (Figure 2.53). If the small soft iron piece is also removed (Figure 2.52b) and we assume,

1. there is no leakage or fringing flux in the magnetic circuit.
2. the permeability of the soft iron is infinite.

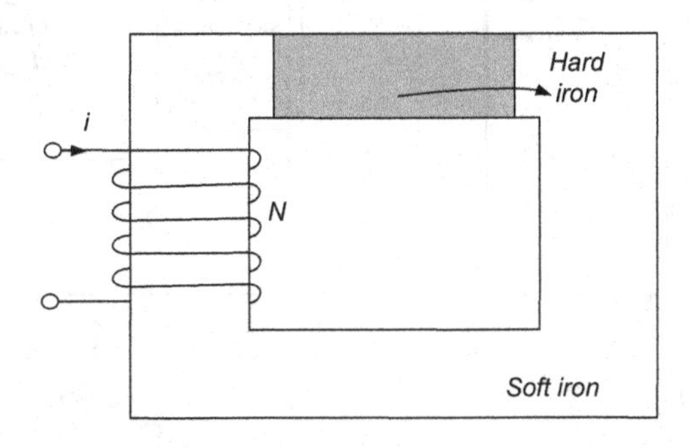

FIGURE 2.50 Making of a permanent magnet.

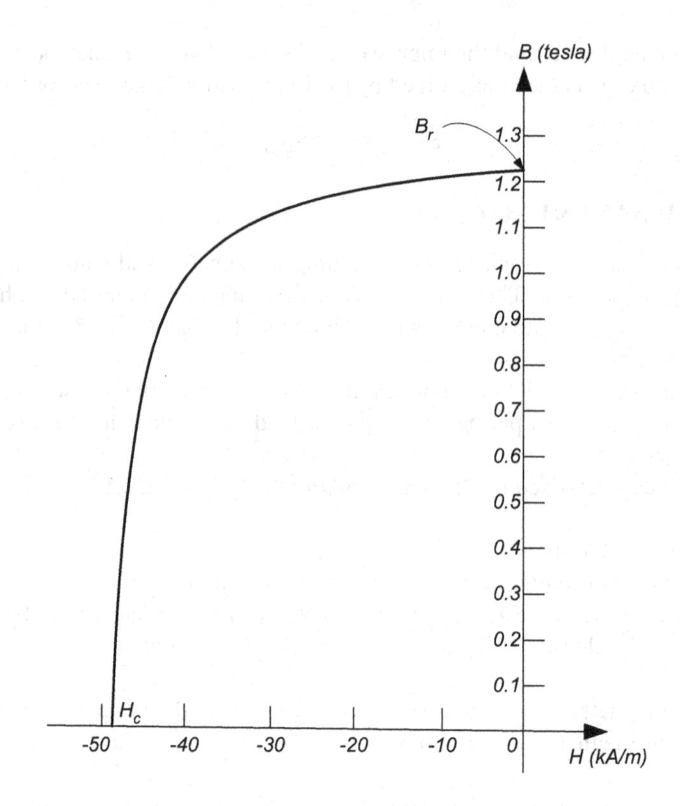

FIGURE 2.51 Permanent magnet B-H characteristics (Alnico).

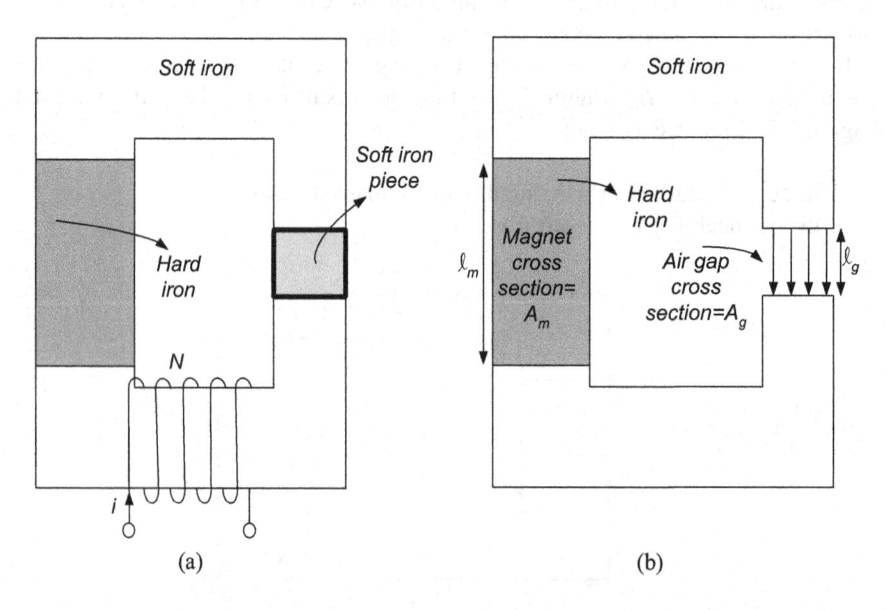

FIGURE 2.52 Permanent magnet (a) with external emf (b) emf and hard iron removed.

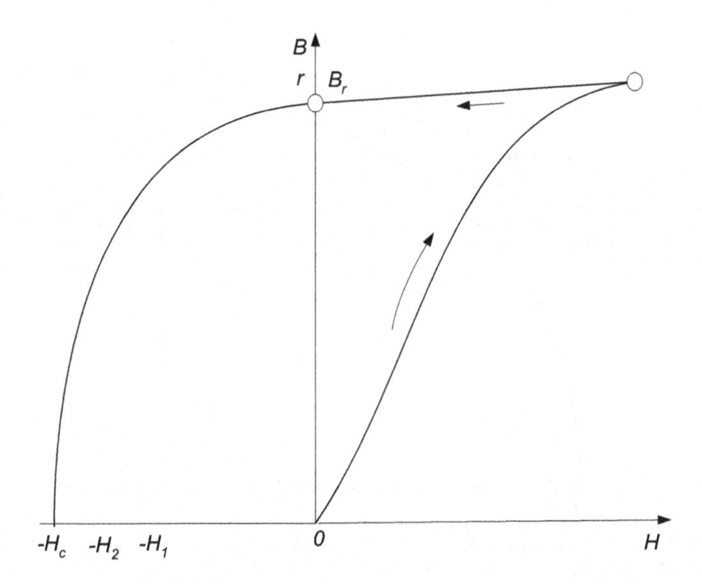

FIGURE 2.53 External mmf removed.

From Ampère's circuital law,

$$H_m l_m + H_g l_g = 0 \quad \rightarrow H_g = \frac{-H_m l_m}{l_g}$$

Since the flux in the magnetic circuit must remain a constant, we must have

$$\Phi = B_m A_m = B_g A_g \rightarrow \quad B_m = \frac{B_g A_g}{A_m}$$

$$B_g = \mu_0 H_g \quad \rightarrow B_m = \mu_0 H_g \times \frac{A_g}{A_m} = \frac{-H_m l_m}{l_g} \times \frac{\mu_0 A_g}{A_m}$$

$$B_m = -\mu_0 \left(\frac{l_m}{l_g} \times \frac{A_g}{A_m} \right) H_m$$

Notice that above equation represents a straight line going through the origin called the "shear line" (see Figure 2.54).

The intersection of the shear line with the demagnetisation curve at point "s" (Figure 2.54) determines the operating point and hence the operating values of B and H of the hard iron material with the "soft iron piece" and "mmf" removed.

We see that by opening the air gap, hard iron is demagnetised to point "s", reducing its remanent magnetisation to point B_s which is less than B_r. We also see that if the hard iron has low coercive force (H_c), for the same geometry, its operating point will lower to a point t (Figure 2.54) with a remanent magnetisation B_t which is much

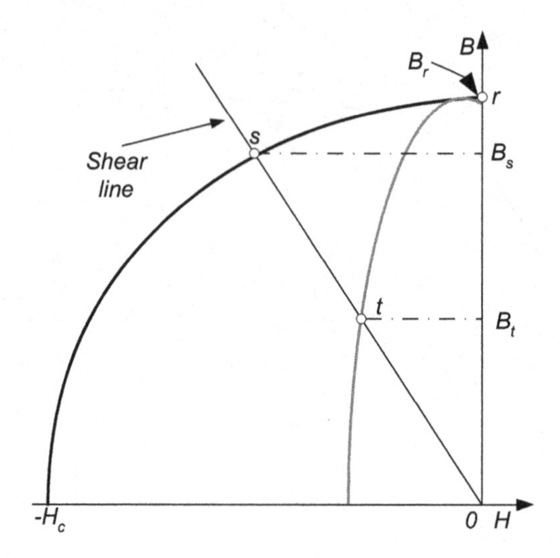

FIGURE 2.54 B-H Curve with "shear line."

less than B_r and B_s. Therefore, we can conclude that good permanent magnets should have high coercivity.

If we calculate the volume of the hard iron piece,

$$V_m = A_m l_m = \frac{B_g A_g}{B_m} \times \frac{H_g l_g}{H_m} = \frac{B_g^2 V_g}{\mu_0 B_m H_m}$$

Thus, to establish a flux density B_g in the air gap of volume V_g, a minimum volume of the hard iron is required if the final operating point is located such that the $B_m H_m$ product is a maximum. This quantity $B_m H_m$ is known as the "energy product" of the hard iron. The unit of the energy product is J/m³. We can write the above equation as

$$V_m = \frac{V_g}{B_m H_m}\left[\frac{B_g^2}{\mu_0}\right] = \frac{2 \times \text{Energy deposited in the air gap volume}}{B_m H_m}$$

$$\text{Energy deposited in the air gap volume} = \frac{B_m H_m V_m}{2}$$

To obtain maximum energy in an air gap with a known volume of hard iron, the energy product of the hard iron (permanent magnet) must be a maximum. The higher this product, the smaller need be the volume of the magnet.

It was only in the early twentieth century that high carbon steels and then tungsten/chromium containing steels replaced lodestone as the best available permanent magnet material. These steels are mechanically very hard and are the origin of the term hard magnetic. These magnets had an energy product of approximately 8 kJ/m³.

Exercise 2.11:

The air gap dimensions in the magnetic circuit shown in Figure 2.55 are $A_g = 1.0\,\text{cm}^2$ and $l_g = 0.1$ cm. Assuming maximum energy product for Alnico 5 is at $B_m = 1.5$ T and $H_m = -50$ kA/m, find the minimum permanent magnet (Alnico 5) volume required to achieve an air gap flux density of 0.9 T.

Solution:

The smallest magnet volume will be achieved with the magnet operating at its point of maximum energy product. At this operating point, $B_m = 1.5$ T and $H_m = -50$ kA/m.

$$A_m = \text{Cross-sectional area of the magnet} = A_g\left(\frac{B_g}{B_m}\right)$$

$$A_m = 10^{-4} \times \frac{0.9}{1.5} = 0.6 \times 10^{-4}\, m^2$$

$$\text{Length of magnet} = l_m = -l_g\left[\frac{H_g}{H_m}\right]$$

$$= -\left(0.1 \times 10^{-2}\right)\left[\frac{0.9}{-4\pi \times 10^{-7} \times \left(-50 \times 10^3\right)}\right] = 1.43\text{ cm}$$

$$\text{Volume of the magnet} = V_m = 1.43 \times 0.6 = 0.858\text{ cm}^3$$

Going back to Figure 2.54, we saw that by opening the air gap, hard iron is demagnetised to point *s*, and its remanent magnetisation reduced to B_s.

If the "soft iron piece" is now reinserted (remagnetisation), the operating point moves up the recoil line sq to point *q* and not to *r* where it is effectively weakened (Figure 2.56). As a result, hard magnetic materials often do not operate stably with

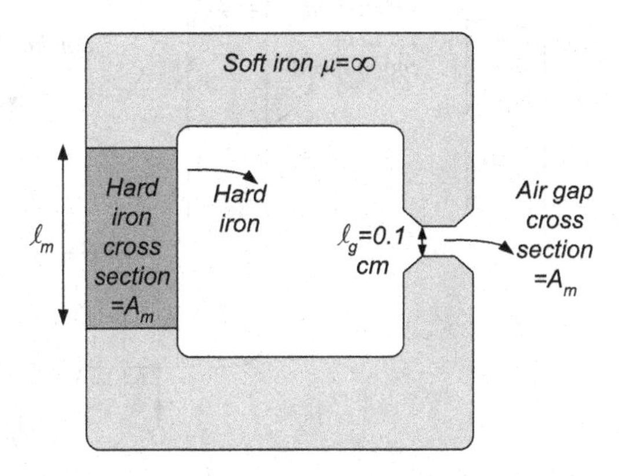

FIGURE 2.55

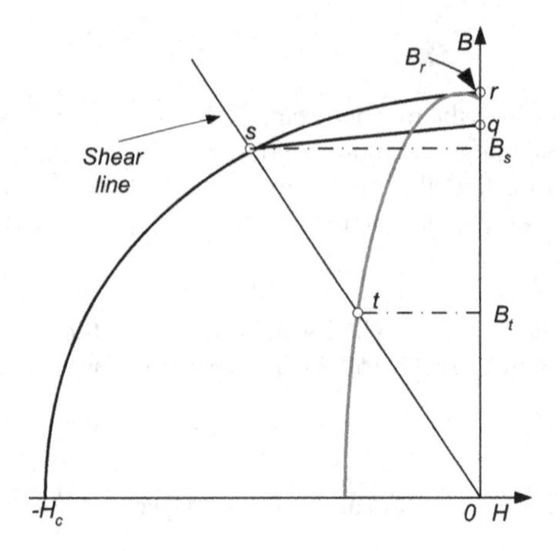

FIGURE 2.56 B-H curve with soft iron piece reinserted.

varying mmf and geometry. There is the risk of demagnetisation with improper operation. An advantage of materials such as samarium–cobalt is that their recoil lines closely match their magnetisation characteristic.

They also have a very high coercivity which means they are not easily demagnetised. Therefore, demagnetisation effects are significantly reduced in these materials. However, manufacturing samarium–cobalt magnets require elaborate techniques like induction melting, sintering, solution treatment, and tempering.

The magnetic circuit of Figure 2.57 can be used to magnetise hard magnetic materials. This process requires a large excitation be applied to the winding and then reduced to zero, leaving the material at a remanent magnetisation B_r (point r).

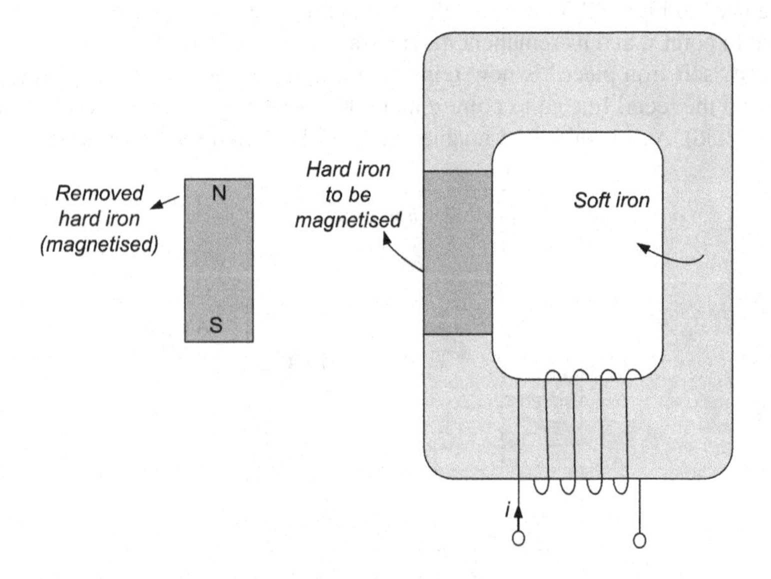

FIGURE 2.57 Making permanent magnet from hard iron.

If now the hard iron material is removed (Figure 2.57) from the core, this would be equivalent to opening a large air gap in the magnetic circuit, demagnetising the hard iron material to a lower value.

Exercise 2.12:

The dimensions of the magnetic circuit of Figure 2.58 are as follows: $l_g = 0.5\,\text{cm}$, $A_g = A_c = A_m = 5\,\text{cm}^2$, $l_c = 40\,\text{cm}$, and $\mu_r = 400$ for steel.

What is the minimum length of the magnet required for the maximum energy product shown in Figure 2.59?

Solution:

$$H_m l_m + H_c l_c + H_g l_g + H_c l_c = 0$$

$$H_m = -\frac{\left[2H_c l_c + H_g l_g\right]}{l_m}$$

Since the cross-sectional areas are equal,

$$B_m = B_c = B_g; \quad H_g = \frac{B_m}{\mu_0}; \quad H_c = \frac{B_m}{\mu_c}$$

$$H_m = -\frac{2B_m l_c}{\mu_c l_m} - \frac{B_m l_g}{\mu_0 l_m} = -\frac{B_m}{\mu_0 l_m}\left[\frac{2l_c}{400} + l_g\right]$$

$$-39\times 10^3 = -\frac{0.62}{4\pi\times 10^{-7}\times l_m}\left[\frac{2\times 0.4}{400} + 0.5\times 10^{-2}\right]$$

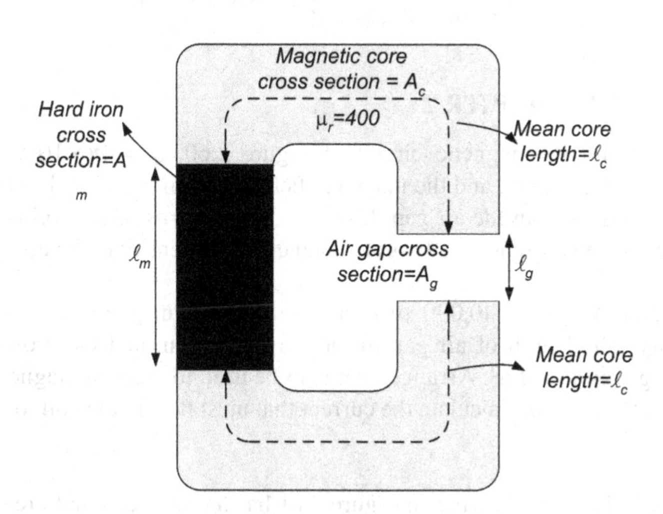

FIGURE 2.58

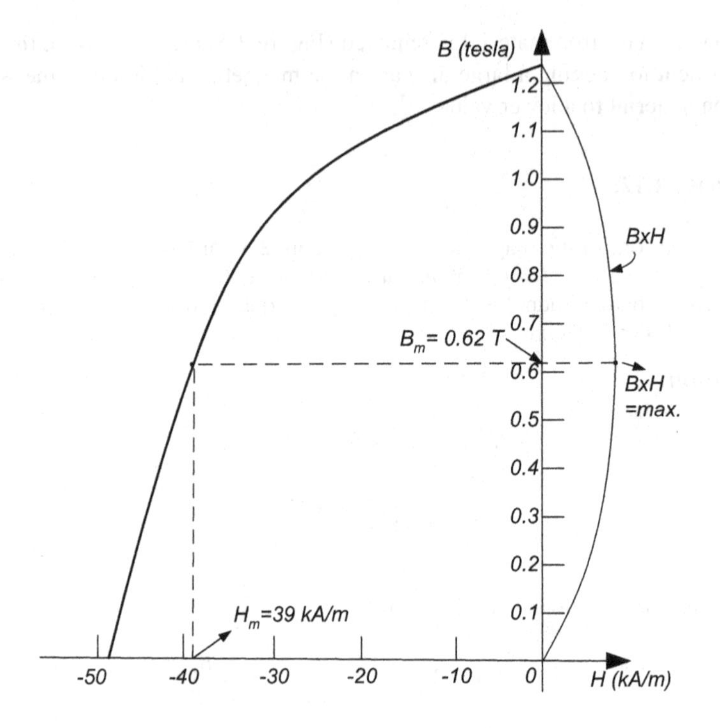

FIGURE 2.59

$$= -\frac{0.62}{4\pi \times 10^{-7} \times l_m}[0.002 + 0.005] = -\frac{0.62}{4\pi \times 10^{-7} \times l_m} \times 0.007$$

$$l_m = \frac{-0.62 \times 0.007}{-39 \times 10^3 \times 4\pi \times 10^{-7}} = 8.85 \times 10^{-2} \text{ m}$$

PROBLEMS ON CHAPTER 2

Problem 2.1: In the magnetic circuit of Figure 2.60, $\mu_0 = 4\pi \times 10^{-7}$, I = 5 A, N = 1,000 turns, g = 1 cm, and the magnetic field in the air gaps is 0.1 Wb. Find the magnetic field density in the air gap. If the number of turns of the coil is 500, how much would be the current for a twofold magnetic field density in the air gap?

Problem 2.2: A coil of 10,000 turns is wound on a magnetic core of infinite permeability. The length of air gap in the core is 0.4 cm, and the cross-sectional area of the gap is 400 mm^2. Arrangement is to be used to create a magnetic field of 0.02 Wbs in the air gap. Calculate the current that must flow in the coil to create this magnetic field.

Problem 2.3: The toroidal ring in Figure 2.61 has a cross-sectional area of 6 cm^2, and its mean core length is 50 cm. Assuming the relative permeability of the core as

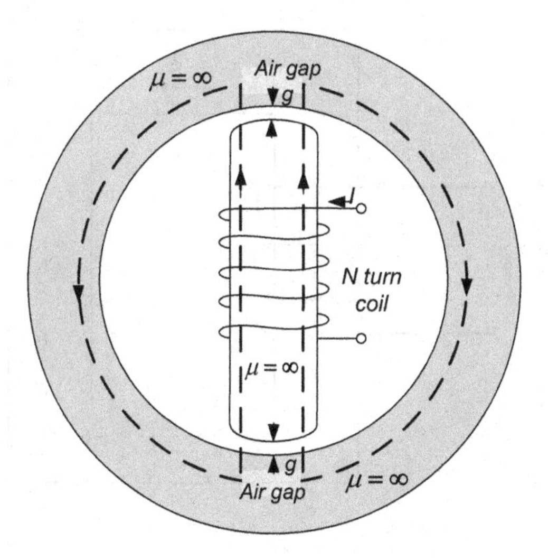

FIGURE 2.60

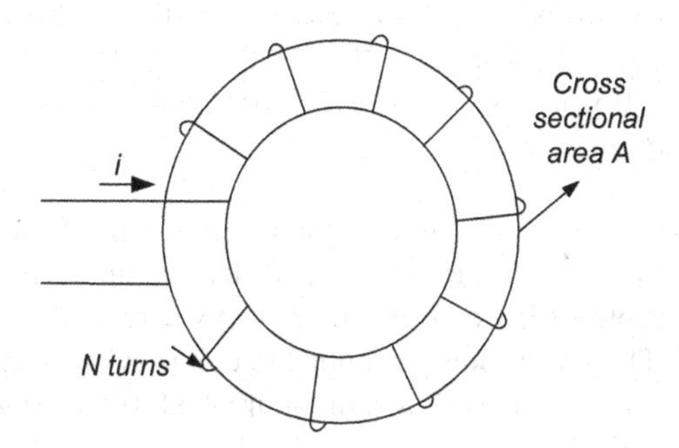

FIGURE 2.61

500, calculate the reluctance of the toroidal ring. If we wind a coil of 1,500 turns on this ring and make 5 A flow in the coil, what would be the magnetic field density in the ring?

Problem 2.4: The magnetic core in Figure 2.62 has a mean core length of 80 cm, and its uniform cross-sectional area is $10\,cm^2$. The relative permeability of the core is 1,500, and an AC voltage source $e(t) = 60\cos(120t)V$ is connected across the coil of $N = 100$ turns. Obtain an expression for the flux density in the core and the current in the coil.

Problem 2.5: Two AC sources of frequencies f_1 and f_2 with rms values of 200 and 300 V, respectively, are separately connected to a coil wound on iron core and the hysteresis losses of the core are measured. If the ratio of the measured hysteresis losses is $P_{h1}/P_{h2} = 0.5$, find the ratio f_1/f_2 and the eddy current losses ratio of P_{e1}/P_{e2}. For hysteresis loss consider, n = Steinmetz constant = 2.

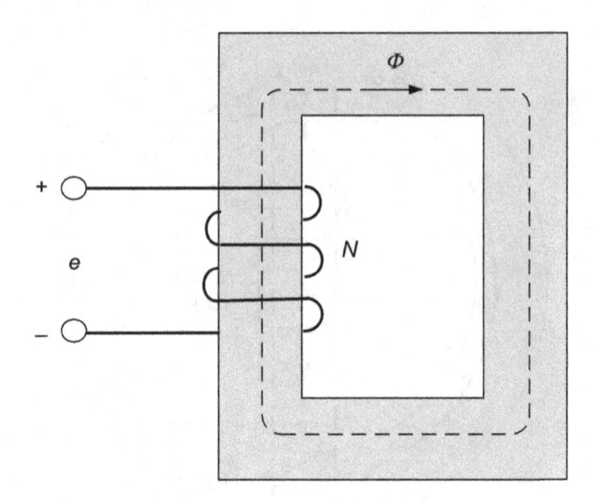

FIGURE 2.62

Problem 2.6: The permanent magnet shown in Figure 2.63 is made of alnico 5. The air gap cross section is $A_g = 3\,\text{cm}^2$, and its length is $l_g = 0.5\,\text{cm}$. The energy product $H_m B_m$ occurs for $B_m = 0.95$ T, $H_m = -42$ kA/m. To create a magnetic field density of 2 T in the air gap, what must be the minimum magnet volume, length, and cross-sectional area?

Problem 2.7: The core of the relay system in Figure 2.64 has a magnetic permeability of 15×10^{-4} Henries/meter and has a uniform cross section of $6\,\text{cm} \times 10\,\text{cm}$. The depth of the core is 10 cm everywhere, and the number turns of the coil is $N = 300$. The DC excitation applied to coil provides a current just enough to keep the air gap at $g = 5\,\text{mm}$ as shown in Figure 2.64. If the air gap magnetic field density B_g is 1.2 T, find

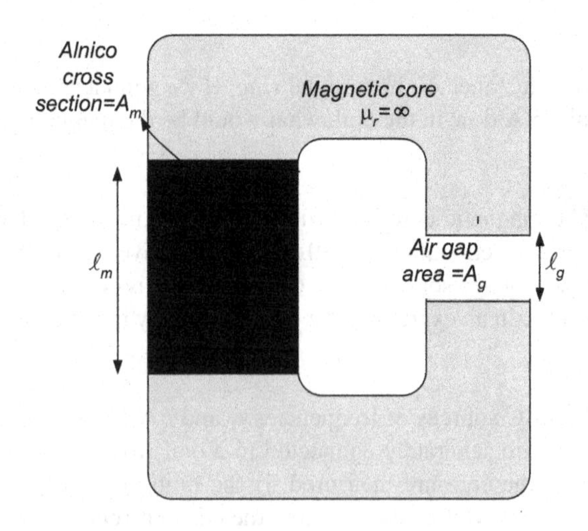

FIGURE 2.63

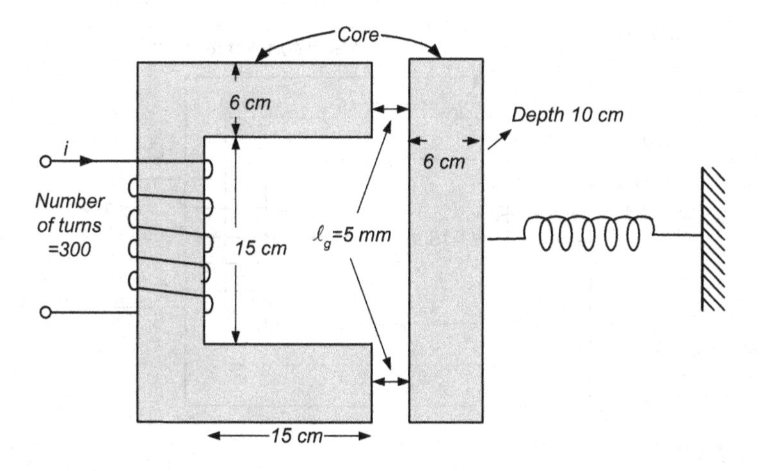

FIGURE 2.64

a. the current taken from the DC source
b. the stored magnetic energy in the system.

Problem 2.8: The two coils in Figure 2.65 are wound so as to help each other to increase the total mmf in the magnetic circuit. Their self-inductances are $L_1 = 10$ mH and $L_2 = 0.1$ mH with a coupling coefficient of $k = 0.8$. If $i_1 = 15\cos(2000\pi t)$ mA and $i_2 = 0$, find the expressions for the voltages $v_1(t)$ and $v_2(t)$.

Problem 2.9: If the material of the core in Figure 2.66 has a relative permeability of 10,000,

a. what are the self- and mutual inductances of the two coils ?
b. if sinusoidal currents are flowing in the two coils as shown in the figure, how much are the two magnetic fluxes linking coils 1 and 2?
c. find the induced voltages that appear on coils 1 and 2?

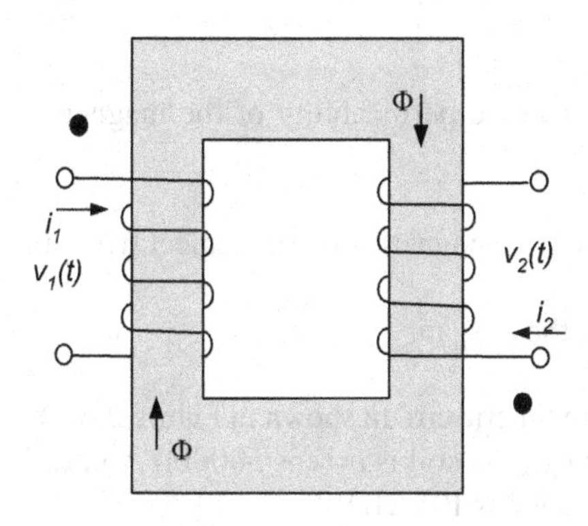

FIGURE 2.65

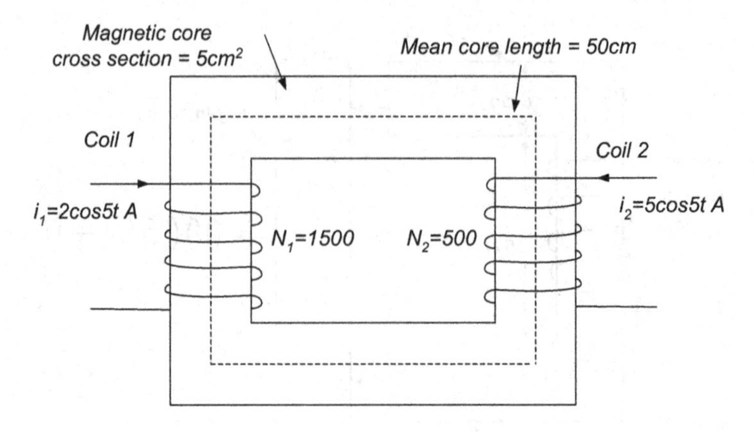

FIGURE 2.66

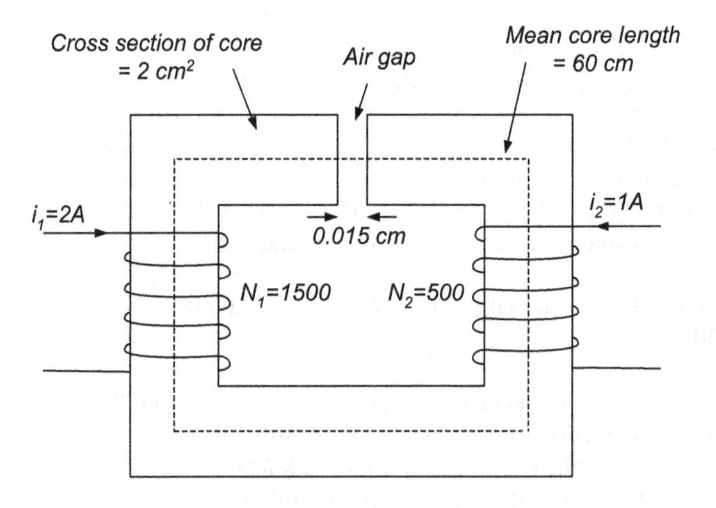

FIGURE 2.67

Problem 2.10: If the relative permeability of the magnetic core in Figure 2.67 is 10,000, find the following:

a. total reluctance of the magnetic circuit (neglect fringing of magnetic flux in the air gap)
b. magnetic flux in the air gap.

Problem 2.11: In the magnetic circuit shown in Figure 2.68, the number of turns N is 1,000, the current through the coil is $i = \cos(50000t)$ Amps, the cross-sectional area is $1\,\text{cm}^2$, and the reluctance is 10^7 AT/Wb.

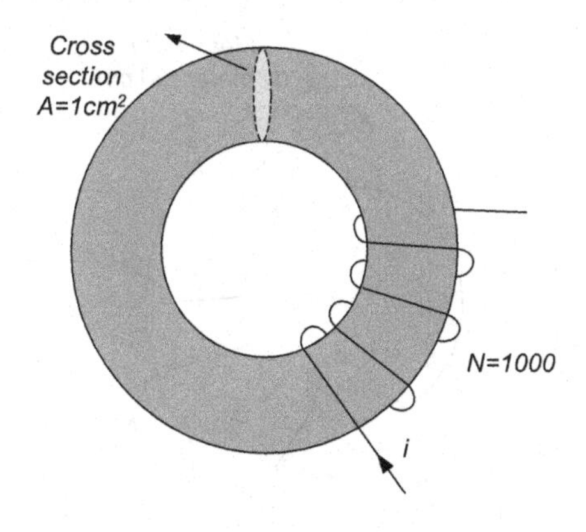

FIGURE 2.68

a. calculate the magnetic field density in the core.
b. if the characteristic constant of the core (k_h) is 1.5, Steinmetz exponent n is 2, and the hysteresis loss in the core is 1 W, what is the volume of the toroid?
c. if the eddy current constant k_e is 0.01, what is the eddy current loss of the core?

Problem 2.12:

a. In the magnetic circuit in Figure 2.69, the coil has 500 number of turns (N) and a voltage of $e(t) = 5\cos(2t)$ volt is applied across its terminals. What is the time-varying function of the magnetic field $Ø(t)$ in the magnetic core?
b. if the area of the hysteresis loop is 10,000 J/m^3 and the volume of the core is 200 cm^3, find the hysteresis energy loss as heat in π seconds.
c. what is the hysteresis loss in watts of this core?

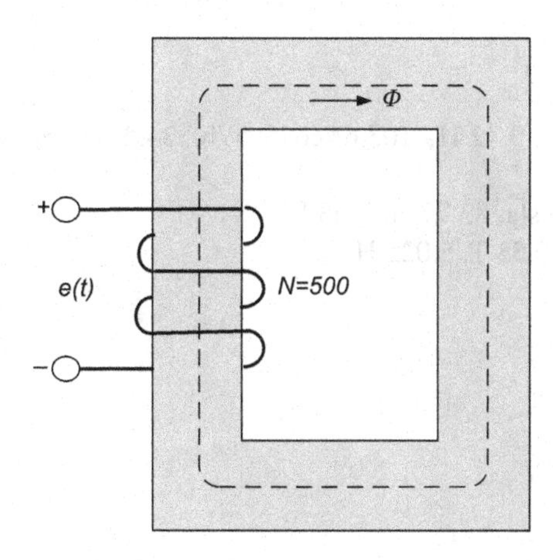

FIGURE 2.69

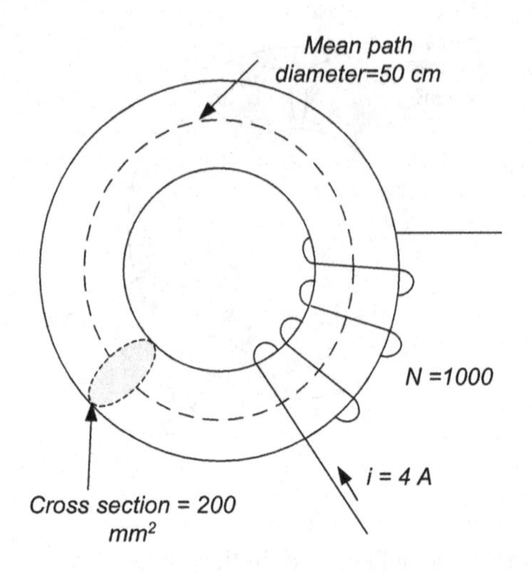

FIGURE 2.70

Problem 2.13: In the magnetic circuit of Figure 2.70, if the number of turns $N =$ 1,000, current $i = 4$ A, mean core length $l = 50\,\text{cm}$, cross-sectional area $A = 200\,\text{mm}^2$, and $\mu_r = 10{,}000$, find

a. magnetic field intensity H
b. magnetic field density B
c. self-inductance value of the magnetic circuit.

Answers to odd questions:

2.1: 19.8 A
2.3: 9.42 T
2.5: 44.4;
2.7: 34.07 A, 36.8 J
2.9: 28.27 H, 3.14 H, 9.42 H, $103.64\cos 5t$ Wb, $34.54\cos 5t$ Wb, $-518.2\sin 5t$ V, $-172.7\sin 5t$ V
2.11: $\cos(50000\text{t})$ Tesla, $83.77\ \text{cm}^3$, 53 W
2.13: 8,000 At/m, 100.53 T, 5.025 H

3 Principles of Electromechanical Energy Conversion

3.1 INTRODUCTION

Electromechanical devices convert electrical energy into mechanical energy and vice versa. Energy conversion takes place through the medium of magnetic field. Electromechanical energy conversion is a reversible process except for the losses in the system. We can categorise the electromechanical energy conversion devices in the following manner:

1. Transducers: These are the first category of devices that involve low-energy conversion signals from electrical to mechanical or vice versa. Microphones, sensors, and loudspeakers are typical examples of this category of devices which operate generally under linear conditions.
2. Force-producing devices: These are the second category of devices that consist of force- or torque-producing devices with limited mechanical motion. Examples are solenoids, relays, and electromagnets.
3. Continuous energy conversion equipment: Examples are motors and generators that are used for bulk energy conversion and utilisation.

In this chapter, we first discuss transducers and force-producing devices and then continuous energy conversion devices. In doing these, we try to keep as short and as less time consuming as possible to understand the basics of the subject.

3.2 FORCES AND TORQUES IN MAGNETIC FIELDS

According to Lorentz's force law, in pure magnetic field systems, the force on a moving particle of charge q is indicated by the following vector cross product:

$$F = q\left(\vec{v} \times \vec{B}\right)$$

where v is the velocity in m/s of the particle of charge q moving in a magnetic field density of B Tesla (Figure 3.1). The magnitude of this vector cross product is

$$F = qvB\sin\theta = q\frac{l}{t}B\sin\theta = ilB\sin\theta$$

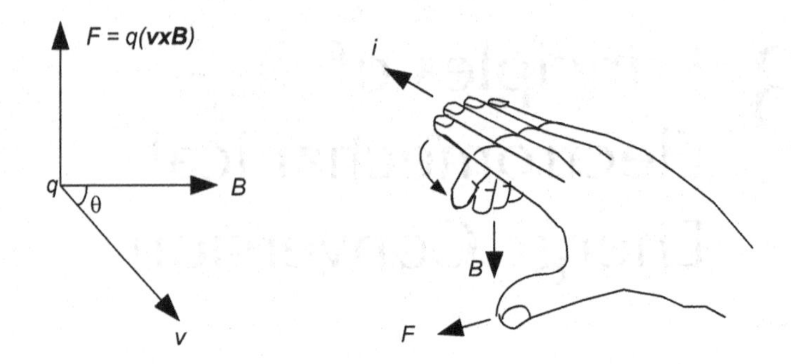

FIGURE 3.1 Direction of force (right-hand rule II).

The direction of the force is the advance of right-handed screw rotated from the first vector to second.

The direction of the force can be determined from the right-hand rule II (Figure 3.1), which states that

- point the fingers of your right hand in the direction of the current.
- then bend them in the direction of the magnetic flux.
- your thumb indicates the direction of the force (this is consistent with the definition of a vector product).

Exercise 3.1:

A 6 m long wire carries a current of 100 A in a magnetic field density of 0.8 T as shown in Figure 3.2. The angle between the magnetic field and the current flow is 110°. What is the force on the wire, and what is its direction?

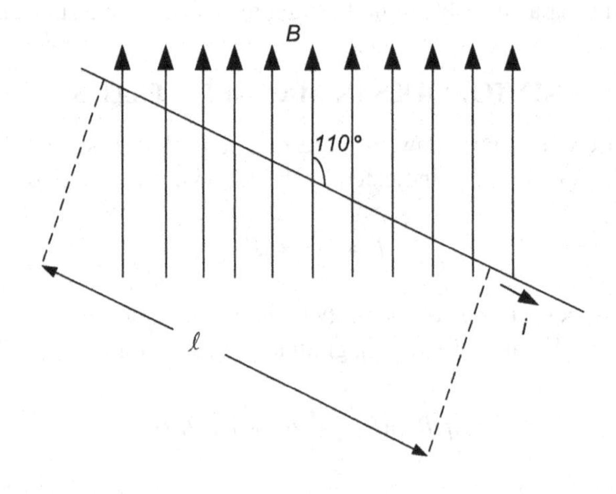

FIGURE 3.2

Solution:

$$F = Bil\sin\theta = 0.8\times100\times6\times\sin110^{\circ} = 451\,\text{N}$$

According to right-hand rule II, the direction of the force would be out of the paper with 90°.

3.3 FORCE BETWEEN TWO CURRENT-CARRYING WIRES

As illustrated in Figure 3.3, when a current-carrying conductor is brought near another current-carrying conductor, a magnetic force is experienced between the wires. If we have

- two parallel wires of length l separated by a distance d
- currents i_1 and i_2 flowing in opposite directions in these parallel wires

from our previous chapters, we know that a magnetic field intensity H due to current i_2 on conductor carrying current i_1 at distance d would be (Figures 3.3 and 3.4)

$$|H_d| = \frac{i_2}{2\pi d}$$

The force exerted on i_1 will be in an outward direction (right-hand rule II) (Figure 3.3) with a magnitude of

$$F = Bil\sin\theta = i_1 \times \mu_0 \frac{i_2}{2\pi d} \times l = \mu_0 \frac{i_1 i_2 l}{2\pi d} \quad N$$

where μ_0 represents the magnetic permeability of the media which happens to be air in our example.

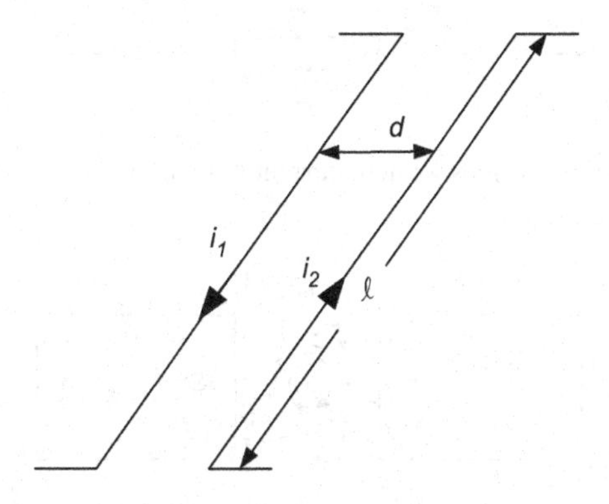

FIGURE 3.3 Force on parallel current carrying wires.

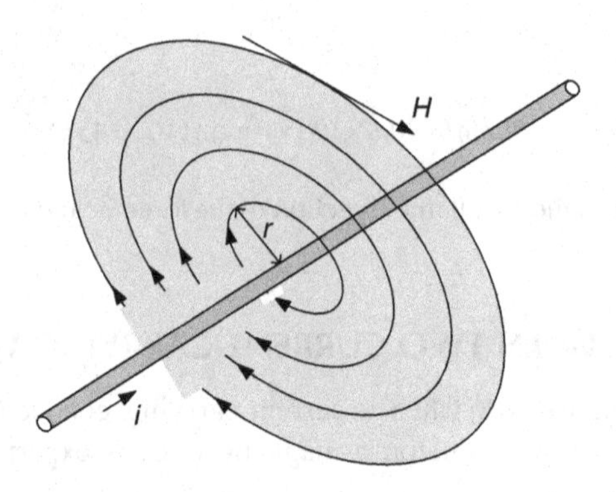

FIGURE 3.4 Magnetic flux lines around current carrying conductor.

3.4 FORCE ON A CURRENT-CARRYING WIRE IN A MAGNETIC FIELD

Now we will look at the magnetic force from a different point of view of magnetic flux lines. Let us consider two permanent magnets with their N and S poles facing each other as shown in Figure 3.5. Let us also consider a piece of wire running midway and parallel to the pole faces and right through and perpendicular to the paper. Assuming the wire is thin, the magnetic field and flux lines between the two pole faces will be as plotted in Figure 3.5.

If a current flows in the wire and into the paper (Figure 3.6), it will cause a magnetic field of its own which will help the magnetic flux lines in the upper part and oppose the magnetic flux lines in the lower part. As a result the distorted magnetic flux lines will be as illustrated in Figure 3.6. The distorted lines of flux can be

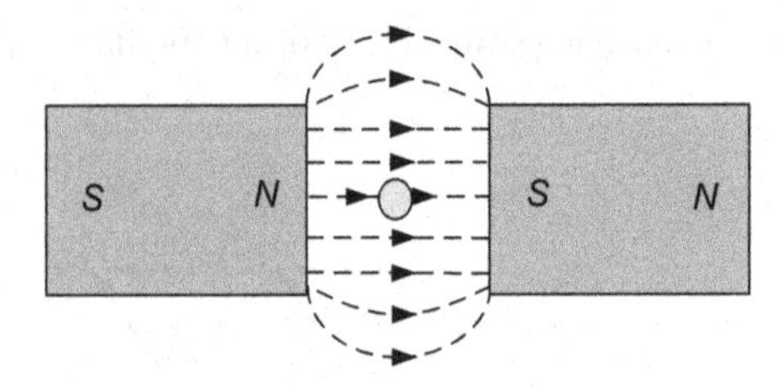

FIGURE 3.5 Magnetic flux lines without current flowing.

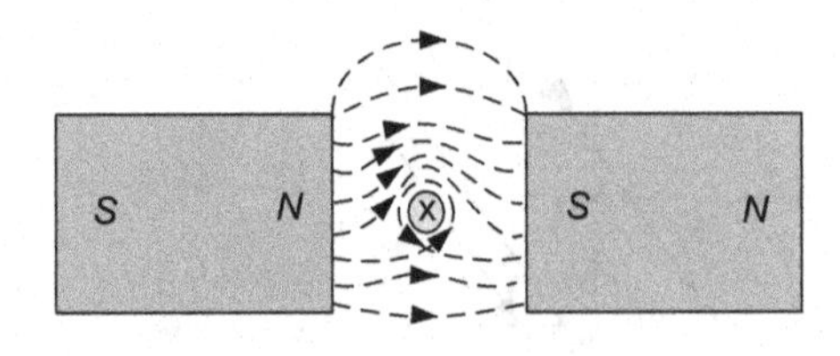

FIGURE 3.6 Magnetic flux lines with current flowing.

visualised as acting like stretched elastic cords. They will try to return to the shortest paths between pole faces, thereby exerting a force pushing the conductor out of the way. The resultant effect is to strengthen the magnetic field on the upper side and weaken that on the lower side, thus giving the distribution shown in Figure 3.6. This is a good way to visualise how the forces appear on the current carrying wires in magnetic fields.

Exercise 3.2:

The two conductors in Figure 3.7 are 2 cm apart, and both carry 2000 A of current in opposite directions as shown in the figure. What is the force on each conductor if both wires are 1 m long?

Solution:

$$F = \mu_0 \frac{i_1 i_2 l}{2\pi d} = 4\pi \times 10^{-7} \times \frac{2000 \times 2000 \times 1}{2\pi \times 0.02} = 40\,\text{N}$$

Exercise 3.3:

Figure 3.8 shows a single winding coil of length l = 40 cm on a wooden rotor placed in a uniform magnetic field of density B. If the current flowing in the coil is 15 A,

a. find the force (F) exerted on the wires.
b. find the torque on the coil. In which direction the rotor will rotate?

Solution:

According to right-hand rule II, the force exerted on the wires will be as shown in Figure 3.9.

$$\text{Force on the wires, } F = Bil \sin 90^\circ = 1 \times 15 \times 0.4 \sin 90^\circ = 6\,\text{N}$$

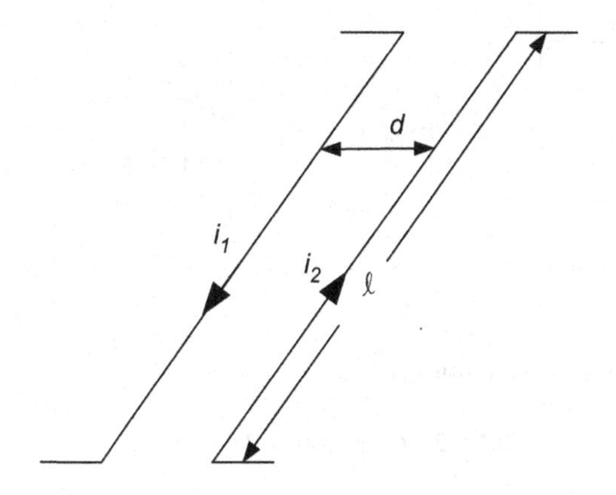

FIGURE 3.7

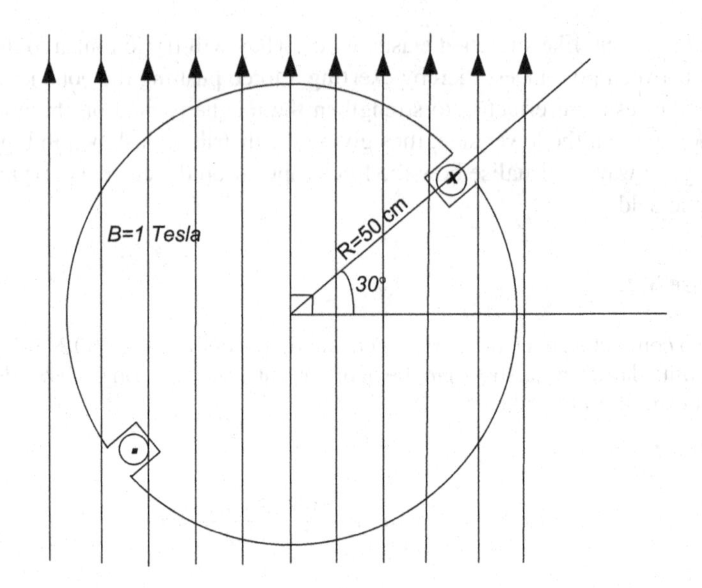

FIGURE 3.8

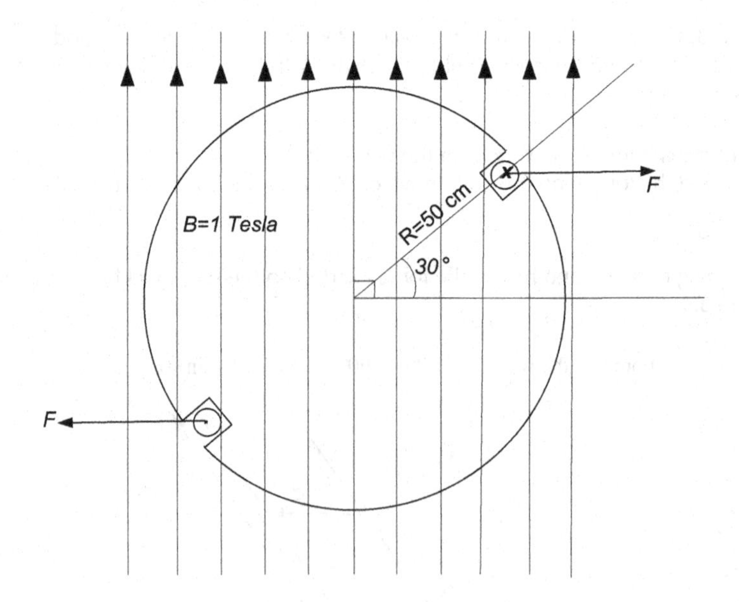

FIGURE 3.9

The torque exerted on the rotor is

$$T = 2 \times 6 \times \cos 60° \times 0.5 = 3 \text{ N-m}$$

Rotor will rotate in clockwise direction.

3.5 GENERATED VOLTAGE IN MAGNETIC SYSTEMS

In previous chapters, we stated Faraday's law as the voltage induced (e) in an N-turn coil when the flux Φ linking or threading through it changes.

$$e = N\frac{d\Phi}{dt} = \frac{d\lambda}{dt}$$

Now consider the magnetic circuit of Figure 3.10 where a changing voltage $v(t)$ is applied to a coil of N turns wound on a magnetic material to produce a changing magnetic flux of $\Phi(t)$ passing through the coil. According to Faraday's law, changing magnetic flux $\Phi(t)$ will induce a voltage $e(t)$ on the coil:

$$e(t) = N\frac{d\Phi(t)}{dt} = \frac{d\lambda(t)}{dt}\ \text{V}$$

According to Lenz's law, the direction of the induced voltage $e(t)$ on the coil will be such as to cause a current flow in the "closed coil" which will produce a flux opposing the original flux in the core. In more general terms, $\Phi(t)$ in the above expression is defined with the surface integral of the dot product $\vec{B}.\overrightarrow{ds}$:

$$\Phi(t) = \int \vec{B}.\overrightarrow{ds} = \text{Total flux passing through the coil}$$

where "s" is the area enclosed by the coil. Therefore for our case, in scalar terms,

$$\Phi(t) = B.s.\sin\theta$$

Here, θ is the angle between the plane of the area s and the magnetic flux lines. The induced voltage according to Faraday's law is also known as the "transformer voltage." If we consider the static single loop and a changing magnetic flux going

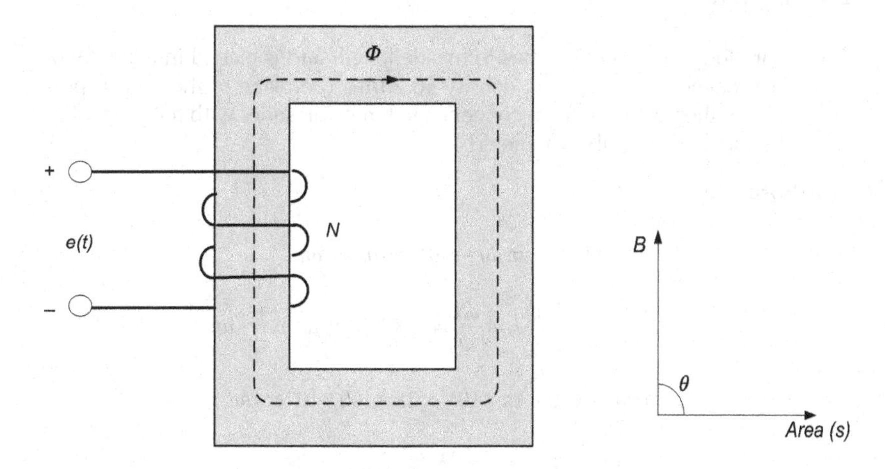

FIGURE 3.10 Induced voltage in a coil with changing magnetic field.

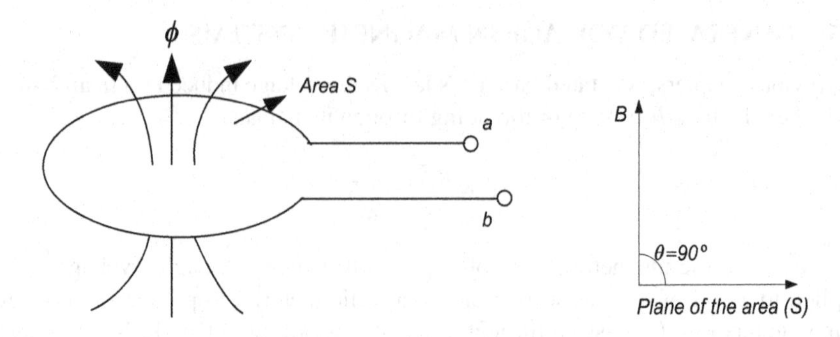

FIGURE 3.11 Generation of induced voltage.

through it as shown in Figure 3.11, according to Faraday's law the induced transformer voltage would be

$$e_{\text{transformer}} = \frac{d\lambda}{dt} = N\frac{d\Phi}{dt} = N.A.(\sin\theta).\left[\frac{dB}{dt}\right]$$

Here,

$$\theta = \text{Angle between the plane of the area enclosed } (A) \text{ and flux } (\Phi)$$

In accordance with Lenz's law, the direction of this induced voltage would be so as to make a current flow from *a* to *b* outwards to oppose the original flux change that is creating this induced voltage.

We can apply the "right-hand rule" to this loop to find the direction of the current due to the induced voltage that would have to flow from *a* to *b* (outwards) [If $\Phi(t)$ is increasing, the induced voltage is positive].

Exercise 3.4:

The square loop in Figure 3.12 has 5 cm side length and is placed in a 100 MHz sinusoidal magnetic flux of peak density 50 AT/m. The plane of the loop is perpendicular to the flux lines. If we connect a voltmeter in series with the loop what would be the reading of the voltmeter?

Solution:

$$H = 50\sin\omega t \rightarrow B = 50\mu_0 \sin\omega t$$

$$e = \frac{d\Phi}{dt} = \frac{d(B.A)}{dt} = A\frac{dB}{dt} = 0.0025\times 50\mu_0.\omega.\cos\omega t$$

$$= 0.0025\times 50\times 4\pi\times 10^{-7}\times 2\pi\times 100\times 10^{6}\cos\omega t$$

$$e_{\text{rms}} = \frac{10\pi^2}{\sqrt{2}} = 70\,\text{V}$$

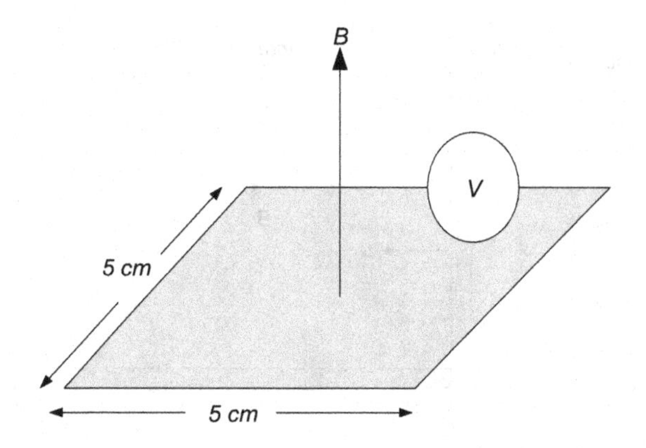

FIGURE 3.12

3.6 VOLTAGE INDUCED ON A MOVING CONDUCTOR

For a straight conductor with length l moving with velocity v in a uniform magnetic field B which is not changing with time (Figure 3.13), the induced motion voltage on the conductor e_{motional} would be

$$\text{Induced voltage on the conductor, } e_m = \frac{d\Phi}{dt}$$

If the distance covered by the conductor in Δt seconds is Δx, then

$$e_m = \frac{d\Phi}{dt} = \frac{Bldx.\sin\theta}{dt} = v.l.B.\sin\theta$$

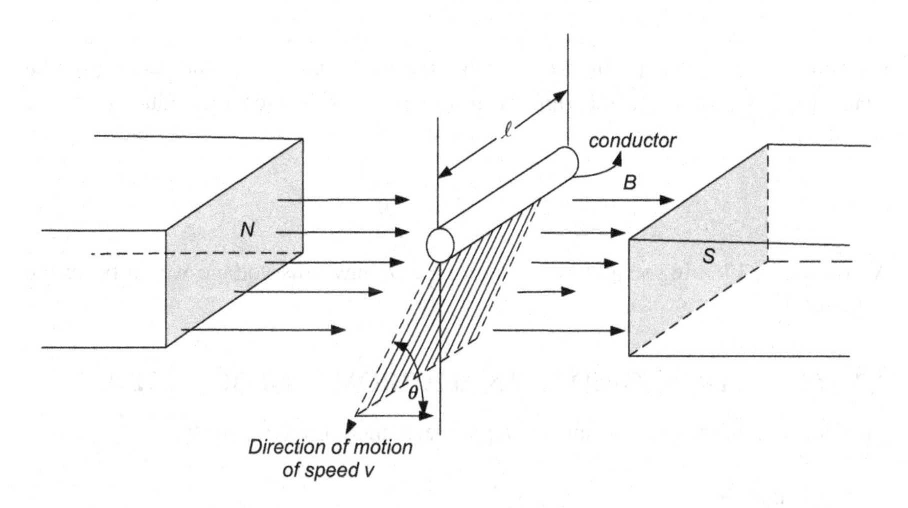

FIGURE 3.13 Moving conductor at an angle in a magnetic field.

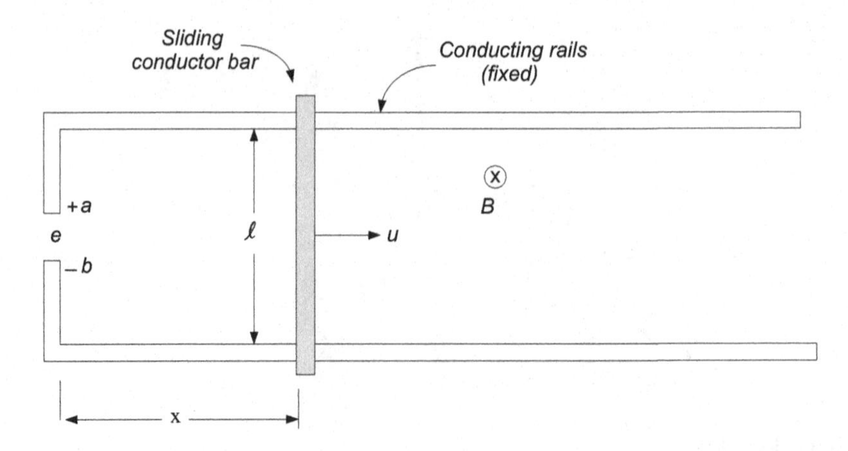

FIGURE 3.14 Moving conductive bar on rails in a magnetic field.

Here, θ is the angle between velocity v and magnetic field B (from v to B). Direction of this induced voltage can be found by Lenz's law or by the right-hand rule of Fleming which can be summarised as

"If the thumb, forefinger, and middle finger of the right hand are held at right angles to one another with the

- thumb pointing in the direction of the motion of the conductor the forefinger pointing in the direction of the magnetic field then the middle finger will point in the direction of the induced current. If the angle θ is 90°, then the voltage induced is a maximum:

$$e_m = \frac{d\Phi}{dt} = \frac{Bldx.\sin\theta}{dt} = v.l.B$$

If in Figure 3.14 magnetic field B is in the direction shown (into the paper) and the conductor l is moving to the right with a velocity of u, then the induced voltage would be

$$e_m = \frac{d\lambda}{dt} = u.l.B$$

According to Fleming's right-hand rule or Lenz' law, this voltage would be in the direction V_{ab}.

3.7 FORCE DEVELOPED IN AN ELECTROMAGNETIC SYSTEM

An electromagnetic system can develop a mechanical force in two ways:

1. by alignment
2. by interaction.

3.7.1 Force of Alignment

To be able to understand the force of alignment, consider the two illustrations shown in Figure 3.15 and 3.16, in which two poles are opposite to each other; each is made of a ferromagnetic material. In Figure 3.16 a flux passes from one to the other and the two magnetised surfaces are attracted towards one another.

Magnetic attraction forces will try to bring the poles together since this decreases the reluctance of the air gap in the magnetic circuit. Hence, this will increase the flux and consequently the stored energy.

In the second case shown in Figure 3.16, forces try to achieve greater stored magnetic energy by two component actions:

1. by attraction of the poles
2. by aligning the poles laterally.

If the poles move laterally, the cross-sectional area of the air gap is increased and the reluctance is reduced. This will again increase the flux and consequently the stored

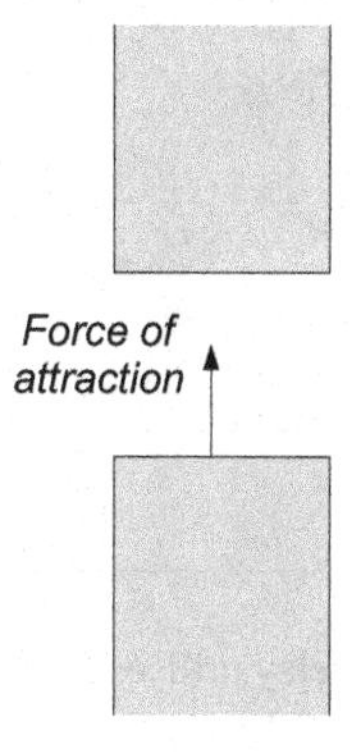

FIGURE 3.15 Force of attraction.

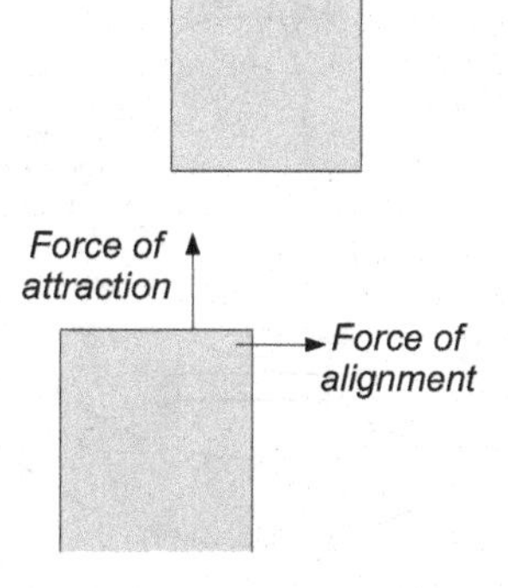

FIGURE 3.16 Force of attraction and alignment.

magnetic energy as before. In both of the above cases, forces (force of alignment) occur so as to maximise the magnetic energy stored in the magnetic circuit.

The arrangement (relay) shown in Figure 3.17 is an example of an electromechanical energy conversion device which utilises force of alignment giving rise to linear motion. When the coil is energised, a flux is set up in the relay core and air gap. The air gap surfaces become magnetised and are attracted. Hence, armature plate is pulled in the direction indicated.

This relay is used extensively in old type of telephone exchanges as a simple switch. Recently, they have been replaced by digital switches.

The force of alignment can also be used to produce rotary motion as in the reluctance motor shown in Figure 3.18. The rotating part (piece) experiences a torque due to the magnetised rotor and pole surfaces attempting to align themselves.

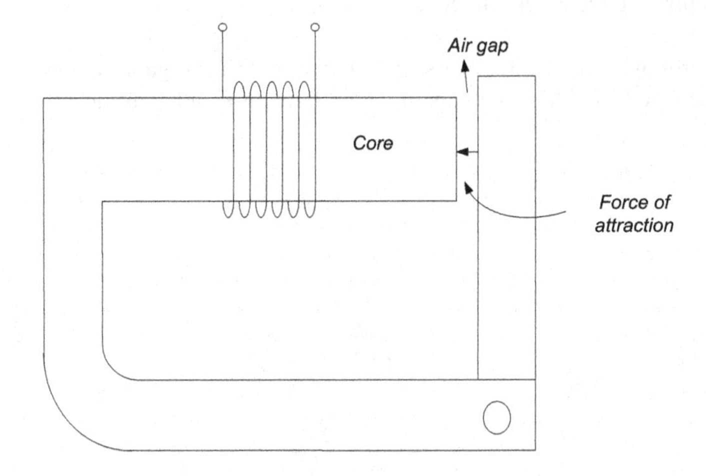

FIGURE 3.17 Telephone exchange relay utilises force of alignment.

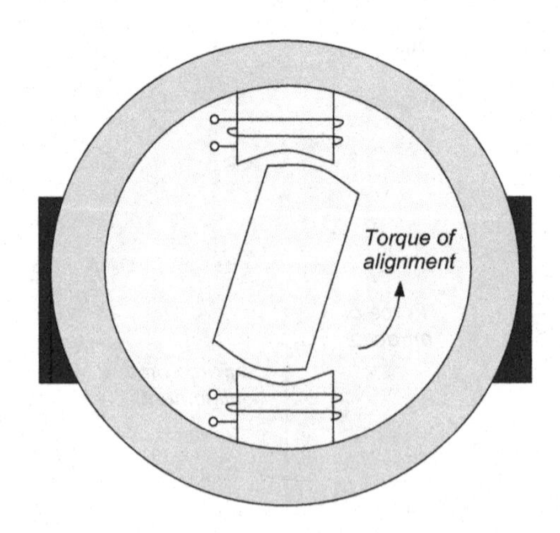

FIGURE 3.18 Reluctance motor utilises force of alignment.

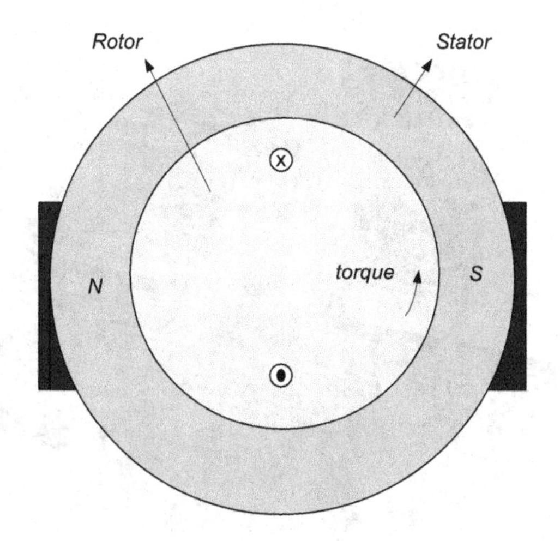

FIGURE 3.19 Simple electrical machine.

3.7.2 Force of Interaction

Force of interaction is utilised in the principles of operation of the electrical machines. The simple machine shown in Figure 3.19 illustrates the principle involved in force of interaction. By passing a current through the coil, a force is experienced on each of the coil sides resulting in a torque about the axis of rotation. A practical machine requires many turns of conductors in order to develop a sufficient torque. Depending on how the conductors are arranged, various electrical machines are created.

3.8 BASIC STRUCTURE OF ROTATING ELECTRICAL MACHINES

Structurewise electrical machines have two major parts, stator and rotor, separated by an air gap, as shown in Figures 3.20 and 3.21.

Stator: This part of the machine does not move and normally is the outer frame of the machine.
Rotor: This part of the machine is free to move and normally is the inner part of the machine.

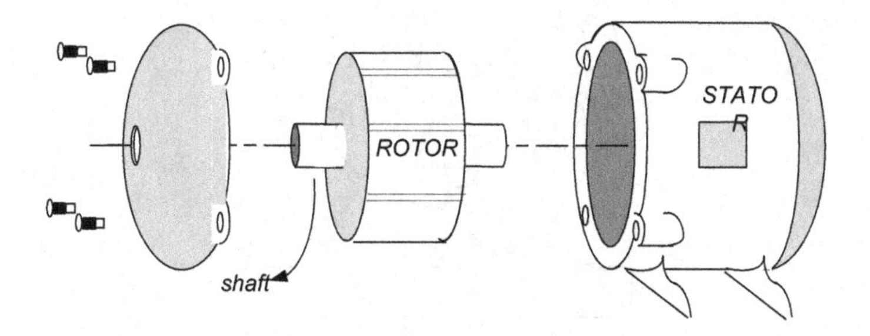

FIGURE 3.20 Typical electrical machine.

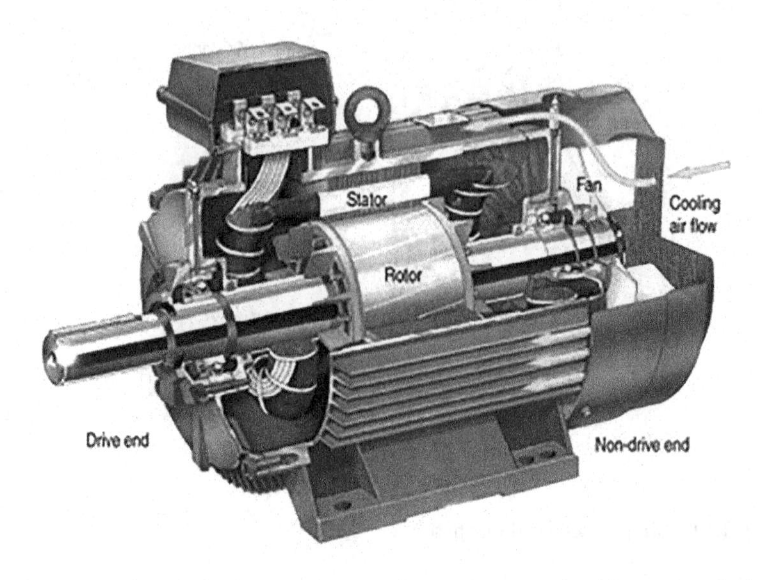

FIGURE 3.21 Electrical machine, 4 W induction motor (GE Power Systems, U.S.A).

Magnetic materials of electrical machines require materials of high permeability and low resistivity. Proper mechanical strength is another requirement. A complete 4 kW electrical machine is illustrated in Figure 3.21.

In electrical machines,

- the coil in which the magnetic field is created is called the "field winding."
- the coil to which the voltage is applied (or received) is called the "armature winding."

Whether the field or the armature winding is on the stator or rotor winding varies according to the type of the machine.

Machine windings are inserted in the slots of the machine (Figure 3.22) or wound on the protruding parts of the relevant structure (Figure 3.27).

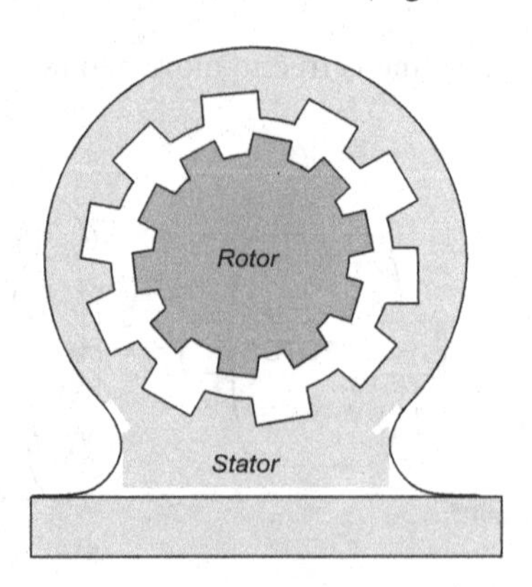

FIGURE 3.22 Rotor, stator, and slots of electrical machines.

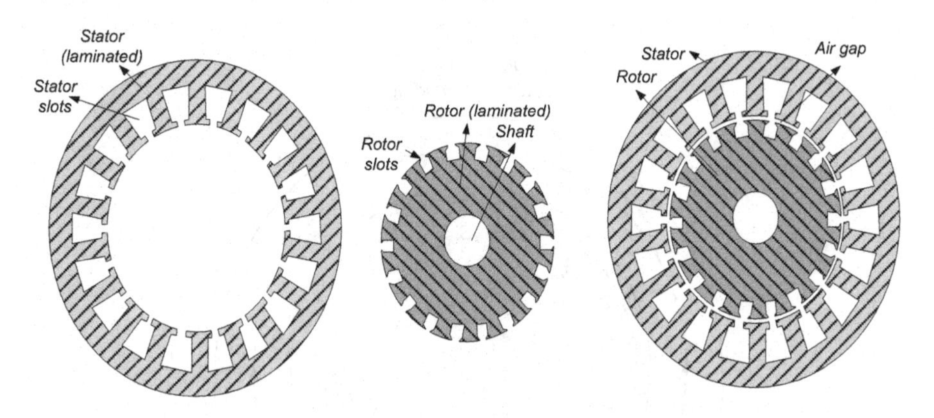

FIGURE 3.23 Stator and rotor laminations.

The part of the machine (stator and/or rotor) in which AC voltages and currents are involved, must be made in laminated form for smaller iron losses (Figure 3.23). Laminated structure is not required if the part is not subjected to AC currents.

3.8.1 Structure Types of Rotating Electrical Machines

In "salient pole structure" coil is wound around protruding poles (Figures 3.24, 3.27). Rotating machines which have a salient pole rotor or stator are called salient pole machines. In salient pole machines, there is a nonuniform air gap.

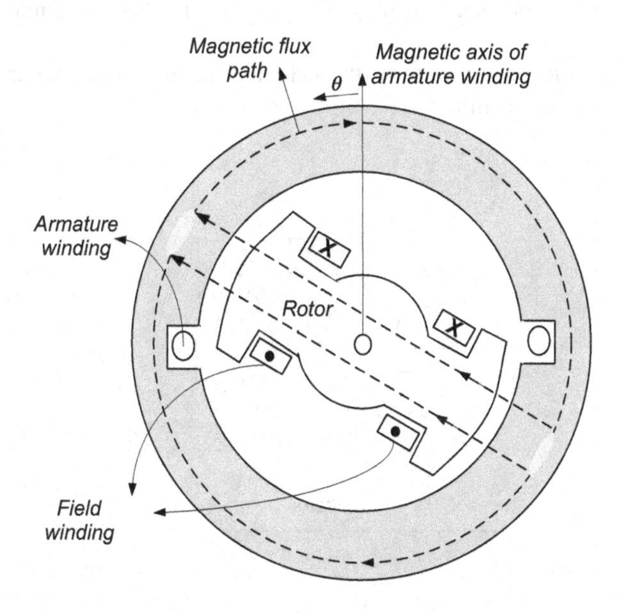

FIGURE 3.24 Two-pole single-phase salient pole (rotor) synchronous machine.

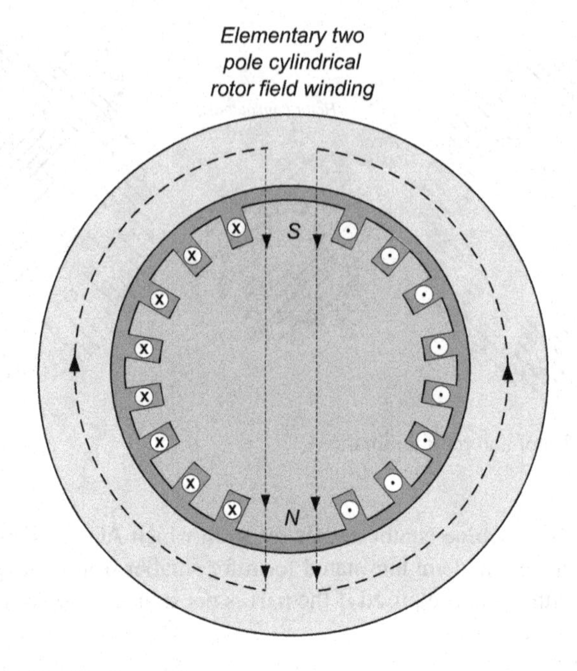

FIGURE 3.25 Cylindrical machine.

In cylindrical (nonsalient) pole structure, coil is wound in slots cut into a cylindrical magnetic structure (Figure 3.26).

Rotating machines which have a cylindrical rotor and stator (Figure 3.25) are called "cylindrical" or "nonsalient pole" machines. In these machines, there is a uniform air gap.

Depending on the machine type, the field and the armature coils can be on the stator or rotor of the machine.

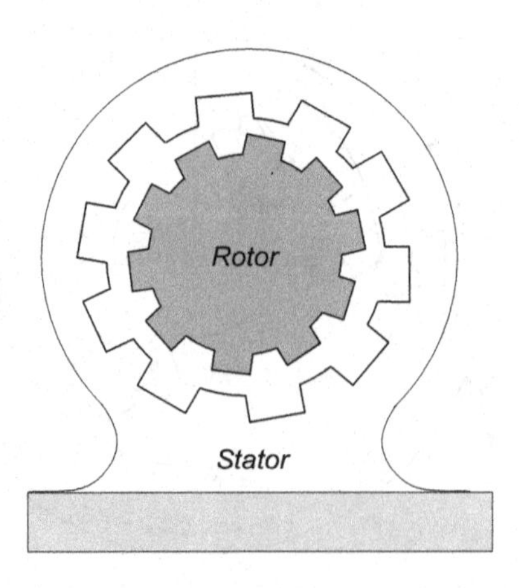

FIGURE 3.26 Cylindrical machine with slots in rotor and stator.

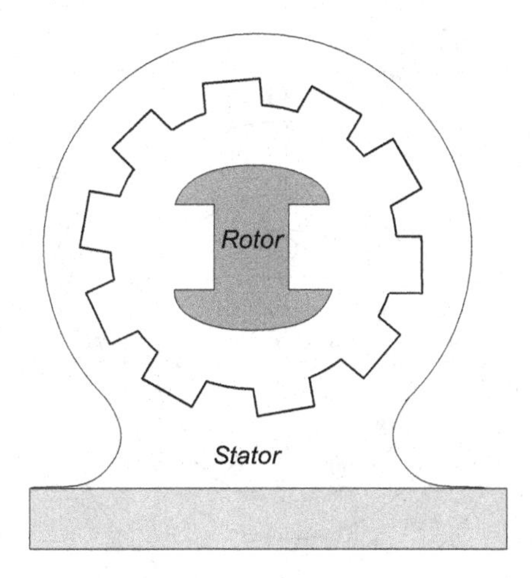

FIGURE 3.27 Salient pole machine with slots in stator.

a. DC machines: Rotor is cylindrical and statoris salient pole (Figure 3.28)
    - field is on stator.
    - armature is on rotor.
b. Induction machines: Both rotor and stator are cylindrical (Figure 3.29).
    - Field winding is on the stator.
    - Armature winding is on the rotor.
c. Synchronous machines: Stator is cylindrical, and rotor is cylindrical or salient pole (Figure 3.30).
    - Field is on rotor.
    - Armature is on stator.

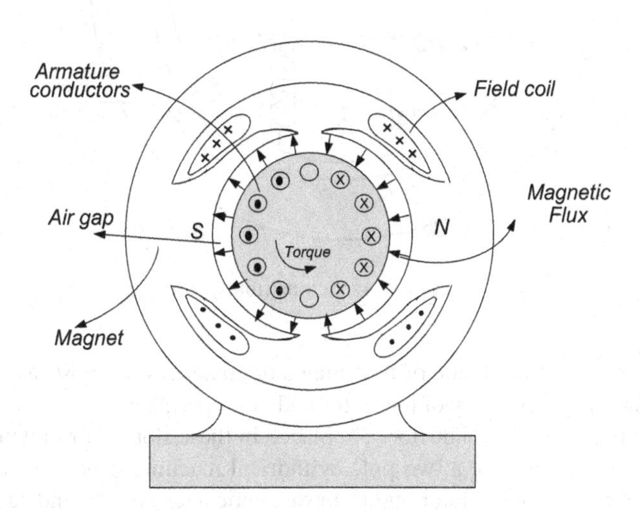

FIGURE 3.28 Salient pole DC machine (two-pole).

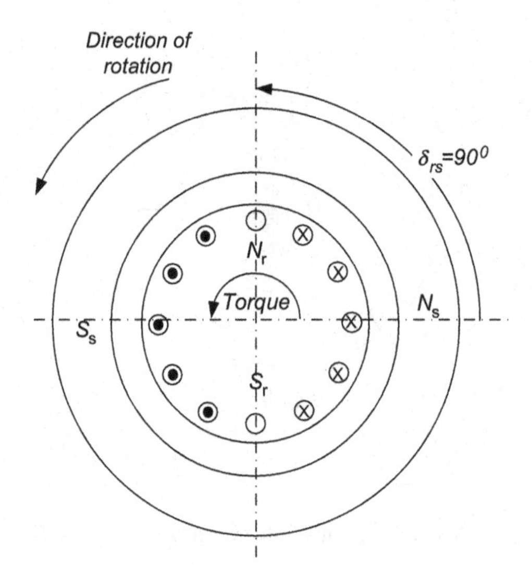

FIGURE 3.29 Induction motor structure.

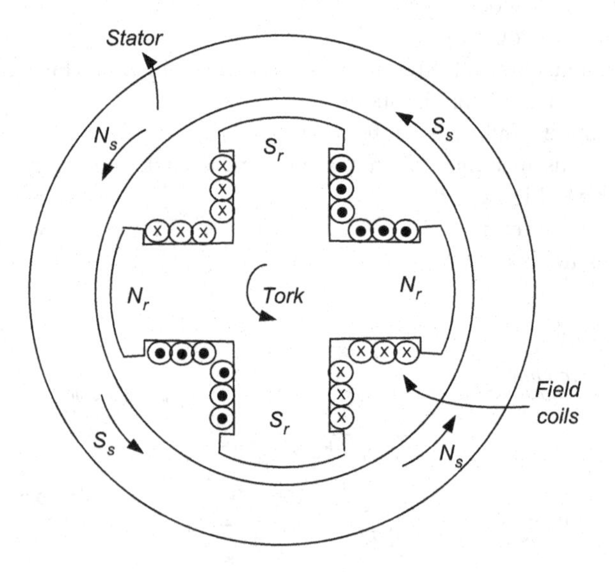

FIGURE 3.30 Salient pole synchronous machine (four-pole).

Both stator and rotor are made of ferromagnetic materials. In most machines, slots are cut on the inner periphery of the stator and outer periphery of the rotor structure, as shown in Figure 3.31. Conductors are placed in these slots to form windings.

In Figure 3.32A, we see a two-pole cylindrical machine structure showing how the windings are situated. Stator and rotor magnetic axes are also indicated. We see the schematic illustration of the same machine in Figure 3.32B.

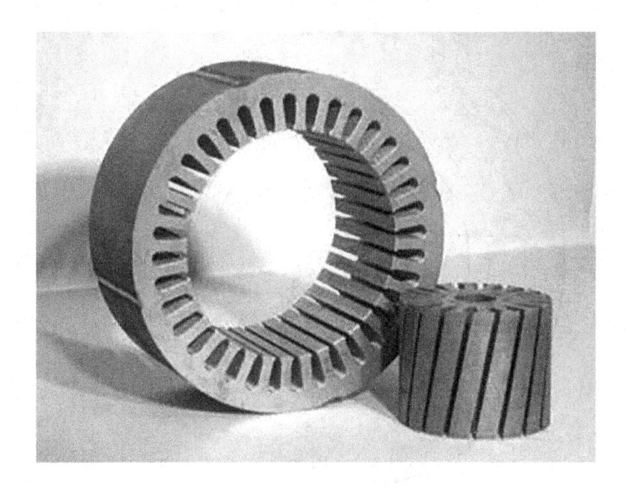

FIGURE 3.31 Slotted structure of rotor and stator.

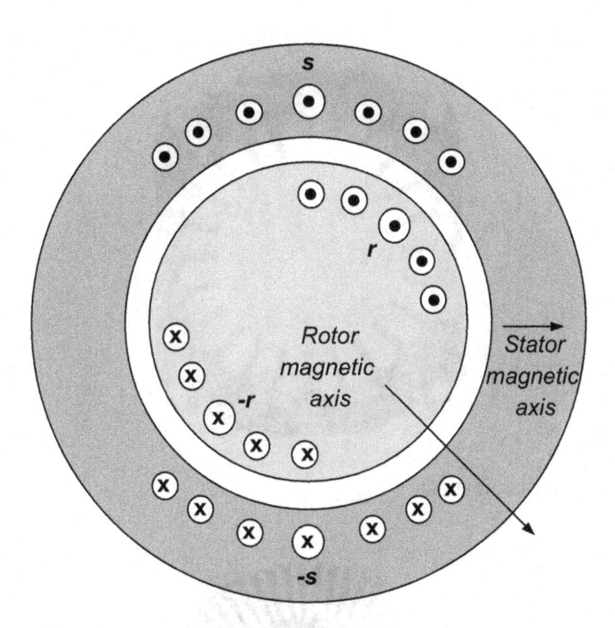

FIGURE 3.32A Two-pole cylindrical machine showing windings.

If the stator or rotor (or both) is subjected to a time-varying magnetic flux, the iron core is laminated to reduce eddy current losses. Eddy currents heat up the core and wastes energy. If laminations (Figures 3.33 and 3.34) are provided, the area is reduced and hence resistance is very high which limits the current to a minimum value.

The winding in which voltage is induced is called the armature winding. The winding through which a current is passed to produce the primary source of flux in the machine is called the field winding. Permanent magnets are used in some

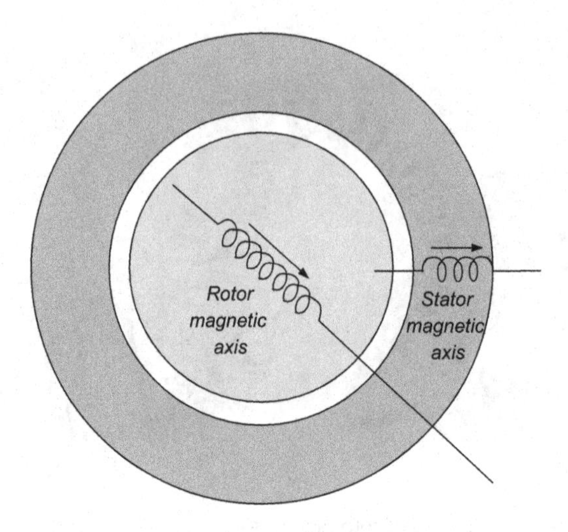

FIGURE 3.32B Two-pole cylindrical machine schematic illustration.

FIGURE 3.33 Stator laminations.

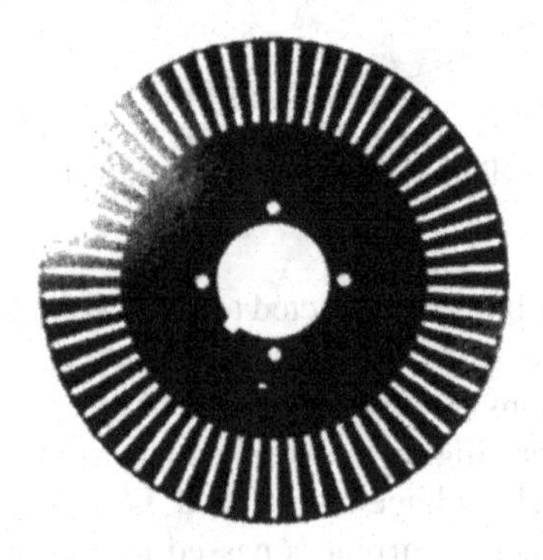

FIGURE 3.34 Rotor laminations.

machines to provide the major source of flux in the machine. The three basic and common rotating electrical machines are

- DC machines
- induction machines
- synchronous machines.

To understand the behaviour of rotating machinery, we must realise that the rotor structure is a magnetic field, and the stator structure is another magnetic field. In a similar manner, a compass needle tries to align with the Earth's magnetic field, these two sets of fields attempt to align, and a torque is therefore produced with their displacement from alignment. Thus, in a motor, the stator magnetic field rotates ahead of that of the rotor, pulling on it and performing work. The opposite is true for a generator in which the rotor does the work on the stator. The following formula

$$F = q.v.B\sin\theta = i.l.B.\sin\theta$$

is the simplest and easiest way to calculate the forces acting on systems. Unfortunately, it is not useful apart from some practical situations. A more general method called "energy method" is used to calculate the net forces in the electromechanical energy conversion devices.

3.9 ENERGY BALANCE METHOD

Basics of the energy balance method rely on the principle of conservation of energy. The simple force-producing device shown in Figure 3.35 is schematically modelled in Figure 3.36 as a lossless, magnetic field-based electromechanical energy conversion device with an electrical input and mechanical output. In between is a lossless magnetic energy storage system which acts as a reservoir between the electrical input and mechanical output terminals.

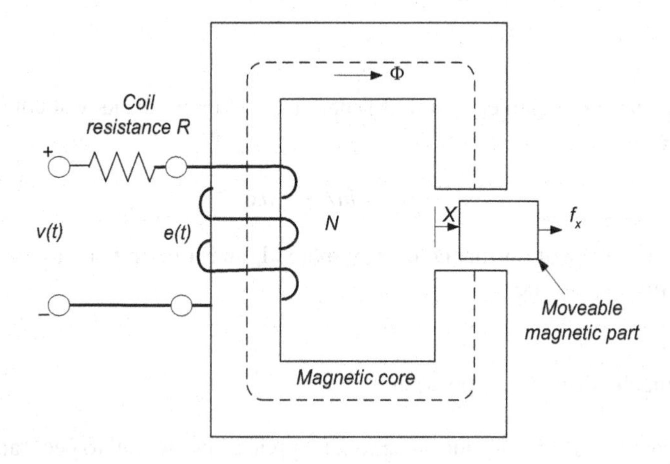

FIGURE 3.35 Magnetic field-based electromechanical energy conversion device.

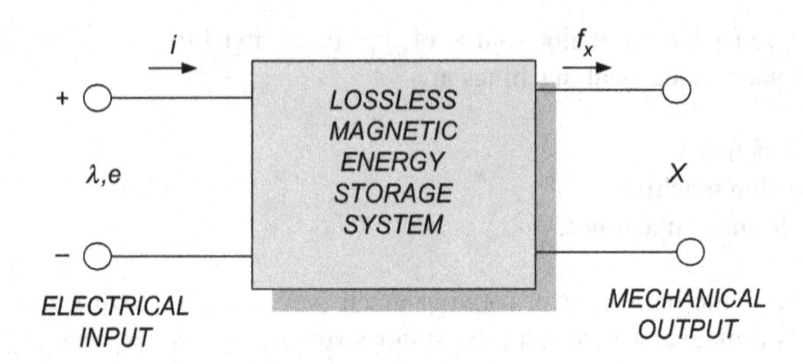

FIGURE 3.36 Model of the device in Figure 3.35.

The electrical input has two variables:

- a voltage e
- a current i.

The mechanical output also has two variables:

- a force f_x
- a position x.

The electromechanical energy conversion will occur through the medium of the magnetic energy. The time rate of change of the stored energy in the magnetic field (W_f) is equal to the electric power input minus the mechanical power output.

$$\frac{dW_f}{dt} = ei - f_x \frac{dx}{dt} \tag{3.1}$$

From Faraday's law, we have

$$e = \frac{d\lambda}{dt} \tag{3.2}$$

Inserting equation (3.2) in equation (3.1) and multiplying both sides of equation (3.1) with dt, we get

$$dW_f = id\lambda - f_x dx \tag{3.3}$$

This is the basic equation for the energy method, and it permits us to solve for the force as a function of the

- flux λ
- the mechanical output position x.

The equation above holds for motor action. It applies equally well to generator action; right-hand terms would then have negative values.

3.10 ENERGY IN SINGLY-EXCITED MAGNETIC FIELD SYSTEMS

3.10.1 Derivation of Force from Energy

So far, we have mainly dealt with fixed-geometry magnetic circuits. The electromechanical energy conversion device shown in Figure 3.37 constitutes a lossless magnetic energy storage system with its magnetic core and armature, and it has an air gap which can change in dimensions only in the direction of displacement x. The system can schematically be shown as in Figure 3.36. We will assume that magnetic energy stored in the system is totally in the air gap. The air gap in the electromechanical energy conversion device therefore acts like a buffer energy between the electrical and mechanical systems.

Since the system is lossless, the magnetic energy stored in the system (W_f) is a state function of the two independent variables λ and x. As a property of state functions, the value of W_f is determined by the final values of x and λ, and it is irrelevant of how they are brought to these values. The case resembles the potential energy of a particle which is independent of the path through which the particle has come to its final height $(PE = mgh)$. Therefore as illustrated in Figure 3.38, in order to reach to a certain value of stored magnetic energy $W_{f1}(x_1, \lambda_1)$ in the system we can choose any path which starts from the origin in Figure 3.38 and ends at point $W_{f1}(x_1, \lambda_1)$. Three such paths $a\,(a = a_1 + a_2)$, b, and c are shown in Figure 3.38.

The magnetic energy stored (W_f) in a lossless electromechanical energy conversion system such as the one in Figure 3.37 can be obtained by the integration of the energy balance equation (3.3).

$$dW_f = i d\lambda - f_x dx \rightarrow W_f = \int i d\lambda - \int f_x dx$$

Because of the properties of the state functions, the result of the above integral is irrelevant of how we arrived at this value of W_f. We can go along any of the paths (a, b, or c) shown in Figure 3.38. They all ought to give the same result.

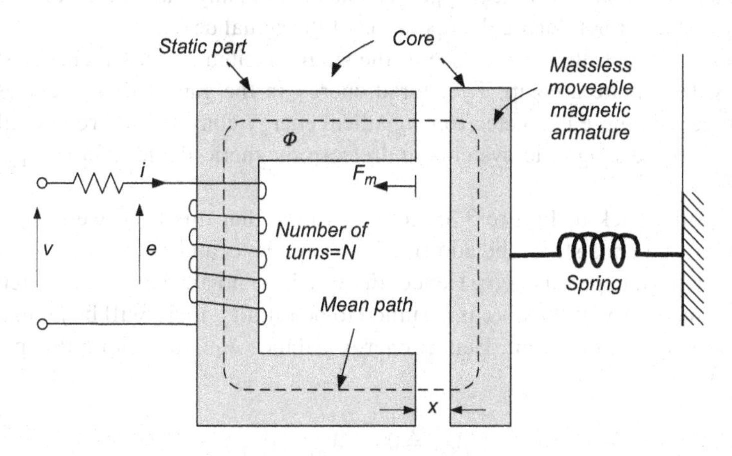

FIGURE 3.37 Singly excited electromechanical system.

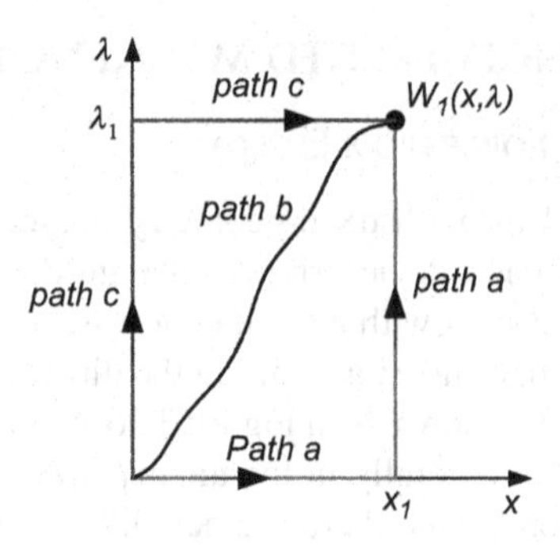

FIGURE 3.38 Energy stored in magnetic field W_f (state function).

Let us assume that we have chosen path "a" which consists of the addition of two paths a_1 and a_2 $(a = a_1 + a_2)$. Along path a_1, we have kept λ as zero. Hence, there will be no magnetic field, and obviously there will be no force f_x, and therefore there will be no magnetic energy stored in the system. Hence, energy balance equation along this path (a_1) reduces to

$$dW_{fa_1} = 0 \rightarrow W_{fa_1} = 0$$

So no contribution to stored magnetic energy comes from this path a_1. In the second part of the path a (a_2), we keep x constant at x_1 and increase λ until we reach to $\lambda = \lambda_1$. In this case, the term dx will be zero. Therefore, the energy balance equation reduces to

$$dW_f = id\lambda \rightarrow i = \left.\frac{\partial W_f}{\partial \lambda}\right|_x \tag{3.4}$$

where the partial derivative is taken while holding x constant. We must keep in mind that the above is a mathematical approach and is not in any way related to whether x is held constant or not during the operation of the actual device.

In actual of fact, this is the case of the singly excited magnetic energy storage system with a static air gap. The stored energy in the gap of these devices does not change, and there is no electromechanical energy conversion. We have already dealt with these magnetic systems and electromagnetic devices in our previous chapters.

Now going back to Figure 3.38, let us assume that this time we have chosen path "c" which consists of the addition of two paths c_1 and c_2 $(c = c_1 + c_2)$. Along path c_1, we have kept x as zero. Hence, there will be no air gap in the system, and as the permeability of the core is assumed to be infinity, there will be no magnetic energy stored in the system. Hence, energy balance equation along this path (c_1) reduces to

$$dW_{fc_1} = 0 \rightarrow W_{fc_1} = 0$$

So no contribution to stored magnetic energy comes from this path c_1. In the second part of the path c (c_2), we keep λ constant at λ_1 and increase x until we reach to $x = x_1$. In this case, the term $d\lambda$ will be zero. Therefore, the energy balance equation reduces to

$$dW_f = -fdx \rightarrow f = -\left.\frac{\partial W_f(\lambda, x)}{\partial x}\right|_{\lambda} \tag{3.5}$$

where the partial derivative is taken while holding λ constant. If $W_f(x, \lambda)$ is known as a function of x and λ, force f_x can be calculated from the above equation. This is the equation we need for determining the force in an electromechanical energy conversion configuration. We must keep in mind that the above is purely a mathematical approach and is not in any way related to whether λ is held constant or not during the operation of the actual device.

For a linear system where $L(x)$ is the self-inductance of the configuration, we have

$$\lambda = i.L(x)$$

Integrating equation (3.4) while keeping x a constant and inserting the above linear relation in the result, we get

$$W_f = \int_0^{\lambda} id\lambda = \int_0^{\lambda} \frac{\lambda}{L(x)} d\lambda = \frac{\lambda^2}{2L(x)} = \frac{1}{2} i^2.L(x) \tag{3.6}$$

Now let us insert this in equation (3.5),

$$f = -\left.\frac{\partial W_f(\lambda, x)}{\partial x}\right|_{\lambda} = -\frac{\partial}{\partial x}\left[\frac{\lambda^2}{2L(x)}\right]_{\lambda} = \frac{\lambda^2}{2L(x)^2}\frac{dL(x)}{dx}$$

Using this expression, we get

$$\lambda = L(x)i \rightarrow f_x = \frac{i^2}{2}\frac{dL(x)}{dx} \tag{3.7}$$

3.10.2 Derivation of Force from Coenergy

According to the energy equation (3.3) for a particular value of the air gap length shown in Figure 3.37, the energy stored in the magnetic field W_f is given by

$$dW_f = id\lambda$$

which means that the area between the λ axis and the $\lambda - i$ characteristic represents the energy stored in the magnetic field in the air gap. The area between the i axis and the $\lambda - i$ characteristic is known as the coenergy W_f' and is defined as

$$dW_f' = \int_0^i \lambda di$$

From Figure 3.39,

$$W_f' + W_f = \lambda i$$

Coenergy has no physical significance. However, it can be used to derive expressions for force (or torque) developed in an electromagnetic system. Let us remember that

if $\lambda - i$ curve is nonlinear $\rightarrow W_f' > W_f$
if this curve is linear $\rightarrow W_f' = W_f$

Now we shall find the expression for force f_x from another approach. Without changing anything else, let us slowly push the movable part in Figure 3.37 from position x_1 to another position x_2 so that air gap decreases. The current i before and after the movement will be the same. Since W_f' is a state function, what happens to current i during the motion has no importance. We can assume the movable part has moved slowly and the current i has remained the same. The λ-i characteristics of the system for these two positions are shown in Figure 3.40 as a and b.

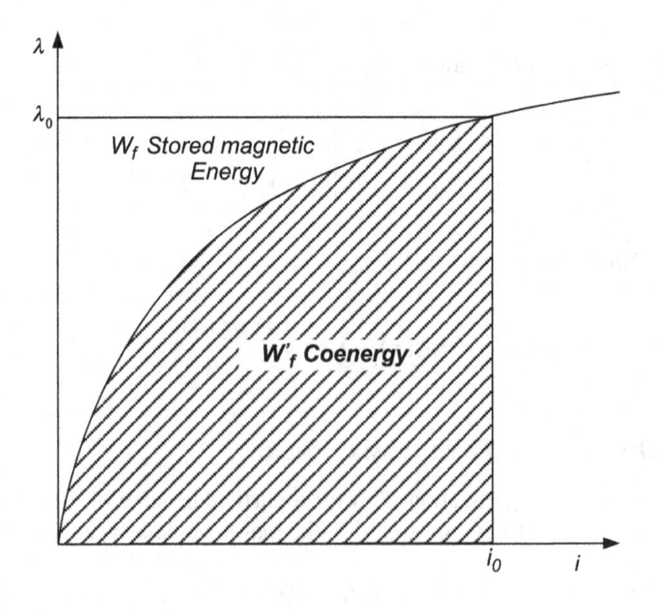

FIGURE 3.39 Stored magnetic energy and co-energy.

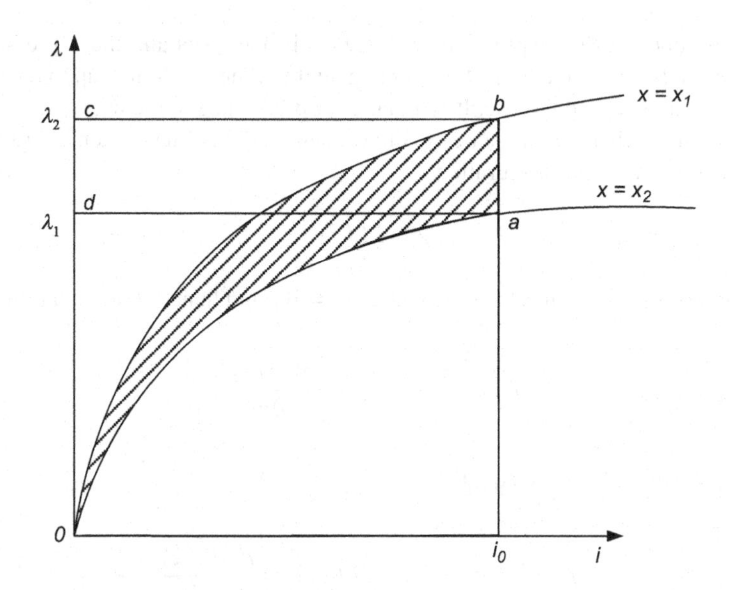

FIGURE 3.40 Energy and coenergy of magnetic fields.

The operating point has therefore moved upward from point a to b as shown in Figure 3.40. During the motion,

$$dW_e = \int ei.dt = \int_{\lambda_1}^{\lambda_2} i.d\lambda = \text{alan } abcd$$

$$dW_f = \text{alan } 0bc - 0ad$$

$$dW_{\text{mech}} = dW_e - dW_f$$

$$= \text{area } abcd + \text{area } 0ad - \text{area } 0bc$$

$$= \text{area } 0ab$$

If the motion has occurred under constant-current conditions, the mechanical work done is represented by the shaded area, which, in fact, is the increase in the coenergy (or decrease in the magnetic energy):

$$dW_{mech} = dW_f'$$

If f_x is the mechanical force causing the differential displacement dx,

$$f_x dx = dW_{\text{mech}} = dW'$$

$$f_x = \left. \frac{\partial W_f'(i, x)}{\partial x} \right|_{i=\text{constant}} \tag{3.8}$$

We have found another expression for force f_x. In fact, this and the expression we found before for the same force f_x ought to give the same result. It is up to the user to decide which expression to use. It is a matter of preference of the user.

If we assume that the magnetic system in Figure 3.37 is linear, then if $L(x)$ is the self-inductance of the configuration,

$$\lambda = i.L(x)$$

Because the system is linear, we also have $W_f' = W_f$ and hence from equation (3.7),

$$f_x = \left.\frac{\partial W_f'(i, x)}{\partial x}\right|_i = \left.\frac{\partial W_f(i, x)}{\partial x}\right|_i$$

Inserting (3.6) in the above equation,

$$f_x = \left.\frac{\partial W_f}{\partial x}\right|_i = \frac{\partial}{\partial x}\left[\frac{i^2}{2}L(x)\right]_i = \frac{i^2}{2}\frac{dL(x)}{dx} \tag{3.9}$$

which is as expected the same as the result we obtained in equation (3.7) in the above paragraphs.

3.11 FORCE OF ALIGNMENT BETWEEN PARALLEL MAGNETISED SURFACES

From here onwards, we will be using the following relationship for calculating the force in electromechanical energy conversion arrangements:

$$f_x = \left.\frac{\partial W_f}{\partial x}\right|_i$$

The force of alignment between the core members of the arrangement shown in Figure 3.41 is

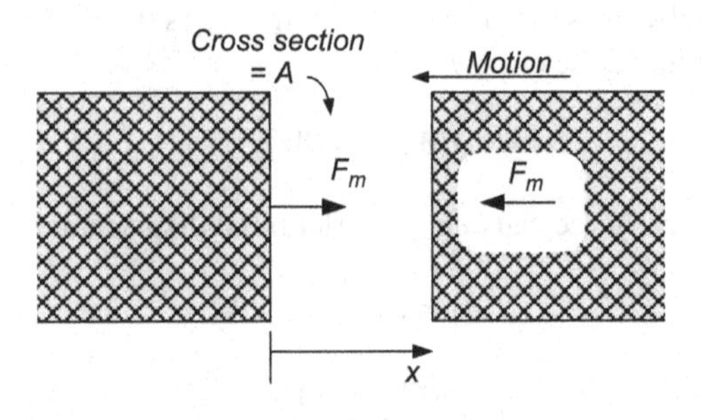

FIGURE 3.41 Force of alignment.

$$\text{Force of alignment, } F = \frac{dW_f}{dx}$$

$$W_f = (\text{Energy density}) \times (\text{Volume of the air gap}) = w_f \times V$$

$$= w_f \times (A.x) = \frac{B^2}{2\mu_0} \times (A.x)$$

As a result,

$$F = \frac{dW_f}{dx} = \frac{B^2 A}{2\mu_0}$$

This force will be in the direction to decrease the reluctance of the system by decreasing the distance (shortening the flux path) between the magnetic poles so that flux can travel the path of least resistance.

3.12 LATERAL FORCE OF ALIGNMENT BETWEEN PARALLEL MAGNETISED SURFACES

Ignoring the effect of leakage flux, let the cross-sectional area of the gap be xl and the gap length be l_g. The air gap volume is given by (Figure 3.42)

$$\text{Volume} = V = ll_g x$$

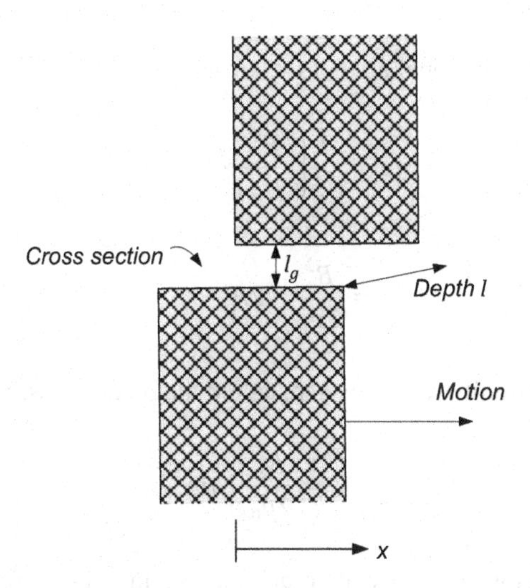

FIGURE 3.42 Lateral force of alignment.

$$W_f = w_f l l_g x = \frac{B^2}{2\mu_0} l l_g x$$

$$F = \frac{dW_f}{dx} = \frac{B^2}{2\mu_0} l l_g$$

3.13 ALTERNATIVE FORCE CALCULATION IN A LINEAR SYSTEM

In Figure 3.37, if the reluctance of the magnetic core path is negligible compared to that of the air gap path, the λ-i relation becomes linear. This was extensively examined in 3.10. For this idealised system where $L(x)$ is the inductance of the coil, whose value depends on the air gap length, we have found the force equation as follows (equations 3.7 and 3.9),

$$f_m = \frac{1}{2} i^2 \frac{dL(x)}{dx}$$

We can yet approach the linear system from another point of view to calculate the force produced in the air gap. For the system in Figure 3.37, if the reluctance of the magnetic core path is neglected, the field energy is

$$W_f = \frac{B_g^2}{2\mu_0} \times \text{Volume of air gap}$$

$$W_f = \frac{B_g^2}{2\mu_0} \times A_g.2g$$

where A_g is the cross-sectional area of the air gap.

$$f_m = \frac{\partial}{\partial g}\left(\frac{B_g^2}{2\mu_0} \times A_g.2g\right)$$

$$= \frac{B_g^2}{2\mu_0} \times 2A_g$$

Hence, the force per unit area of air gap, called magnetic pressure F_m, is

$$F_m = \frac{B_g^2}{2\mu_0} \text{ N/m}^2 \tag{3.10}$$

In most linear systems, the above approach may seem to be more practical to calculate the force in the air gap.

Exercise 3.5:

The magnetic circuit in Figure 3.43 has N (number of turns) = 1,000, i (current) = 1 A, air gap width = 1.0 cm, air gap depth = 2 cm, and air gap length l_g = 2 mm. Assuming the permeability of the core is infinity and the leakage and fringing flux effects are negligible, calculate

a. the force of attraction on the two faces of the air gap.
b. energy stored in the air gap.

Solution:

a. Ampere's circuital law, $\rightarrow Hl = Ni$

$$\rightarrow \left(\frac{B_g}{\mu_0}\right) l_g = Ni \rightarrow B_g = \frac{Ni\mu_0}{l_g}$$

$$\text{Force} = f_m = \frac{B_g^2}{2\mu_0} \times A_g = \frac{\left(\frac{Ni\mu_0}{l_g}\right)^2 \times A_g}{2\mu_0} = \frac{N^2 i^2 \mu_0}{2l_g^2} A_g$$

$$= \frac{1000^2 \times 1^2 \times 4\pi \times 10^{-7}}{2 \times 2^2 \times 10^{-6}} \times 2 \times 1 \times 10^{-4} = 31.4\,\text{N}$$

b.

$$W_f = \frac{B_g^2}{2\mu_0} \times \text{air gap volume} = \frac{B_g^2}{2\mu_0} \times A_g \times l_g = 31.4 \times 2 \times 10^{-3}\,\text{J} = 0.0628\,\text{J}$$

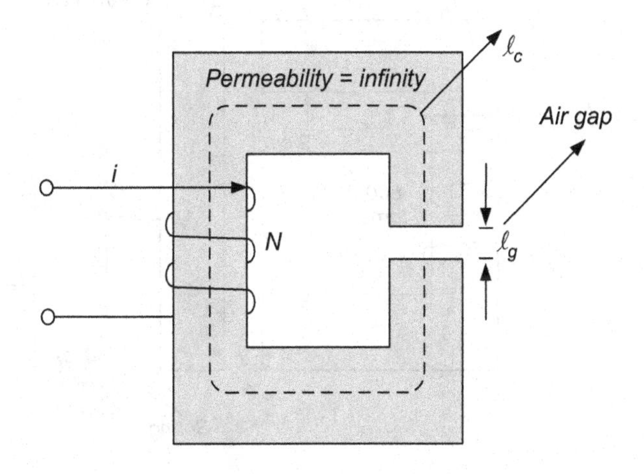

FIGURE 3.43

Exercise 3.6:

The electromechanical energy conversion device shown in Figure 3.44 is in equilibrium when the gap distance g is at 0.5 mm and the current intake is $i = 1$ A. The torque produced by the system on the arm is 30 nm. If the cross-sectional area of the pole faces is $4 \times 4\,\text{cm}^2$ and the reluctance of the core can be neglected, calculate the length l of the arm.

Solution:

Let us write the Ampere's circuital law by ignoring the core,

$$Ni = H_g \times 2g = \frac{B_g}{\mu_0} \times 2g$$

$$\rightarrow B_g = \frac{Ni\mu_0}{2g}$$

$$W = \text{Magnetic energy stored in the air gap} = \frac{B_g^{\ 2}}{2\mu_0} \times \left(\text{Volume of the air gap}\right)$$

$$F = \frac{dW}{dx} = \frac{B_g^{\ 2}}{2\mu_0} \times 2A = \left(\frac{Ni\mu_0}{2g}\right)^2 \times \frac{A}{\mu_0}$$

$$= \frac{500^2 \times 1^2 \times 4\pi \times 10^{-7} \times 4 \times 4 \times 10^{-4}}{4 \times 0.5^2 \times 10^{-6}}$$

$$= 502.4\,\text{N-m}$$

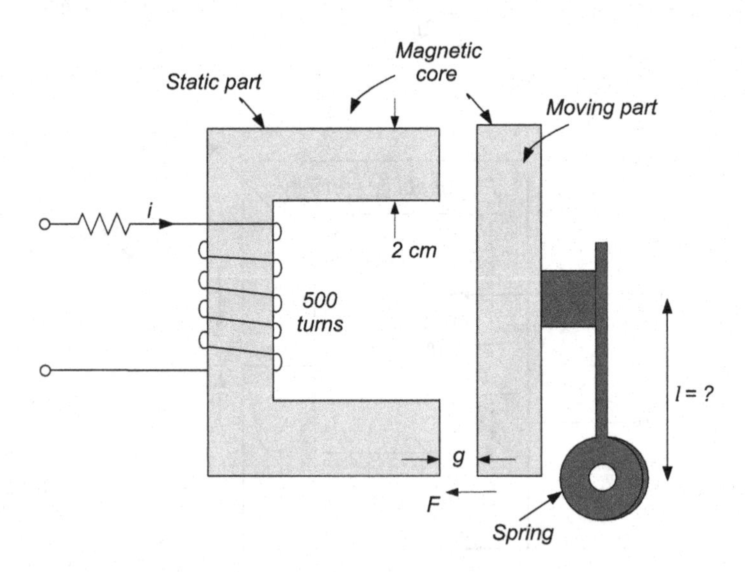

FIGURE 3.44

$$\text{Torque} = 30 = F \times l$$

$$l = \frac{3000}{160\pi} = 5.97\,\text{cm}$$

PROBLEMS ON CHAPTER 3

Problem 3.1: In the magnetic circuit in Figure 3.45, the density of the freely hanging lower part of the core is 20 g/cm^3. The length and the cross-sectional area of the same part are 75 cm and $6 \times 5\,\text{cm}^2$, respectively. The air gap cross-sectional area is $4 \times 6\,\text{cm}^2$, and its depth is 1 cm. If the number of turns of the coil is $N = 200$ and the reluctance of the core and the fringing of the flux in the air gap are negligible, calculate

a. the current intake of the coil.
b. magnetic energy stored in the system.

Problem 3.2: The reluctance of the magnetic material of the relay and the piston in Figure 3.46 can be assumed to be negligible. If the coil has 100 turns and $g = 1.0\,\text{mm}$, $m = 15\,\text{cm}$, $l = 10\,\text{cm}$, and $i = 8$ A, assuming no fringing of the magnetic flux in the air gap, find inductance L of the magnetic circuit as a function of position x.

When $x = 4$ cm what should be the length l of the core in order to store a magnetic energy of 2.7 J? In case (b), what is the force on the piston?

Problem 3.3: In Figure 3.47, the iron block which has a density of 8 g/cm^3 and a cross-sectional area of 50 cm^2 is freely hanging at a distance of 0.1 cm above a magnetic core which is resting on a table. Neglecting the reluctance of the iron block and the core as well as the fringing of the magnetic flux in the air gap and assuming $i = 1$ A, find (a) the stored magnetic energy in the system and (b) the upward force developed on the spring.

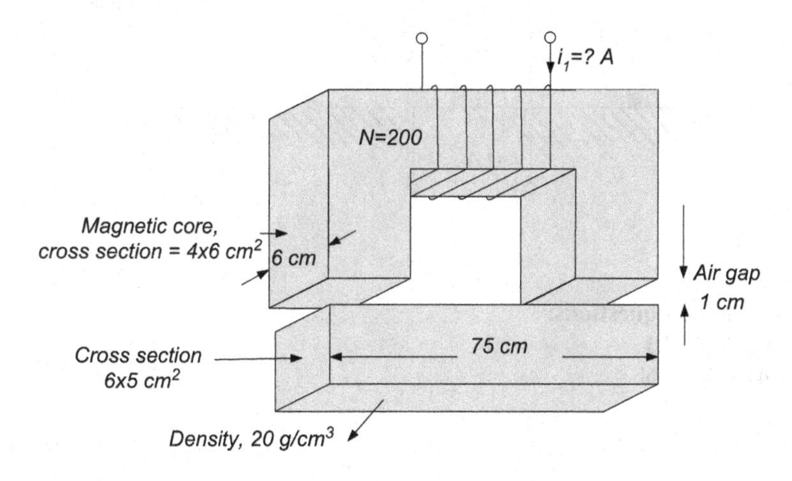

FIGURE 3.45

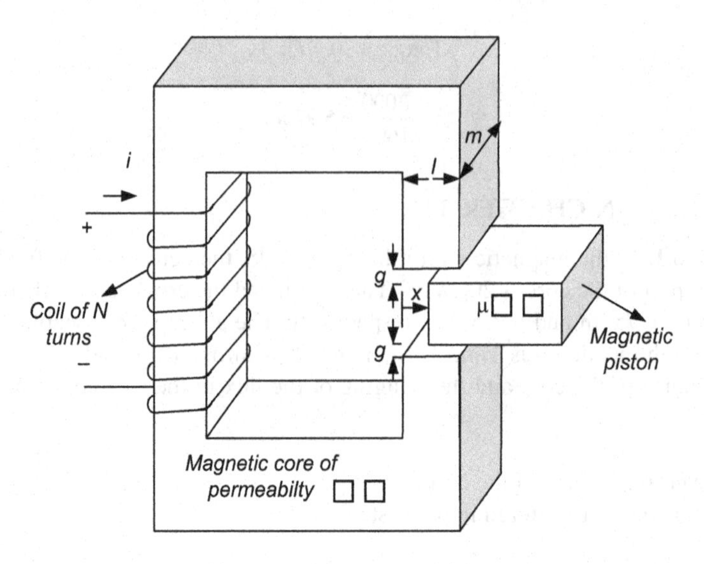

FIGURE 3.46

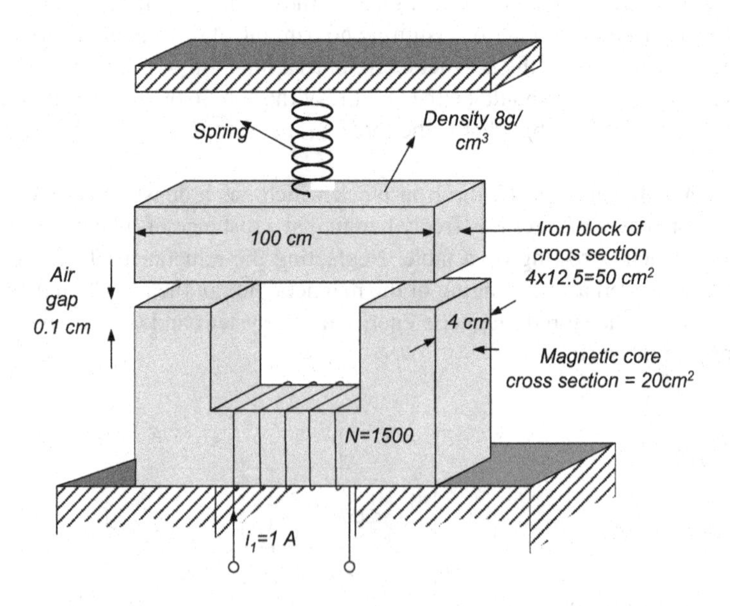

FIGURE 3.47

Answers to odd questions:

3.1: 38.2 A, 4.41 J

3.3: 1.41 J, 1798 N.

4 Rotating Electrical Machines (General)

4.1 ROTATING ELECTRICAL MACHINES

Most of the electromechanical energy converters and especially the high-power ones produce rotational motion. In these types of energy converters, the rotor is mounted on a shaft and is free to rotate between the poles of the stator. Let us take a general case in which both stator and rotor have windings carrying currents i_s and i_r, as shown in Figure 4.1. The current is fed into the rotor circuit through fixed brushes and rotor-mounted slip rings.

As we saw before, W_f, the magnetic energy deposited in the system, is a state function and how it comes to a certain value is of no consequence. So let us assume the system is static, there is no mechanical output, and W_f is the stored magnetic field energy within the system. Consequently,

$$dW_f = e_s i_s dt + e_r i_r dt$$

$$= i_s d\lambda_s + i_r d\lambda_r \tag{4.1}$$

For a linear magnetic system,

$$\lambda_s = L_{ss} i_s + L_{sr} i_r \tag{4.2}$$

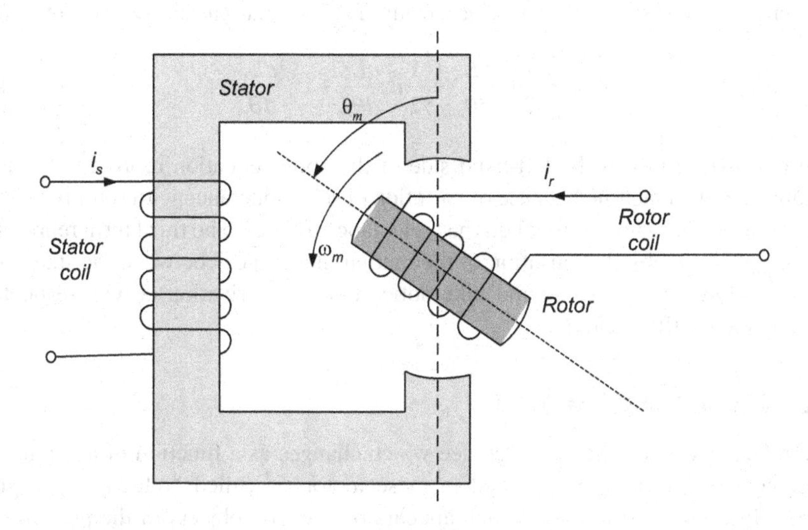

FIGURE 4.1 General structure of a rotating machine.

$$\lambda_r = L_{rs} i_s + L_{rr} i_r \tag{4.3}$$

Here,

L_{ss} is the self-inductance of the stator winding.
L_{rr} is the self-inductance of the rotor winding.
L_{rs} and L_{sr} are mutual inductances of the rotor and stator windings.

From equations (4.1), (4.2), and (4.3),

$$dW_f = i_s d\left(L_{ss} i_s + L_{sr} i_r\right) + i_r d\left(L_{sr} i_s + L_{rr} i_r\right)$$

$$= L_{ss} i_s di_s + L_{rr} i_r di_r + L_{sr} d\left(i_s i_r\right)$$

Total field energy will be

$$W_f = L_{ss} \int_0^{i_s} i_s di_s + L_{rr} \int_0^{i_r} i_r di_r + L_{sr} \int_0^{i_s i_r} d\left(i_s i_r\right)$$

$$= \frac{1}{2} L_{ss} i_s^2 + \frac{1}{2} L_{rr} i_r^2 + L_{sr} i_s i_r$$

Similar to the force equation we obtained in previous chapters, in rotational electromechanical energy conversion systems, the torque developed would be

$$T = \left.\frac{\partial W_f'(i, \theta_m)}{\partial \theta_m}\right|_{i=\text{constant}}$$

In a linear magnetic system, the stored magnetic energy W_f would be equal to the coenergy of the system W_f'. Therefore, from the two equations above, we can write

$$T = \frac{1}{2} i_s^2 \frac{dL_{ss}}{d\theta_m} + \frac{1}{2} i_r^2 \frac{dL_{rr}}{d\theta_m} + i_s i_r \frac{dL_{sr}}{d\theta_m} \tag{4.4}$$

The first two terms on the left-hand side of the above equation represent "torques produced in the machine because of variation of self-inductance with rotor position." This component of torque is called the "reluctance torque." The third term represents "torque produced by the variation of the mutual inductance" between the stator and rotor windings. Here, we should take notice that the derivation is with respect to mechanical (shaft) angle θ_m.

4.2 RELUCTANCE MOTOR

The reluctance motor has a reluctance which changes as a function of angular displacement θ. The torque produced in these motors is called "reluctance torque." This is the same as the force which appears on the iron objects in the presence of an external magnetic field. The external field induces an internal field in the iron which interacts with the external field to produce the reluctance torque of this type of motors. Owing to its constant speed operation, it is commonly used in electric clocks, record players, and other precise timing devices. This motor is usually of a single-phase type and is available in the small power range (Figure 4.2).

In essence, reluctance motor is a synchronous motor. The difference is that a reluctance motor does not have a field winding on the rotor. Figure 4.3 shows an elementary, single-phase, two-pole reluctance motor. We assume that all the reluctance of the magnetic circuit is in the air gap. In other words, the permeability of the core material is considered to be infinity.

When the angular displacement θ between stator magnetic axes and the rotor magnetic axis is zero, the effective air gap length (g) is a minimum and the cross-sectional area of the poles A is a maximum. Hence at this instant, the reluctance of the magnetic circuit

$$\mathcal{R} = \frac{2g}{\mu_0 A}$$

is a minimum. Consequently, the inductance of the magnetic circuit,

$$L = \frac{N^2}{\mathcal{R}}$$

would be a maximum. When the magnetic axes of the rotor and the stator are at right angles to each other, the reluctance $\mathcal{R}$ is maximum, leading to a minimum inductance L. Consequently, we can say that the change in air gap g causes the inductance change sinusoidally. Hence, we can assume that

$$L = a + b\cos 2\theta$$

FIGURE 4.2 Reluctance motors. (Courtesy of Switched Reluctance Drives Ltd.)

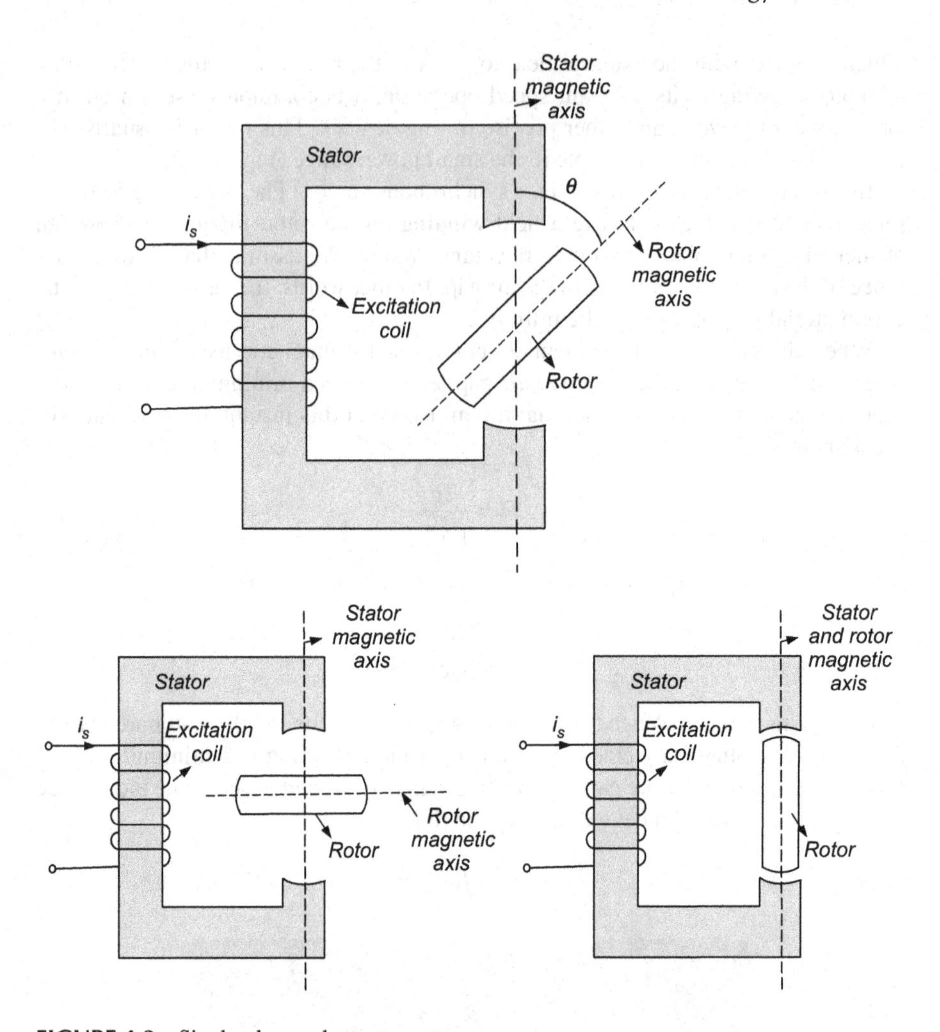

FIGURE 4.3 Single-phase reluctance motor.

When the rotor is rotating with an angular velocity of ω_m, the inductance of the magnetic assembly varies between its maximum and minimum values as shown in Figures 4.4 and 4.3.

$$\theta = 0,\ \pi,\ 2\pi,\ 3\pi\ldots\ldots \rightarrow L = a + b\cos(2\times 0) = a + b = L_{\max} = L_d$$

$$\theta = \frac{\pi}{2},\ \frac{3\pi}{2},\ \frac{5\pi}{2}\ldots\ldots \rightarrow L = a + b\cos\left(\frac{2\pi}{2}\right) = a - b = L_{\min} = L_q$$

Hence,

$$a = \frac{(L_{\max} + L_{\min})}{2}\ ;\ b = \frac{(L_{\max} - L_{\min})}{2}$$

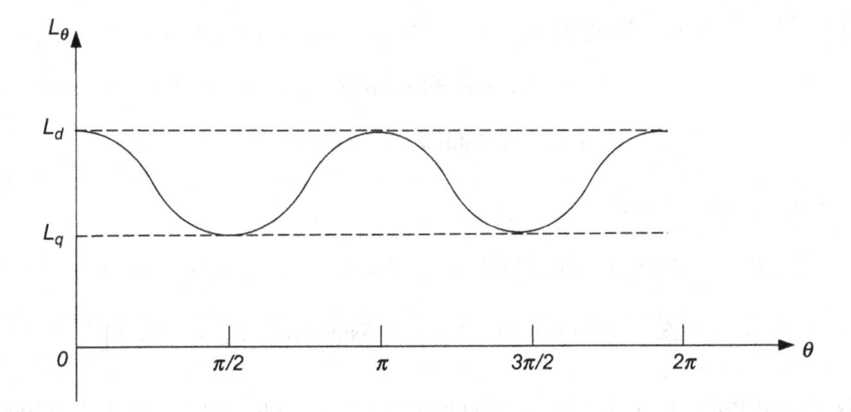

FIGURE 4.4 Variation of inductance of a reluctance motor with displacement.

We can therefore express inductance L as a function of θ as follows:

$$L(\theta) = 0.5(L_d + L_q) + 0.5(L_d - L_q)\cos 2\theta$$

On the other hand, we know from our previous chapters (equation 3.9) that the developed torque is given by

$$T_e = \frac{1}{2} i^2 \left.\frac{\partial L}{\partial \theta}\right|_{i\ \text{constant}}$$

Therefore, the torque on the rotor is

$$T_e = -\frac{1}{2} i^2 (L_d - L_q)\sin 2\theta$$

We can express θ as follows:

$$\theta = \omega_m t + \delta_{rs}$$

In the above equation, θ is the angle of the rotor magnetic axis relative to stator magnetic axis. As a result, the torque equation becomes

$$T_e = -\frac{1}{2} i^2 (L_d - L_q)\sin\left[2(\omega_m t + \delta_{rs})\right]$$

The current in the coil is sinusoidal, so for current i, we can write

$$i = I_m \cos\omega t$$

The torque developed by the rotating system is

$$T_e = -0.5 I_m^2 (L_d - L_q)\cos^2(\omega t)\sin(2\omega_m t + 2\delta_{rs})$$

Using the following trigonometric identities,

$$2\cos^2\alpha = 1 + \cos 2\alpha$$

$$2\sin\alpha\cos\beta = \sin(\alpha+\beta) + \sin(\alpha-\beta)$$

the torque equation becomes

$$T_e = -0.25(L_d - L_q)I_m^2\big[\sin(2\omega_m t + 2\delta_{rs})$$

$$+ 0.5\sin\{2(\omega+\omega_m)t + 2\delta_{rs}\} - 0.5\sin\{2(\omega-\omega_m)t - 2\delta_{rs}\}\big]$$

By examination of the above torque equation, we can conclude that average torque only exists if the mechanical angular speed ω_m is equal to the electrical angular speed ω ($\omega_m = \omega$) and δ_{rs} is not equal to $n\pi$ ($\delta_{rs} \neq n\pi$) where ($n = 0, 1, 2,\ldots$). In other words, average torque exists only for the following conditions:

$$\omega_m = \omega$$

$$\delta_{rs} \neq n\pi \text{ where } n = 0, 1, 2, \ldots..$$

Therefore in order to develop an average torque the reluctance motor must be rotating in either direction with an angular speed ω_m which is equal to the electrical angular speed ω. This speed is called the synchronous speed ω_s.

When the rotor is rotating at this synchronous speed and if $\delta_{rs} = 0°$, there would be no average torque, and if $\delta_{rs} = 45°$, torque has a maximum value. The negative sign in the above torque equation indicates that the torque is towards the nearest pole. The expression for the developed torque at this synchronous speed then would be

$$T_{ef\,\text{avg}} = -0.125(L_d - L_q)I_m^2\sin 2\delta_{rs}$$

It is clear from this expression that torque is a maximum when $\delta_{rs} = 45°$ (pull-out torque). δ_{rs} is known as the "torque angle" (Figure 4.5).

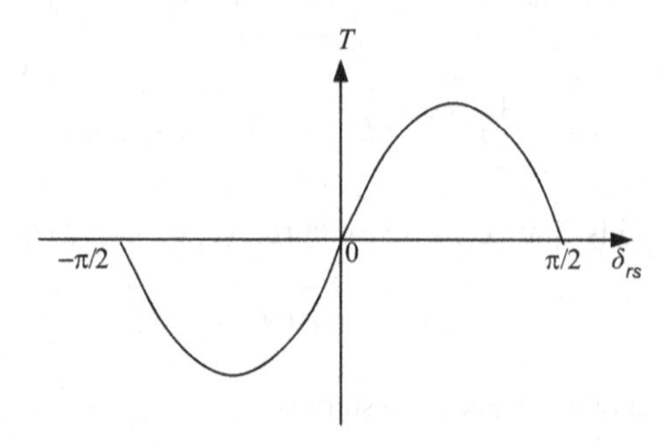

FIGURE 4.5 Variation of torque with torque angle in reluctance motors.

Reluctance motor is very similar to "synchronous motors" that we discuss in Chapter 8. Like synchronous motors, reluctance motors are not self-starting. Precautions must be taken for their "self-starting." Therefore, their rotors are built very similar to "induction motors." The rotor has salient poles for normal steady-state operation and also has a squirrel cage or amortisseur windings for starting (see Chapter 9). Stator can be monophase or three-phase structure.

Exercise 4.1:

A two-pole reluctance motor has a speed of 1,800 rpm. The minimum and maximum values of the inductance are 4 H and 2 H, respectively. If the motor current intake is 20 A (rms), (a) find the frequency of the stator voltage. (b) If torque angle δ_{rs} is 22.5°, what is the average torque developed by the motor?

Solution:

a. Frequency of the stator voltage:

$$f = \frac{1800}{60} = 30 \text{ Hz}$$

b. Average developed torque:

$$T_{ef\,avg} = -0.125(4-2)\left(20\times\sqrt{2}\right)^2 \sin 45° = -141,42 \text{ N}$$

4.3 STEP MOTOR

Step (or stepper) motors are brushless DC motors that move in finite increments (steps) in either direction. Steps are controlled by applying command pulses to the stator windings. They respond to digital information in proportional mechanical movement. Therefore, they are very suitable for digital applications with micro-controllers. They are used in applications where motion and positioning are critical (printers, plotters). Step motors can hold their position and resist rotation. There are three types of step motors.

4.3.1 Permanent Magnet Step Motor

Unlike in reluctance motor, the rotor in this motor is permanently magnetised. If a voltage pulse is applied to coil $a - a_1$, then the rotor would line up in the position shown in Figure 4.6. If coils a and a_1 are deenergised and coils b and b_1 are energised by applying a voltage, then the rotor will rotate 90° and line up with the b and b_1 coils.

The direction it rotates will depend on whether b or b_1 is the north pole. Therefore, to rotate the rotor in the opposite direction, we simply reverse the voltage sequence of the command pulses. Let us continue the process in Figure 4.6 by switching back to coils a and a_1 again but this time voltage being applied in the opposite direction. Each time, the rotor moves by 90°. Desired speeds can be achieved in a very short

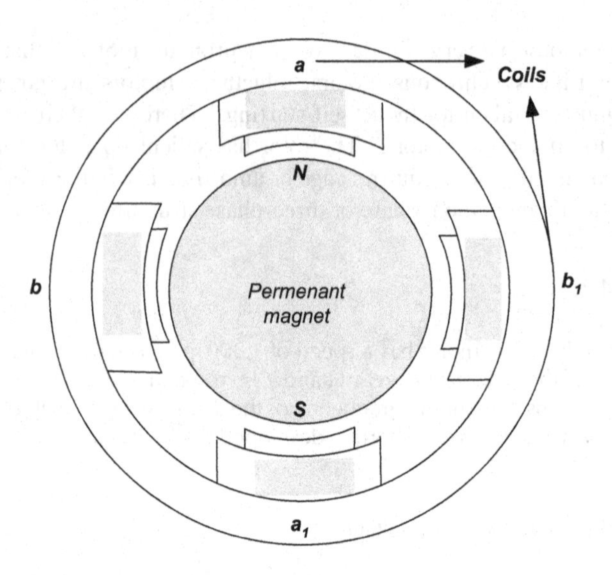

FIGURE 4.6 Permanent magnet step motor.

time. We can decrease the step angle in a permanent magnet step motor by either increasing the number of pole pairs on the rotor or adding more stator coils and phases.

In Figure 4.7, we see a permanent magnet step motor which rotates in steps of 30°. Should the power be removed from a permanent magnet step motor, the motor will remain in its position because the rotor magnetic flux will prefer the low reluctance path of the salient stator.

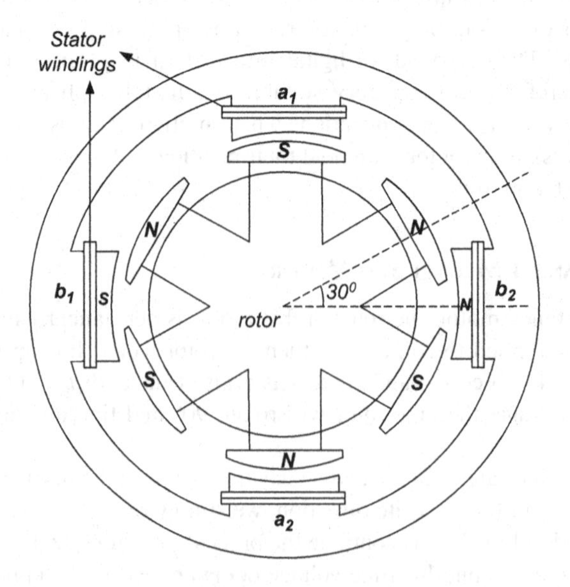

FIGURE 4.7 Step motor with 30° steps.

Permanent magnet step motors are generally preferred when the required power is high. The basic control mechanism is simplified as shown in Figure 4.8. There are two elements in this control mechanism:

- the pulse rate which determines the speed of rotation
- the direction data which determine in which order the poles are excited. This consequently determines the direction of rotation of the rotor.

4.3.2 Variable Reluctance Step Motor

The operation principle of variable reluctance step motors relies on the tendency of the flux lines for low reluctance paths. This motor uses a non-magnetised, soft-iron rotor. The rotor has teeth that are offset from the stator poles to accomplish rotation. When the stator winding is energised, the rotor of the motor is aligned in such a way that the magnetic reluctance is minimum (shortest flux path).

The variable reluctance motor in Figure 4.9 uses a nonsalient (teethed) stator structure and a four-pole rotor which is not a permanent magnet. The stator poles are energised in sequence. If the No. 1 stator poles are excited, the rotor will align itself

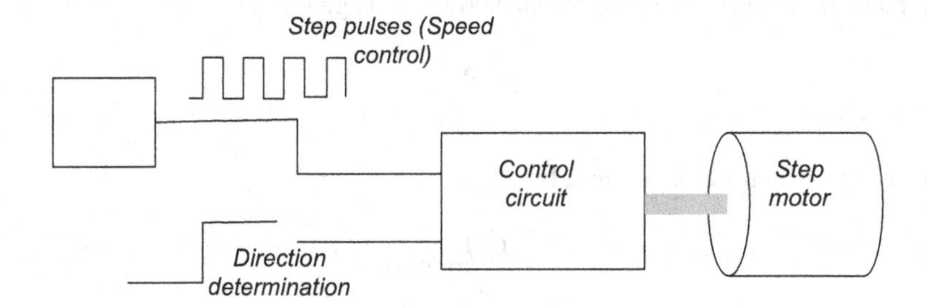

FIGURE 4.8 Simplified control mechanism of step motors.

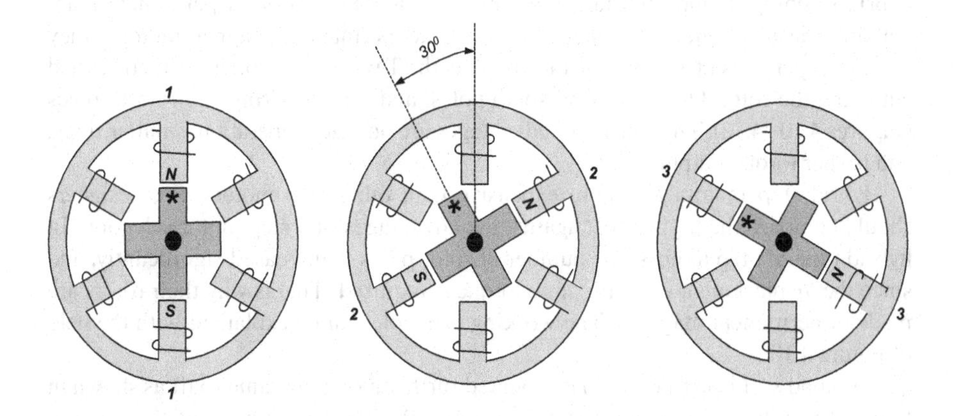

FIGURE 4.9 Variable reluctance step motor.

as shown in the Figure 4.9. The rotor poles at right angles to the marked pair have no function since they are effectively neutral.

Now let us excite the next pair of stator poles. The rotor moves to align the unmarked rotor poles with the No. 2 stator poles. This causes a rotation of 30° anti-clockwise. This arrangement is not dependent on the direction of the excitation current. Instead only the order in which the pole pairs are excited is important. This type of step motors is noisy.

The motor operates much like the permanent magnet stepper motor. Note that the motors' step angle is actually half of what it is with a permanent magnet motor with the same number of stator/windings. The main difference is that the variable reluctance motor has no detent torque when there is no power applied; that is, it is free to rotate. To increase the resolution on this type of motor, typically more teeth are added to the geared rotor.

The step angle of the step motor (θ_s) can be calculated from the number of stator poles (N_s) and the number of rotor poles (N_r) as follows:

$$\theta_s = \frac{|N_s - N_r|}{N_s N_r} \times 360^\circ$$

The number of steps (s) for one complete rotation is given by

$$s = \frac{360^\circ}{\theta_s}$$

The speed of the rotor (n) is given by

$$n = \frac{60f}{s}(\text{rpm})$$

4.3.3 Hybrid Step Motor

In order to obtain smaller angles (1.8° and 2.5°) of alignment, we have to use the hybrid stepping motor. Hybrid stepper motor is a combination of permanent magnet and variable reluctance step motors. Like permanent magnet motors, they contain a permanent magnet in the rotor teeth. Two sets of soft iron teeth called cups ring the rotor. One ring is all south poles, and the other ring is all north poles (Figures 4.10–4.12). Their primary advantages are noiseless operation, smaller steps, and higher torque output.

Hybrid step motors are more expensive than other step motors. So, designers should compare the higher cost against the advantages of a step motor selection. In hybrid type of step motors, the number of rotor poles is increased significantly, and since the displacement is small, more torque is required. That is why their rotors are made of permanent magnets. The working principle can be explained with the help of Figure 4.10.

A second soft iron rotor (cup) is mounted further along the same shaft as shown in Figure 4.12. The surfaces of the cups have typically 50 teeth per cup, and the cups are aligned so that the teeth of the top cup are offset by a certain amount (typically 3.6°)

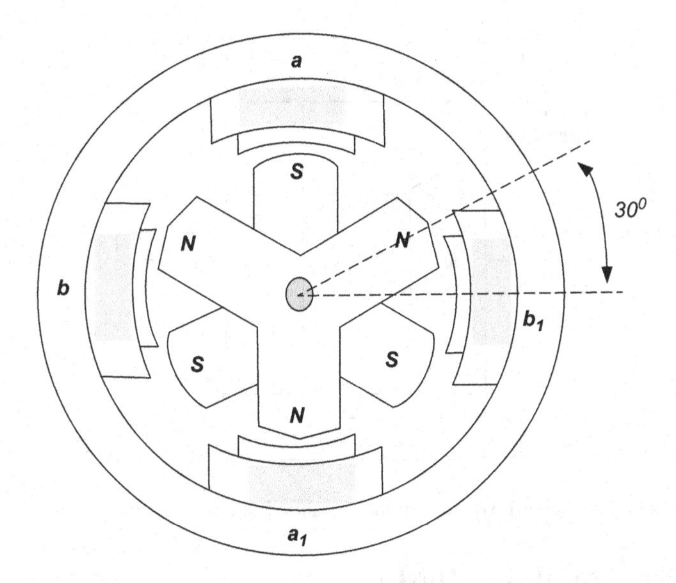

FIGURE 4.10 Hybrid step motor.

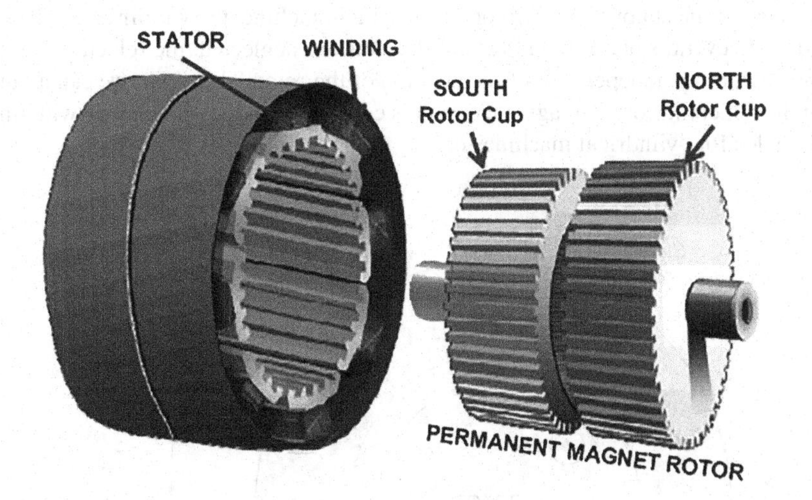

FIGURE 4.11 Hybrid step motors. (Courtesy of Microchip Technology Incorporated.)

from the teeth of the bottom cup. The entire rotor is magnetised axially, with one end polarised north and the other end polarised south.

This rotor and toothed stator construction makes hybrid motors capable of producing higher torque than other step motor types. If power is removed from coils a and a_1 and applied to coils b and b_1, rotor will move into alignment with stator poles b and b_1. The direction of rotation will depend on the polarity of the voltage applied to the coils.

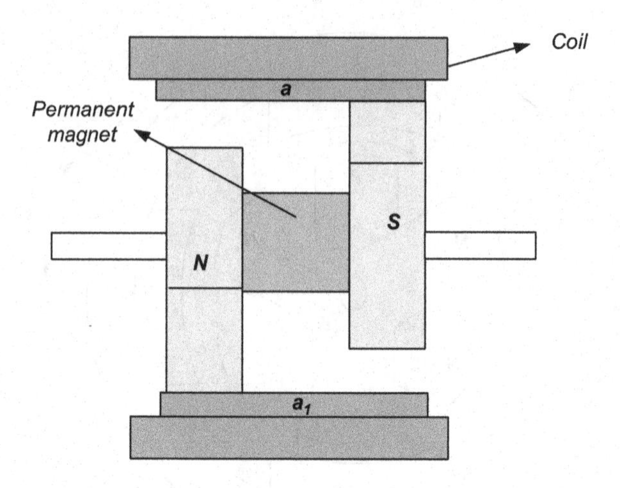

FIGURE 4.12 Side view of hybrid step motors showing cups.

4.4 CYLINDRICAL MACHINES

In Figure 4.13, we see the cross-sectional view of an elementary two-pole cylindrical rotating machine with a uniform air gap. The stator and rotor windings are placed on two slots on the stator and the rotor. In an actual machine, the windings are distributed over several slots. If the effects of the slots are neglected, the reluctance of the magnetic path is independent of the position of the rotor. Therefore, we can assume that the reluctance of the magnetic circuit is constant and does not change with time. This holds for cylindrical machines only.

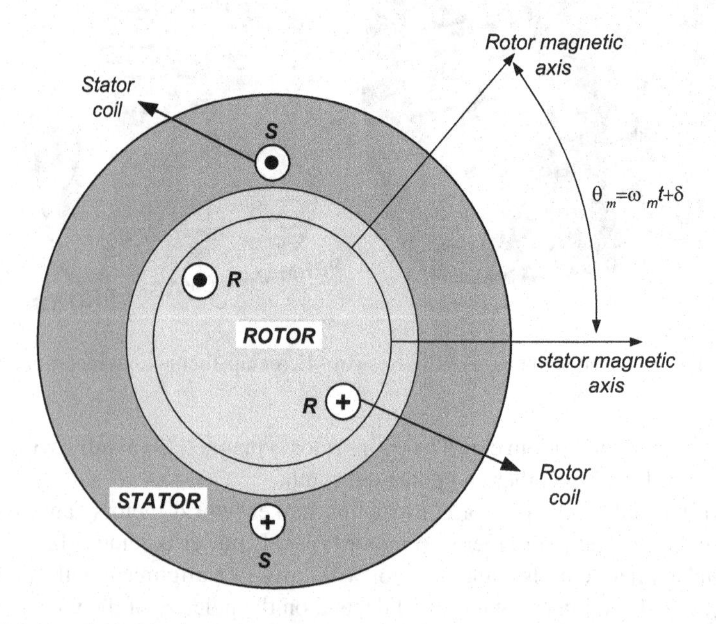

FIGURE 4.13 Cylindrical rotating machine.

Hence, the self-inductances L_{ss} and L_{rr} do not change by time, and therefore, there will be no reluctance torque due to these components in the following torque equation:

$$\text{Torque equation,}\quad T = \frac{1}{2} i_s^2 \frac{dL_{ss}}{d\theta_m} + \frac{1}{2} i_r^2 \frac{dL_{rr}}{d\theta_m} + i_s i_r \frac{dL_{sr}}{d\theta_m} \tag{4.4}$$

However, the mutual inductance L_{sr} will vary with the position of the rotor and therefore will be responsible for the electromagnetic torque in the cylindrical machines (Figure 4.13) as expressed in the following equation:

$$T = i_s i_r \frac{dL_{sr}}{d\theta_m} \tag{4.5}$$

Let us remember that here θ_m represents the shaft angle or the mechanical angle (position of the rotor), and T is the electromechanical torque that the magnetic system generates. The two currents in the stator and the rotor coils, and the mechanical angle θ_m are

$$i_s = I_{sm} \cos \omega_s t$$

$$i_r = I_{rm} \cos(\omega_r t + \alpha)$$

$$\theta_m = \omega_m t + \delta_{rs}$$

Here, δ_{rs} is the initial rotor position angle (angle between rotor and stator magnetic axes), and α is the phase angle of the rotor current i_r with respect to stator current i_s. If M is the maximum value of the mutual inductance L_{sr}, we can write that

$$L_{sr} = M \cos \theta_m$$

For a p-pole machine, θ_m and L_{sr} must be replaced by

$$\theta_e = \text{Equivalent electrical displacement angle} = \frac{p}{2} \theta_m$$

$$L_{sr} = M \cos \theta_e$$

Hence for a two-pole machine, from the above torque equation, we can state the following (equation 4.5):

$$T = i_s i_r \frac{dL_{sr}}{d\theta_m}$$

$$T = (I_{sm} \cos \omega_s t)[I_{rm} \cos(\omega_r t + \alpha)] \frac{d(M \cos \theta_m)}{d\theta_m}$$

$$T = -(I_{sm}\cos\omega_s t)[I_{rm}\cos(\omega_r t + \alpha)]M\sin\theta_m$$

$$T = -(I_{sm}\cos\omega_s t)[I_{rm}\cos(\omega_r t + \alpha)]M\sin(\omega_m t + \delta_{rs}) \tag{4.6}$$

Notice that the minus sign indicates that electromagnetic torque decreases the angle between the two magnetic axes, which means it is acting to accelerate the rotor. This will be the motor action. From this, we can say that in generators, the electromagnetic torque will try to increase this angle and hence act in the opposite direction to the rotor rotation. Now if we make use of the below-given trigonometric identity in the above torque equation:

$$\sin x \cos y = \frac{1}{2}\left[\sin(x+y) + \sin(x-y)\right]$$

$$T = -\frac{I_{sm}I_{rm}M}{2}\cos(\omega_r t + \alpha)\left[\sin(\omega_m t + \delta_{rs} + \omega_s t) + \sin(\omega_m t + \delta_{rs} - \omega_s t)\right]$$

$$= -\frac{I_{sm}I_{rm}M}{2}\left\{\frac{1}{2}\left[\sin(\omega_m t + \delta_{rs} + \omega_s t + \omega_r t + \alpha) + \sin(\omega_m t + \delta_{rs} + \omega_s t - \omega_r t - \alpha)\right]\right.$$

$$\left. + \frac{1}{2}\left[\sin(\omega_m t + \delta_{rs} - \omega_s t + \omega_r t + \alpha) + \sin(\omega_m t + \delta_{rs} - \omega_s t - \omega_r t - \alpha)\right]\right\}$$

$$\begin{aligned} T = -\frac{I_{sm}I_{rm}M}{4}\Big[& \sin\{(\omega_m + (\omega_s + \omega_r))t + \alpha + \delta_{rs}\} \\ & + \sin\{(\omega_m - (\omega_s + \omega_r))t - \alpha + \delta_{rs}\} \\ & + \sin\{(\omega_m + (\omega_s - \omega_r))t - \alpha + \delta_{rs}\} \\ & + \sin\{(\omega_m - (\omega_s - \omega_r))t + \alpha + \delta_{rs}\}\Big] \end{aligned} \tag{4.7}$$

Therefore, the machine will develop a torque only if

$$\omega_m = \mp(\omega_s \mp \omega_r)$$

which means the machine will develop average torque if only it rotates, in either direction, at a speed that is equal to the sum or difference of the angular speeds of the stator and rotor currents. If the machine is a p-pole machine, then we must have

$$L_{sr} = M\cos\left(\frac{p}{2}\theta_m\right)$$

Hence for a p-pole machine, the above torque equation becomes

$$T = -\frac{p}{2}\frac{I_{sm}I_{rm}M}{4}\left[\sin\left\{\left(\frac{p}{2}\omega_m + (\omega_s + \omega_r)\right)t + \alpha + \frac{P}{2}\delta_{rs}\right\}\right.$$

$$+\sin\left\{\left(\frac{p}{2}\omega_m - (\omega_s + \omega_r)\right)t - \alpha + \frac{P}{2}\delta_{rs}\right\}$$

$$+\sin\left\{\left(\frac{p}{2}\omega_m + (\omega_s - \omega_r)\right)t - \alpha + \frac{P}{2}\delta_{rs}\right\}$$

$$\left.+\sin\left\{\left(\frac{p}{2}\omega_m - (\omega_s - \omega_r)\right)t + \alpha + \frac{P}{2}\delta_{rs}\right\}\right] \quad (4.8)$$

Average torque will develop for the following condition:

$$\frac{P}{2}\omega_m = \mp(\omega_s \mp \omega_r)$$

Now let us consider the following cases:

1. $\omega_r = 0$, $\alpha = 0$, $\omega_m = 2\omega_s/p$:
   If the rotor current is a DC I_R and the machine is mechanically brought to its synchronous speed ($\omega_m = 2\omega_s/p$), the developed torque and its average value will be as follows:

$$T = -\frac{p}{2}\frac{I_{sm}I_RM}{2}\left[\sin\left(2\omega_s t + \frac{p}{2}\delta_{rs}\right) + \sin\frac{p}{2}\delta_{rs}\right] \quad (4.9)$$

$$T_{\text{avg}} = -\frac{p}{2}\frac{I_{sm}I_RM}{2}\sin\left(\frac{p}{2}\delta_{rs}\right) \quad (4.10)$$

We can write the same equation as

$$T_{\text{avg}} = -\frac{p}{2}\frac{I_{sm}I_RM}{2}\sin\left(\delta'_{rs}\right)$$

where δ'_{rs} is the phase angle between stator and rotor magnetic axes in electrical degrees.

With single-phase voltage applied to single stator winding, the machine is called a single-phase synchronous machine. This is the case we have examined above. Although it develops an average torque, the instantaneous torque is pulsating (equation 4.9) at twice the stator voltage frequency. The pulsating torque may result in vibration, speed fluctuation, noise, and waste of energy. This is not acceptable in large machines. The pulsating torque can be avoided in a balanced polyphase machine, and all large machines are in fact balanced polyphase machines.

Since $\omega_m = 2\omega_s/p$ does not hold at machine standstill, equation 4.9 cannot be used to find the net starting torque of synchronous machines. Therefore to see if the machine has a net starting torque for standstill, we once again go back to the following instantaneous torque equation:

$$T = \frac{-I_{sm} I_{rm} M}{4}\left[\sin\left\{(\omega_m + (\omega_s + \omega_r))t + \alpha + \frac{p}{2}\delta_{rs}\right\}\right.$$
$$+\sin\left\{(\omega_m - (\omega_s + \omega_r))t - \alpha + \frac{p}{2}\delta_{rs}\right\}$$
$$+\sin\left\{(\omega_m + (\omega_s - \omega_r))t - \alpha + \frac{p}{2}\delta_{rs}\right\}$$
$$\left.+\sin\left\{(\omega_m - (\omega_s - \omega_r))t + \alpha + \frac{p}{2}\delta_{rs}\right\}\right]$$

From the above equation, we see that when the machine is at standstill with $\omega_r = 0$, $\alpha = 0$, and $\omega_m = 0$, there will be no net torque on the shaft. Therefore, the machine is not self-starting.

2. $p\omega_m/2 = \omega_s - \omega_r$

Both stator and rotor windings carry AC at different frequencies, and the motor runs at an asynchronous speed ($\omega_m \neq \omega_s$, $\omega_m \neq \omega_r$). Let us go back to torque equation (equation 4.8) once again:

$$T = -\left(\frac{p}{2}\right)\frac{I_{sm} I_{rm} M}{4}\left[\sin\left\{\left(\frac{p}{2}\omega_m + (\omega_s + \omega_r)\right)t + \alpha + \frac{p}{2}\delta_{rs}\right\}\right.$$
$$+\sin\left\{\left(\frac{p}{2}\omega_m - (\omega_s + \omega_r)\right)t - \alpha + \frac{p}{2}\delta_{rs}\right\}$$
$$+\sin\left\{\left(\frac{p}{2}\omega_m + (\omega_s - \omega_r)\right)t - \alpha + \frac{p}{2}\delta_{rs}\right\}$$
$$\left.+\sin\left\{\left(\frac{p}{2}\omega_m - (\omega_s - \omega_r)\right)t + \alpha + \frac{p}{2}\delta_{rs}\right\}\right]$$

For $p\omega_m/2 = \omega_s - \omega_r$, the above equation reduces to

$$T = -\frac{p}{2}\left(\frac{I_{sm} I_{rm} M}{4}\right)\left[\sin\left(2\omega_s t + \alpha + \frac{p}{2}\delta_{rs}\right) + \sin\left(-2\omega_r t - \alpha + \frac{p}{2}\delta_{rs}\right)\right.$$
$$\left.+\sin\left(2\omega_s t - 2\omega_r t - \alpha + \frac{p}{2}\delta_{rs}\right) + \sin\left(\alpha + \frac{p}{2}\delta_{rs}\right)\right] \quad (4.11)$$

Instantaneous torque pulsates and the mean value of the torque would be

$$T_{\text{average}} = -\frac{p}{2}\left(\frac{I_{sm}I_{rm}M}{4}\right)\sin\left(\alpha + \frac{p}{2}\delta_{rs}\right) \tag{4.12}$$

This is the general principle of how the single-phase induction machines work. An AC excitation is applied to the stator windings and an AC is induced in the rotor windings. At machine standstill, ω_m will be zero ($\omega_m = 0$), and no average torque will be available. Therefore, there will be no net starting torque for the single-phase induction motors. In order that the motor can develop a torque, it must be given an initial impetus and be brought to a speed which is equal to $\omega_m = \omega_s - \omega_r$ when ω_m is other than zero. Even then the pulsating torque will create problems in high-power applications. Therefore for high-power applications, balanced polyphase machines are used.

The mechanism of torque production in electromagnetic systems producing both translational and rotary motions has been discussed. In rotating machines, torque can be produced:

- by variation in the reluctance of the magnetic path (reluctance torque)
- or by variation in the mutual inductance between the windings (torque developed by interaction).

Reluctance machines are simple in construction, but they develop small torque. On the contrary, cylindrical machines are more complex in construction, but produce larger torques. Most electrical machines are therefore of the cylindrical type.

PROBLEMS ON CHAPTER 4

Problem 4.1: Find the instantaneous torque expression of the two-pole reluctance motor shown in Figure 4.14. Assume

- the reluctance of the magnetic core is negligible compared to that of air gap.
- the effective area of each gap is equal to the area of the overlap.

If the number of turns $n = 1{,}000$, coil current $I = 10$ A (rms), radius of the rotor $r = 5$ cm, length of the rotor $l = 50$ cm, displacement angle is 15°, and the air gap length $g = 1$ mm, calculate the value of the average torque.

Problem 4.2: The reluctance of a two-pole reluctance motor varies with displacement angle θ according to the following expression:

$$\mathcal{R}(\theta) = 5000 - 1500\cos 2\theta \text{ H}^{-1}$$

If the number of turns in the coil is $N = 1{,}000$, instantaneous coil current is $i = 10$ A, and the displacement angle θ is 15°, how much is the instantaneous torque developed by the motor?

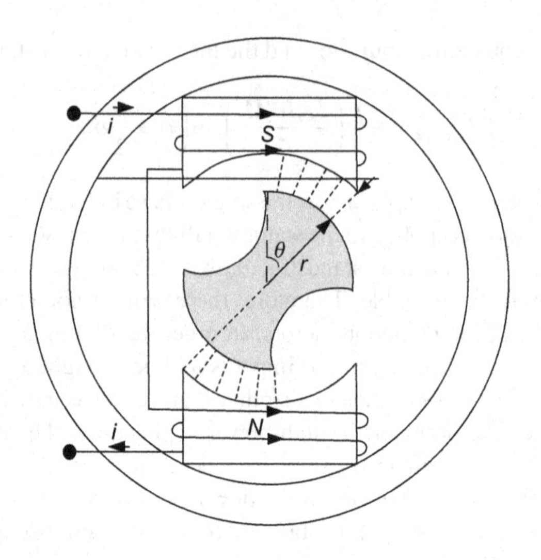

FIGURE 4.14

Problem 4.3: The inductance of a two-pole reluctance motor (Figure 4.15) changes with displacement angle θ as given in the following expression, $L = 0.2 - 0.6\cos 2\theta - 0.4\cos 4\theta$ H

If the stator coil current intake is 5 A(rms) at 50 Hz frequency,

a. Find the rotor angular speed ω_m at which an average torque can be developed.
b. What are the maximum average torque values at each of these speeds?
c. What would be the average torque developed when the speed is zero and the starting angle is $\delta = 45°$

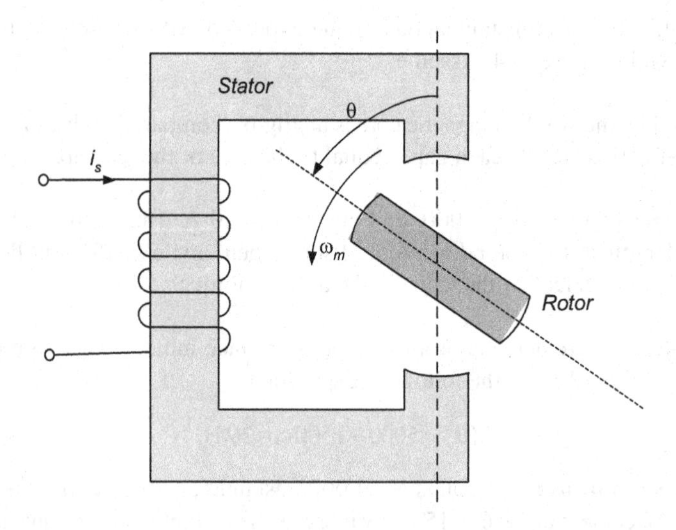

FIGURE 4.15

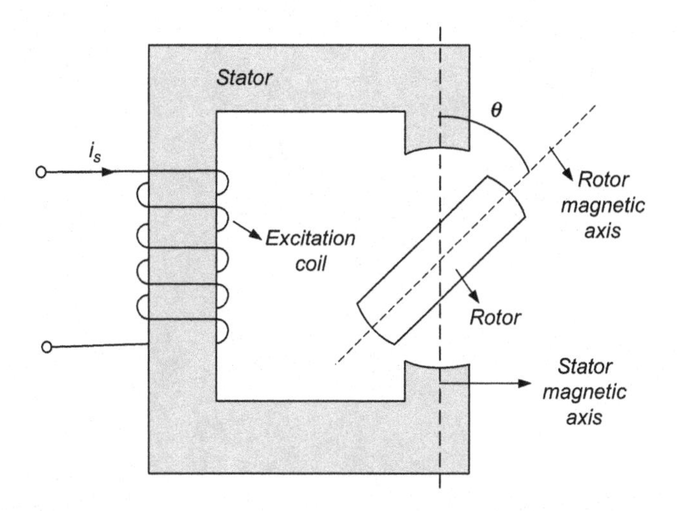

FIGURE 4.16

Problem 4.4: The instantaneous torque equation in the two-pole reluctance motor shown in Figure 4.16 is

$$T_e = -0.25(L_d - L_q)I_m^2$$

$$\left[\sin(2\omega_m t + 2\delta) + 0.5\sin\{2(\omega + \omega_m)t + 2\delta\} - 0.5\sin\{2(\omega - \omega_m)t - 2\delta\}\right]$$

a. Define each term in this equation. Under which condition this expression reduces to an average torque value? What is this average torque expression?
b. When this reluctance motor is rotating at its synchronous speed of 3,600 rpm, the current intake is 15 A(rms) and the average torque is at a maximum value. If L_d and L_q are 5 H and 4 H, respectively, calculate the torque angle δ and the maximum average torque value that can be obtained. What is the frequency of the supply voltage?

Answers to odd questions:
4.1: –102,8 Nm;
4.3: $\omega_m = 100\pi$ rad/s, $\omega_m = 50\pi$ rad/s, $T_{max} = 7.5$ Nm, $T_{max} = 10$ Nm, $T_{avg} = 15$ Nm

5 Electrical Machines

Generator with steam turbine (https://commons.wikimedia.org/)

Barrage and Hydroelectric Power station 2400 MW (https://tr.wikipedia.org/wiki/)

5.1 INTRODUCTION

Continuous electromechanical energy converters are systems that convert electrical and mechanical energies in continuous form from one to the other. As said before when conveying of energy is considered, electrical energy has economical advantages compared to other forms of energy (e.g., heat energy). Therefore, continuous electromechanical energy converters called electrical machines are important. We divide electrical machines into two groups:

1. Motors that convert electrical energy into mechanical energy in continuous form.
2. Generators that convert mechanical energy into electrical energy in continuous form.

An electrical machine can be made to work as a generator or as a motor. Both are inherent in electrical machines. In other words, in these machines, the energy conversion is reversible. In their normal performance,

1. there are voltages that are induced in motors
2. there are forces that are developed in electrical generators

Both motors and generators work on two principles:

1. A force (F) is produced on a current-carrying wire in magnetic field and the wire starts accelerating (conversion of electrical energy to mechanical energy).
2. A voltage is induced on an accelerating wire in a magnetic field and the direction of this induced voltage is opposite to the current that is causing the acceleration of the wire (conversion of mechanical energy to electrical energy).

In this way, the force on the wire decreases and finally becomes zero, and the wire travels at a constant speed. Conversion of electrical energy to mechanical energy and vice versa takes place simultaneously in these machines and therefore they are reversible (Figures 5.1 and 5.2).

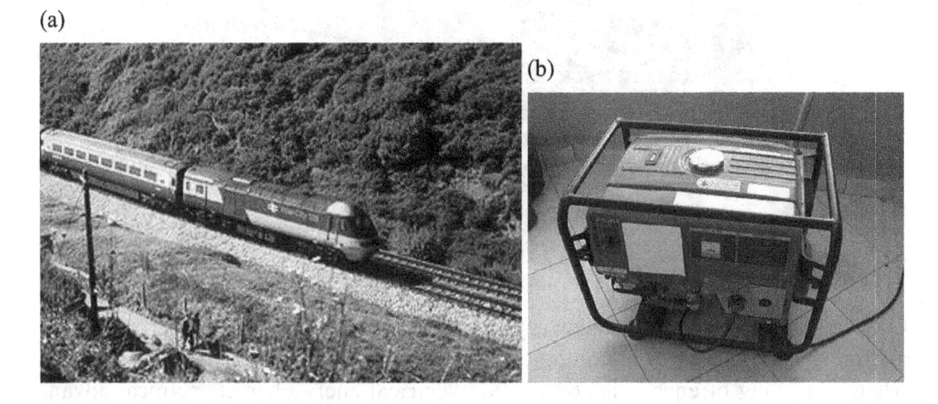

FIGURE 5.1 Generators: (a) diesel generator operated train (https://commons.wikimedia.org) (b) home-type generator.

FIGURE 5.2 1 HP three-phase AC induction motor (https://en.wikipedia.org/).

There is production of force in generators and production of induced voltage in motors. That is why we call them electrical machines. There are two main parts of electrical machines:

1. The winding where we apply (or receive) the voltage (ARMATURE WINDING).
2. The winding that creates the main magnetic field (FIELD WINDING).

In electrical machines, the stationary part is called the stator and the rotating part connected to a shaft that couples the machine to its mechanical load is called the rotor. Depending on the type of the motor, field winding and the armature winding can be in the rotor or in the stator.

In order for a machine to convert electrical energy to mechanical energy or vice versa a magnetic field must exist within its air gap. This field can be set up electromagnetically or by a permanent magnet. It is usually produced electromagnetically by exciting the field windings of the machine. The machine can convert electrical and mechanical energies into one another using the armature winding. Therefore, the armature winding is where the energy conversion takes place. Due to this reason, the power rating of a machine is actually the power rating of the armature section.

5.2 ELECTRIC MOTORS

An electric motor consists mechanically of a stator (a nonrotating part) and a rotor on a shaft which is a rotating part. Slots are cut into the stator, and the rotor and windings are placed into these slots (Figure 5.3). Windings that usually have few turns of

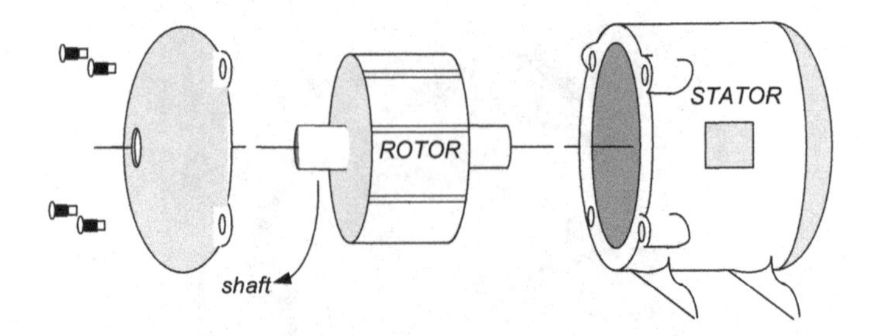

FIGURE 5.3 Main parts of an electric motor.

large wire carrying heavy currents are called the armature windings, and windings that have many turns of wire with smaller diameter are called the field winding. The current in the field winding is independent of the load on the shaft of the motor.

Electric motors are divided into two categories according to how they are powered: alternating current (AC) motors and direct current (DC) motors.

5.2.1 Alternating Current (AC) Motors

AC motors take as their input an AC source, but unlike DC motors, they have no commutation involved, and can be single or multi-phase. Since there is no commutation, AC motors require less maintenance than DC motors. In general, there are two types of AC motors:

1. induction motors (asynchronous motors) that are the most used type because they are rugged, and simple and economical (Figure 5.4)
2. synchronous motors that (assuming frequency of supply is kept constant) run at constant speed regardless of the load on the shaft.

AC power can be single phase or polyphase (usually three phase in practise). Three-phase induction motors are mostly used in industries where requirement is over 5 hp.

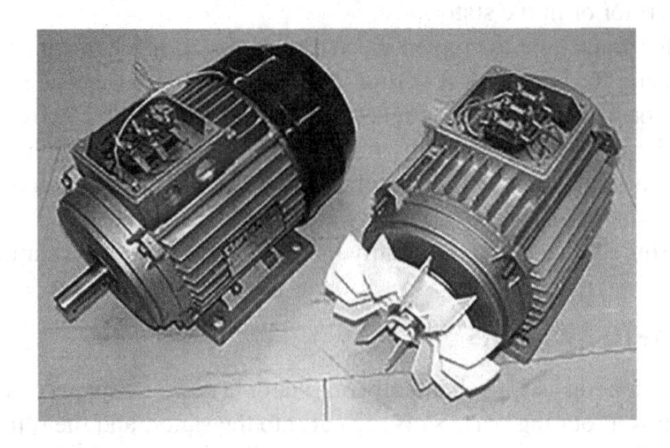

FIGURE 5.4 Three-phase induction motor 0.75 kW, 1420 rpm, 60 Hz, 230–400 V (https://en.wikipedia.org/).

5.2.2 Direct Current (DC) Motors

DC motors used to be much more suitable for speed control applications. However, because of the advancement in solid-state electronics, speed control of AC induction motors by varying the frequency of the source has become equally easy and sometimes economical as well. DC motors are mostly used in vehicles where DC voltage is made available by a 12 V battery. DC motors in common use have brushes and commutators that reverse the connections to the armature conductors.

The main disadvantage of the brushed DC motor is the presence of brushes. These wear down quickly which leads to high maintenance costs. Moreover, due to the sparks that occur during commutation, the motor cannot operate under hazardous conditions.

5.2.3 Efficiency in Motors

If for a motor ω_m is the mechanical angular velocity in radian/s and T_m is the mechanical torque on the shaft in (Nm), then the power output is expressed as

$$P_{\text{out}} = \omega_m T_m$$

The expression for the electrical input differs for three-phase and single-phase AC motors. For a single-phase AC motor,

$$P_{\text{in}} = VI\cos\theta$$

where V and I are the rms values of the source voltage and current, and $\cos\theta$ represents the power factor (Figure 5.5). In the case of a three-phase AC motor, input power takes the form of

$$P_{\text{in}} = \sqrt{3}VI\cos\theta$$

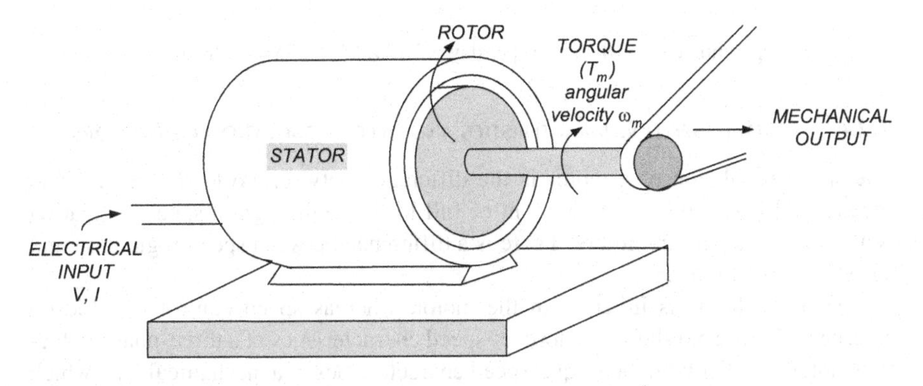

FIGURE 5.5 Electrical motor.

where V and I are rms values of the source line voltage and line current, respectively. Motor efficiency is given by

$$\eta = \frac{P_{out}}{P_{in}}$$

Exercise 5.1:

A single-phase induction motor gives an output power of 2 hp at a speed of 1,750 rpm. If the efficiency of the motor is 70%, and the source voltage and current intake of the motor are 220 V and 15 A, calculate the output torque, and the power factor of the motor. If the cost of electric power is 0.25 euros/kWh, calculate the cost/day of this motor.

Solution:

$$\text{Output power} = 2 \times 746 \text{ W}$$

$$\rightarrow \text{Torque} = \frac{P_{out}}{\omega} = \frac{2 \times 746}{1750 \times 2\pi/60} = 8.141 \text{ Nm}$$

$$\text{Input power, } P_{in} = P_{out}/\text{efficiency}$$

$$= \frac{2 \times 746}{0.7} = 2131.4 \text{ W}$$

$$P_{in} = V \times I \times (\text{Power Factor})$$

$$\rightarrow PF = \frac{P_{in}}{V \times I} = \frac{2 \times 746}{0.7 \times 220 \times 15} = 0.65$$

In these type of motors PF is always lagging.

$$\text{Operating cost of the motor per day} = 2.13 \times 24 \times 0.25 = 12.78 \text{ euros}$$

5.2.4 Torque–Speed Characteristics and Speed Regulation of Motors

The speed regulation of a motor is the difference between no load and full load speeds, and it is expressed in percent of full load speed. A good speed regulation is a desirable feature in motors. There is a difference between speed regulation and speed control of motors.

Speed regulation is inherent in the motor, whereas speed control is achieved externally. Figure 5.6 shows the torque–speed characteristics of a three-phase induction motor together with the torque-speed characteristics of a mechanical load which

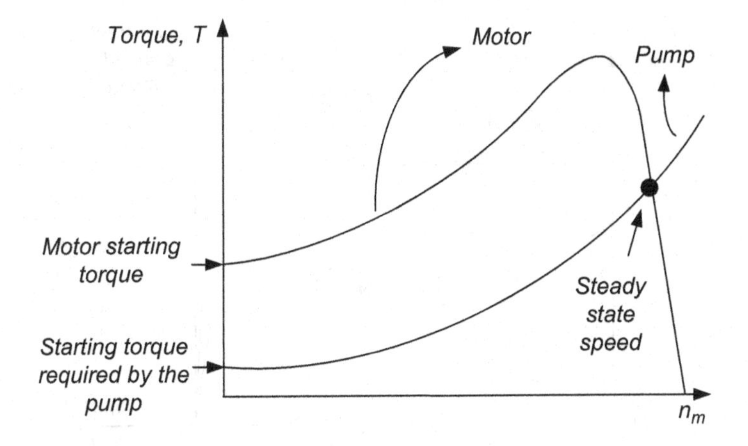

FIGURE 5.6 Induction motor driving a pump.

is supposedly a pump. If this motor is connected to this pump from rest, it would accelerate the pump until it reaches the steady-state speed at which the torque provided by the motor equals the torque required by the pump.

If the same motor is connected to a crane with torque–speed characteristics shown in Figure 5.7, it would not be able to drive it as its starting torque is smaller than the torque necessary for the crane to start up.

Figure 5.8 shows the torque/speed characteristics of several types of electrical motors, and Figure 5.9 shows torque requirements for various mechanical loads.

The designers must tailor their mechanical load requirements (Figure 5.9) against their motor torque/speed characteristic (Figure 5.8).

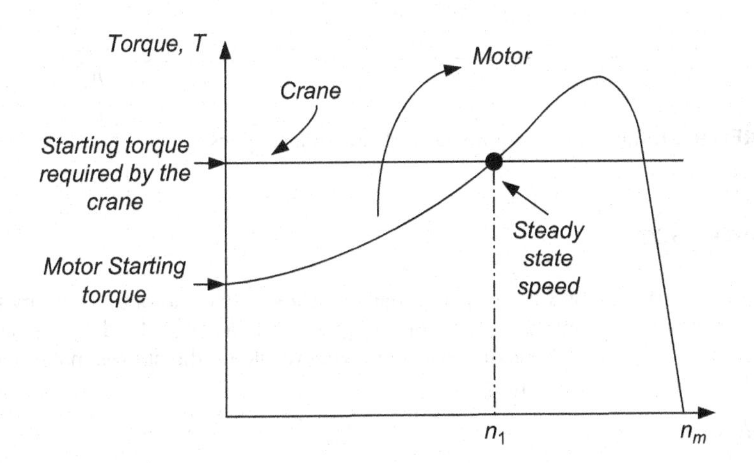

FIGURE 5.7 Induction motor driving a crane.

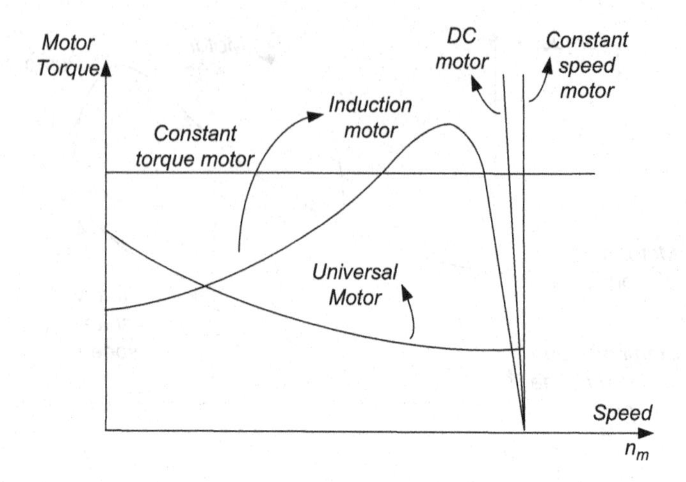

FIGURE 5.8 Torque–speed characteristics of several electrical motors.

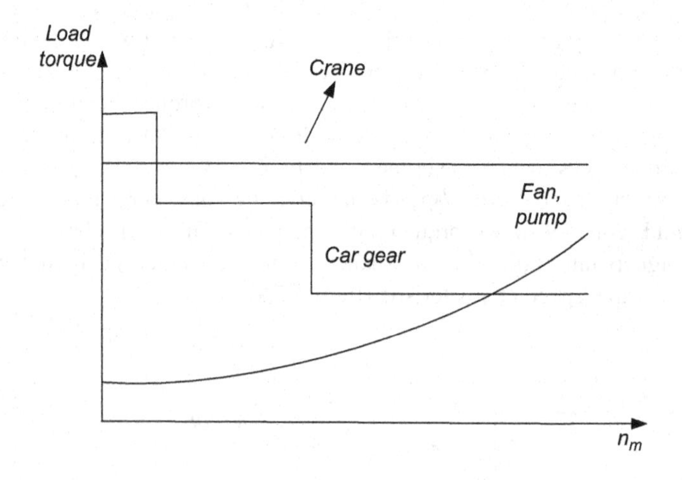

FIGURE 5.9 Torque–speed requirements of several load types.

Exercise 5.2:

Figure 5.10 shows the torque/speed characteristics of two motors. Which motor would come to its stable speed first when connected to a certain load that requires a constant torque of 1 Nm? At this instant, what would be the approximate speed and output power of the motor?

Solution:

From Figure 5.11, it is clear that

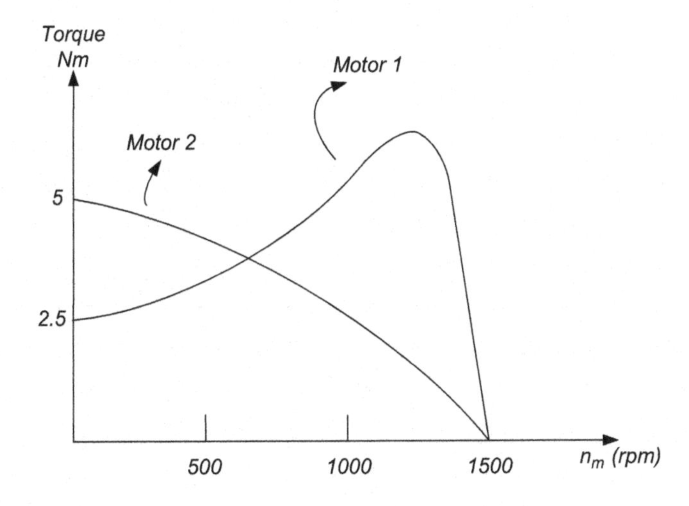

FIGURE 5.10

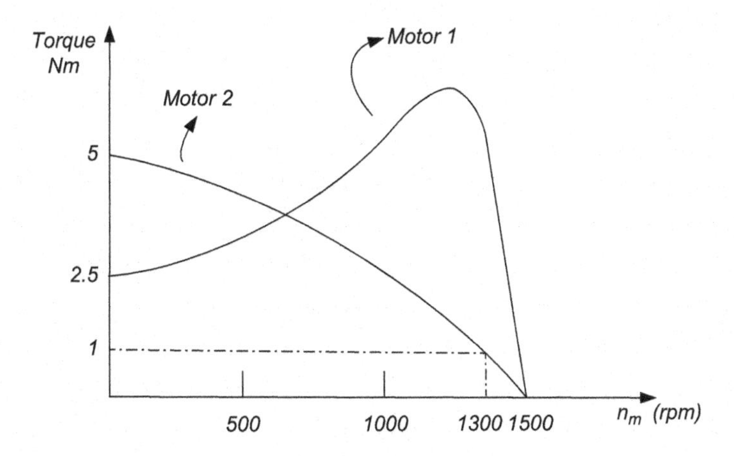

FIGURE 5.11

a. This load would come to its stable operation point at a lower speed with motor 2 than with motor 1.
b. If motor 2 is connected to this load, the stable operation speed of the motor would be approximately ≈ 1,300 rpm (value taken from the graph). The power output of the motor at this instant would be

$$P_{\text{out}} = T \times \omega_m$$

$$\rightarrow P_{\text{out}} = 1 \times \frac{1300 \times 2\pi}{60} = 136 \text{ W}$$

6 DC Machines

Double armature DC Motors (*courtesy of Teco-Westinghouse Company*)

6.1 INTRODUCTION

The two types of DC machines that we will look into are the DC motors and the DC generators. Although their structures are identical, motors have to be more rugged as precautions have to be taken against fire, dust, and water.

Speed regulation which is the change in speed under full load and no load conditions is important for motors. Depending on the connection type, speed regulation of DC motors varies between 5 and 15%. The relative ease of speed control is an important characteristic of DC motors. Speed control can be done by changing the armature voltage, armature resistance, number of poles, or the magnetic field density.

Since Faraday's discovery of force on a current-carrying conductor in 1821, it took a long time (nearly 50 years) to cover the way to DC motors. High-power DC motors are used in vehicles, lifts and cranes, and rolling mills in the iron and steel works (Figure 6.1). However, today following the improvements in electronics, speed control of AC motors has become reasonably feasible as well.

6.2 DC MACHINE PRINCIPLES (LINEAR MACHINE)

Let us set up the system of Figure 6.2 by connecting the power supply V_T to a pair of rails (conductors) through a switch and a resistance R_A and short circuit these rails with a conducting bar of length l which can slide on the rails without any friction. We assume the rails and the bar have zero electrical resistance. If we apply a magnetic

FIGURE 6.1 First rolling mill motor by Westinghouse Company and a rolling mill producing steel slabs by the company Danieli.

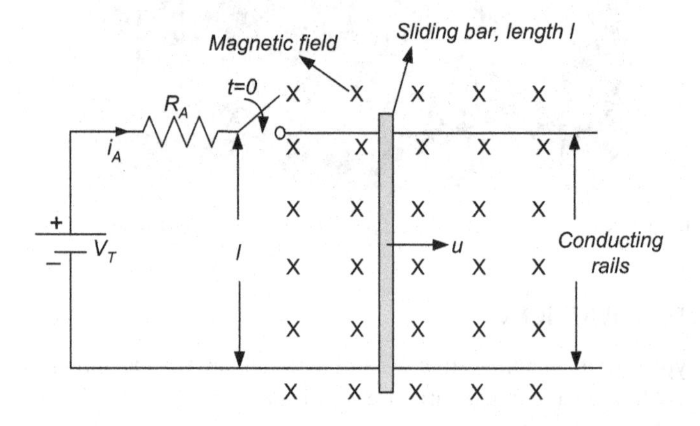

FIGURE 6.2 Linear DC machine.

field density of B perpendicular and into the plane of the rails and close the switch at $t = 0$, a force of

$$F = Bi_A l$$

will move the bar to the right and cause the bar to accelerate.

When the bar reaches the speed u, according to Faraday's and Lenz's laws, a voltage e_A will be induced on the bar and its direction will be opposite in polarity to the source voltage V_T (Figure 6.3).

The equivalent circuit will be as shown in Figure 6.4. If x is the distance covered by the conducting bar in t seconds and u is the speed of the bar, from Faraday's law the induced voltage e_A can be expressed as

$$e_A = \frac{d\lambda}{dt} = \frac{d(B.l.x)}{dt} = Blu$$

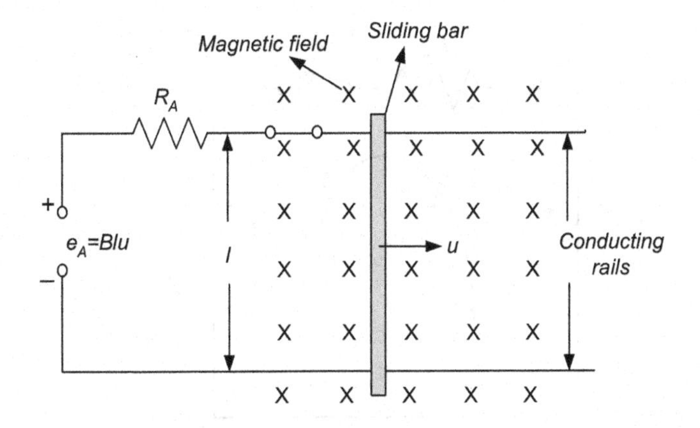

FIGURE 6.3 Induced voltage e due to motion of bar in Figure 6.2.

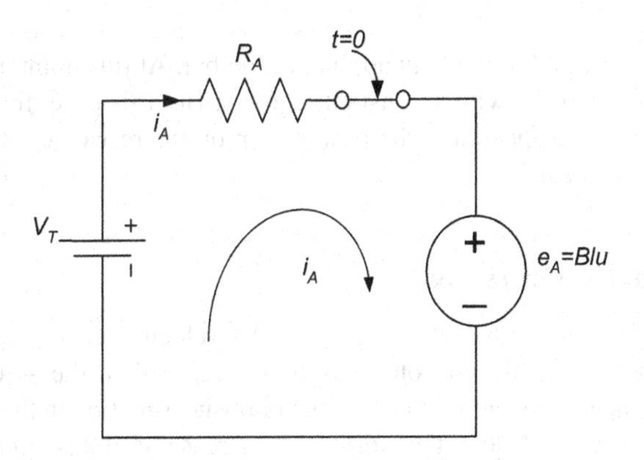

FIGURE 6.4 Equivalent circuit of Figure 6.2.

As the velocity of the bar increases, e_A will increase and the current causing the force F will decrease until e_A becomes equal and opposite to V_T. At this point, i_A and hence the force $f = Bi_Al$ on the bar will be zero and the bar will travel at a velocity of

$$u = \frac{V_T}{Bl}$$

Now let us assume that the bar is travelling with no load at this speed u and let us further visualise the system working as a motor and as a generator as discussed in the next sections.

6.2.1 Motor Operation

If we apply a force (load) to the bar towards left, the velocity of the bar will decrease, and consequently, the induced voltage will decrease and the current responsible for the force towards right will increase. This increase in force towards right will

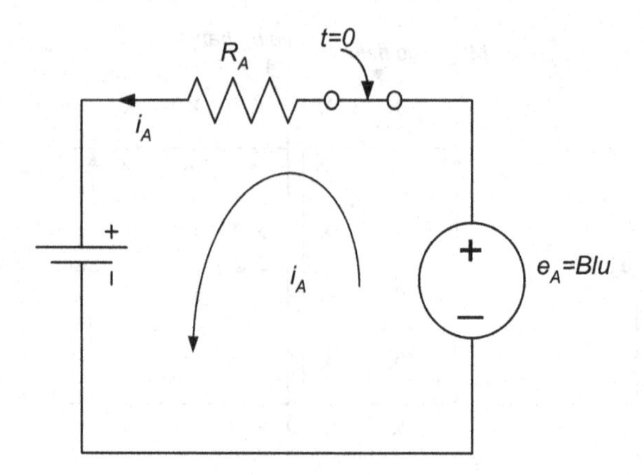

FIGURE 6.5 Equivalent circuit of the generator action in linear DC machine.

continue until it is equal to the load applied to the bar. At this point, the system will reach a stable state again with a constant velocity. But this time, the speed will be lower than the no load speed and the power drop on the resistance R will lower the efficiency of the system.

6.2.2 Generator Operation

If we apply a force towards right to the bar, the velocity of the bar will increase, and consequently, the induced voltage will be higher than the source voltage V_T (Figure 6.5). Hence, a current will flow in the opposite direction in the circuit to exert a force on the bar towards left. This force will increase until it is equal and opposite to the force towards right. When they are equal, there will be no force on the bar and the bar will once again reach a stable state with a constant velocity. But this time, the speed will be higher than the no load speed and the system will be working as a generator supplying power to the source.

Exercise 6.1:

In the linear generator of Figure 6.6, the distance between the conducting rails is 0.5 m, and a magnetic flux density of 1 T is applied in the direction which is perpendicular and into the paper. If the armature is travelling with a speed of u = 10 m/s, find the generated voltage e_A. If a 1 Ω resistor is placed at the end of the rails, what would be the force required to pull the bar at this speed?

Solution:

$$e = Blu = 1 \times 0.5 \times 10 = 5 \text{ V}$$

$$I = \frac{E}{R} = \frac{5}{1} = 5 \text{ A} \rightarrow F = B.i.l = 1 \times 5 \times 0.5 = 2.5 \text{ N}$$

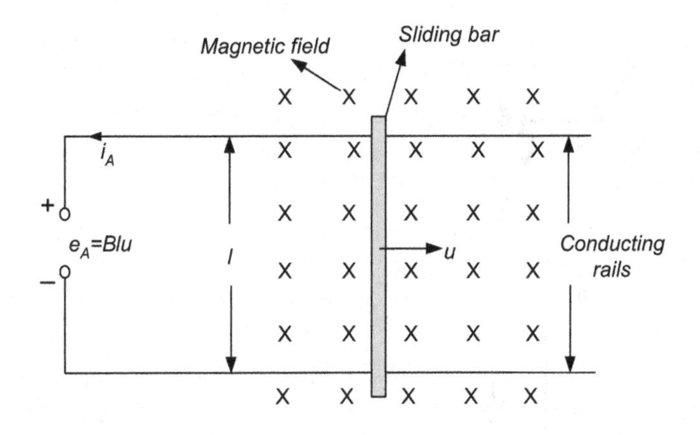

FIGURE 6.6

6.3 ROTATING DC MACHINES

A simple DC machine is illustrated in Figure 6.7. Each end of the coil is connected (fixed) to one half of a ring (segments).

Black wire is fixed to the black segment, and the white wire is fixed to the white segment. Segments do not touch each other. Two brushes ride on the segments as shown in Figure 6.7. The two brushes are used to collect the voltage from these segments.

6.3.1 Principles of DC Generator Operation

If the coil is rotated in the magnetic field of B as shown in Figure 6.8, a voltage of the DC form will be obtained from the brushes. Brushes will short the armature

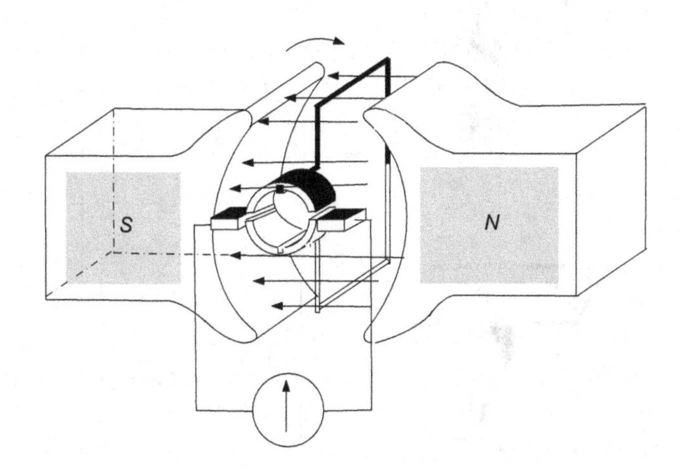

FIGURE 6.7 Simple DC machine.

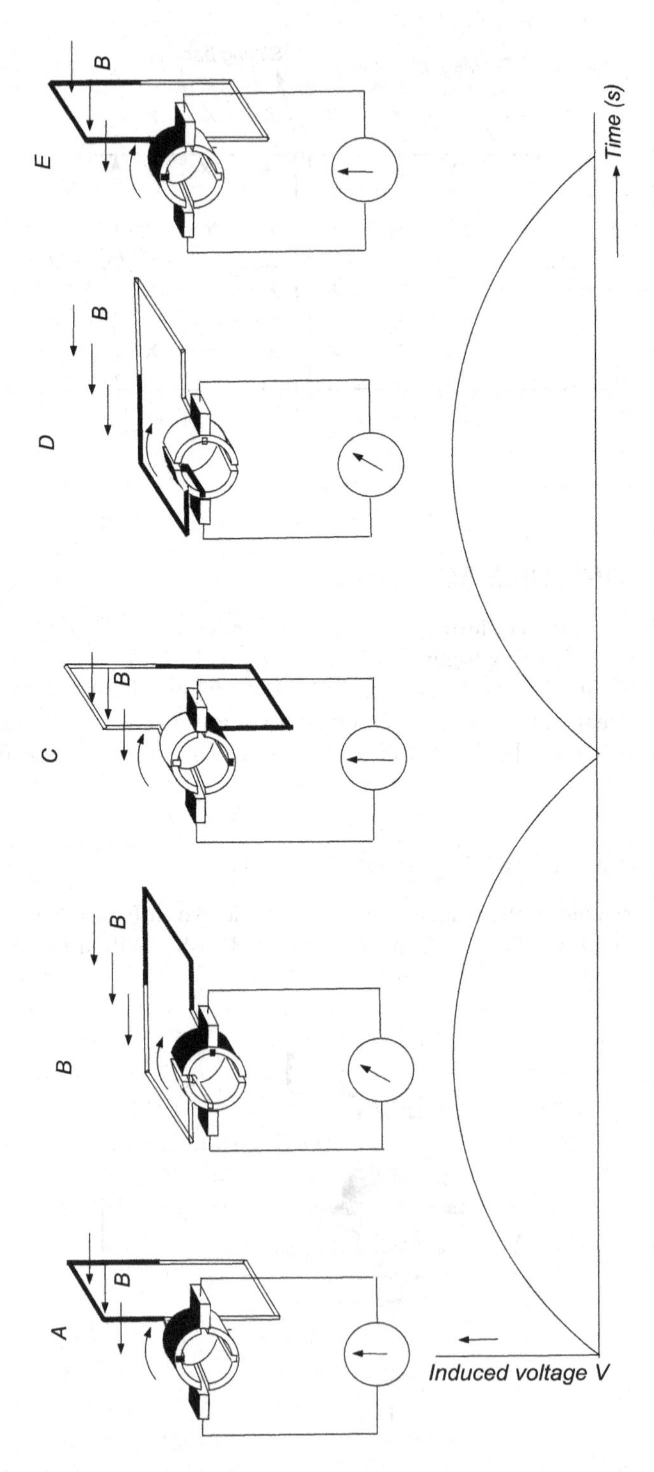

FIGURE 6.8 Generation of DC voltage.

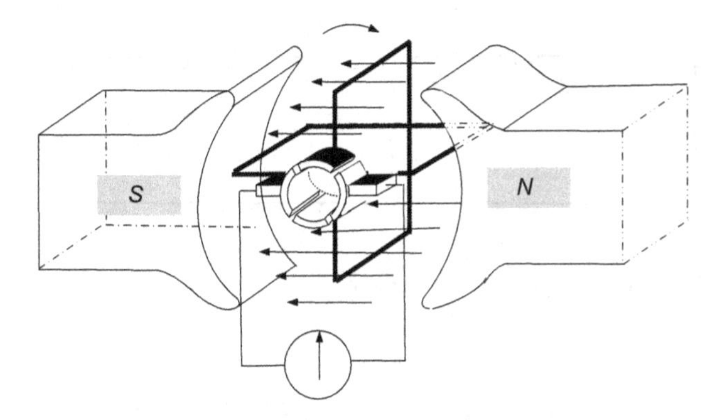

FIGURE 6.9 Simple DC machine with two coils.

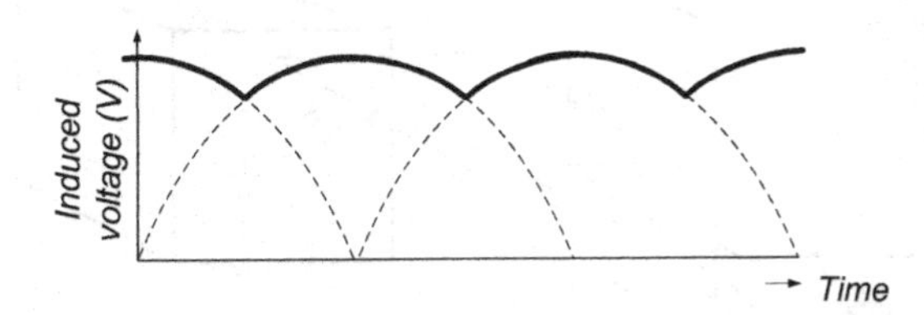

FIGURE 6.10 A simple DC generator output with two coils and four segments.

winding briefly during the switching process. However, this shorting will not cause a problem, because the voltage will be zero when it occurs.

If there are two coils with two commutator segments as shown in Figure 6.9, the voltage obtained from the brushes will be closer to a DC form as shown in Figure 6.10.

Note that the number of commutator segments increases in direct proportion to the number of loops. That is, there are two segments for one loop, four segments for two loops, and eight segments for four loops.

The output voltage of a DC generator is proportional to the

- number of loops
- number of turns per loop
- the total flux per pair of poles in the machine
- the speed of rotation of the armature.

If we think that the coils are spread around an angle of 360° and there are four or more poles, the voltage obtained will be very close to a DC voltage (Figure 6.11).

6.3.2 Principles of DC Motor Operation

Figure 6.12 shows a simple permanent magnet DC machine with a split ring structure. Each end of the coil is fixed to one half of a ring. For motor operation, a DC voltage is applied to the coil through the two brushes which ride on these half rings. The action of changing the connections between the brushes and the coil sides is

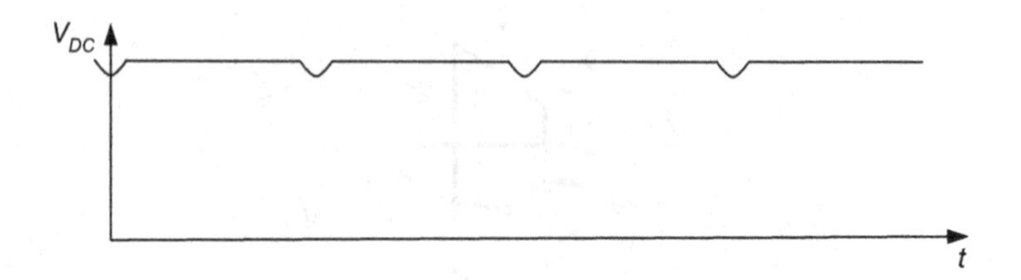

FIGURE 6.11 Voltage output of DC generator.

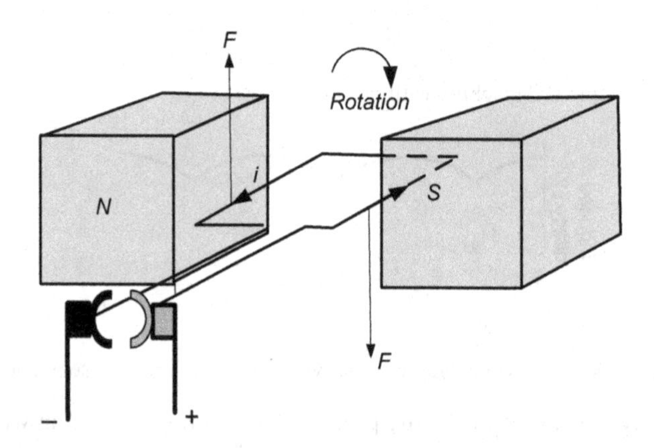

FIGURE 6.12 DC motor operation (maximum torque position-1).

called commutation. Commutation is necessary in order to develop a torque which is always equal and in opposite directions on the two sides of the coil so that rotation of the coil can take place. Through this action, the same polarity of voltage is always applied to the coil side on the right (next to the south pole) and on the left (next to the north pole). This way we obtain a torque which rotates the coil always in the same direction (clockwise in our example).

With the DC voltage applied as shown in Figure 6.12, current is made to flow in the direction as shown in the figure. According to the right-hand rule, an upward force F which is perpendicular to the plane of the coil will be exerted on the left-hand side of the coil which is equal to

$$F = ilB$$

The force on the right-hand side coil (with white segment) will be of the same magnitude but in the opposite direction (downward) due to current flowing in the opposite direction. Hence, the two equal torques in opposite directions on both sides of the coil will make the coil rotate clockwise.

While the coil travels to its second position shown in Figure 6.13, the two equal and opposite forces on the two sides of the coil get weaker because of the decreasing angle between the two magnetic axes but remain always in the same upward and downward directions.

Therefore, the total torque on the coil will weaken but will remain always in the direction to cause the coil to rotate in the clockwise direction. In position shown in Figure 6.13, the current and hence the torque will be zero, but due its momentum, the coil will step towards right and direction of the current flow in the coil sides will reverse due to the change in the polarity of the source voltage (black brush on the white segment and white brush on the black segment) (commutation). In the position shown in Figure 6.13, a short circuit may occur between the brushes, resulting in no current in the coil and causing no significant effect on the performance of the system.

As a result during the motion of the coil sides from the position in Figure 6.13 to the position shown in Figure 6.14, the two forces on the coil sides will reverse and hence the torque on the coil will continue to rotate the coil in the same clockwise direction as shown in Figure 6.14.

The coil will continue its rotation and come to the no torque and no current (neutral) position once again (Figure 6.15). Due to its momentum, it will continue its

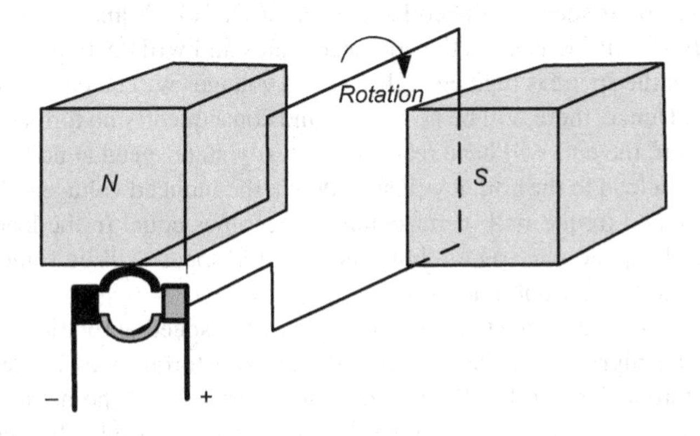

FIGURE 6.13 DC motor operation (minimum torque position-1).

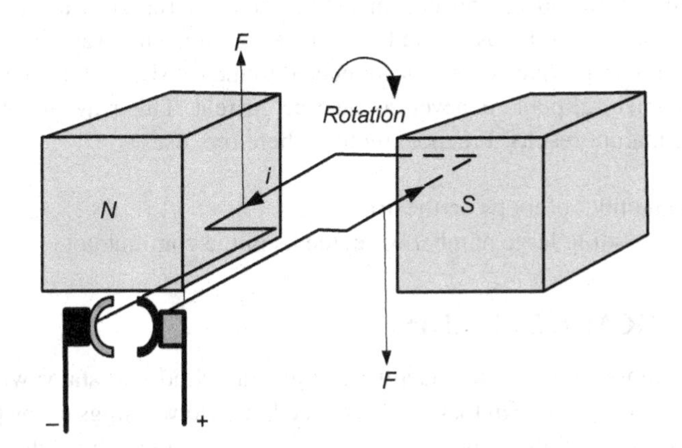

FIGURE 6.14 DC motor operation (maximum torque position-2).

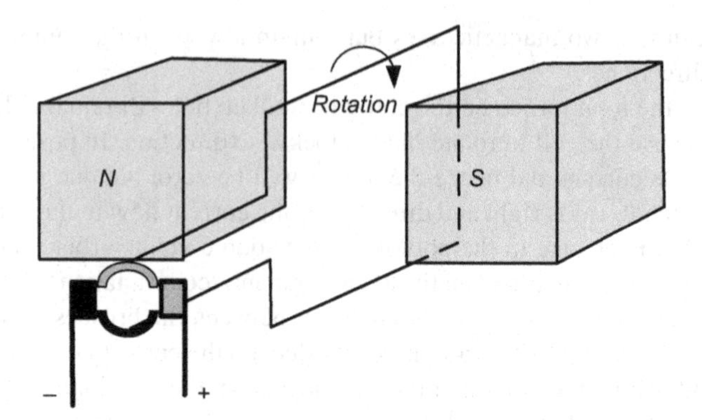

FIGURE 6.15 DC motor operation (minimum torque position-2).

motion and through the similar chain of actions reach to the maximum torque position again (Figure 6.12).

Because the coil moves in a magnetic field, a voltage in the opposite direction to the source will be induced in the coil according to Faraday's and Lenz's laws. This voltage $(= Blu)$ will increase as the coil accelerates and will decrease the torque on the coil. When the speed is high enough, the two voltages will be equal and opposite in directions. Hence, there will be no current and consequently no torque on the coil. At this moment, the coil will have reached its steady-state speed at no load.

If we apply a load to the coil, it will slow down, the induced voltage will decrease, and the developed torque will increase until it becomes equal to the load. The coil runs at a steady speed once again, but this time, the speed will be somehow lower than the no load speed (motor action).

If we apply an extra torque to the coil to increase the speed of rotation, the induced voltage will be higher than the source voltage and a torque will be developed to oppose this torque. The coil will come to a steady speed with no net torque on the coil. This time the speed will be higher than the no load speed. The machine will start working as a generator supplying power to the source.

To develop more torque and eliminate the effect of the zero torque points on armature rotation, more loops are added to the armature. This way only the loop in the best position to produce torque is connected to the brushes. Thus, the loop passing through a neutral plane is never conducting current. This way smoother operation of the armature results. Practical motors, therefore, use

- a large number of loops on the armature
- a corresponding large number of segments in the commutator.

6.4 PRACTICAL DC MACHINES

The most common type of DC machine contains a cylindrical stator with an even number of magnetic poles (p) that are established by field windings (or by permanent magnets). In DC machines, the field winding is wound onto the stator, while the armature winding is located on the rotor (Figure 6.16). Usually, the field winding

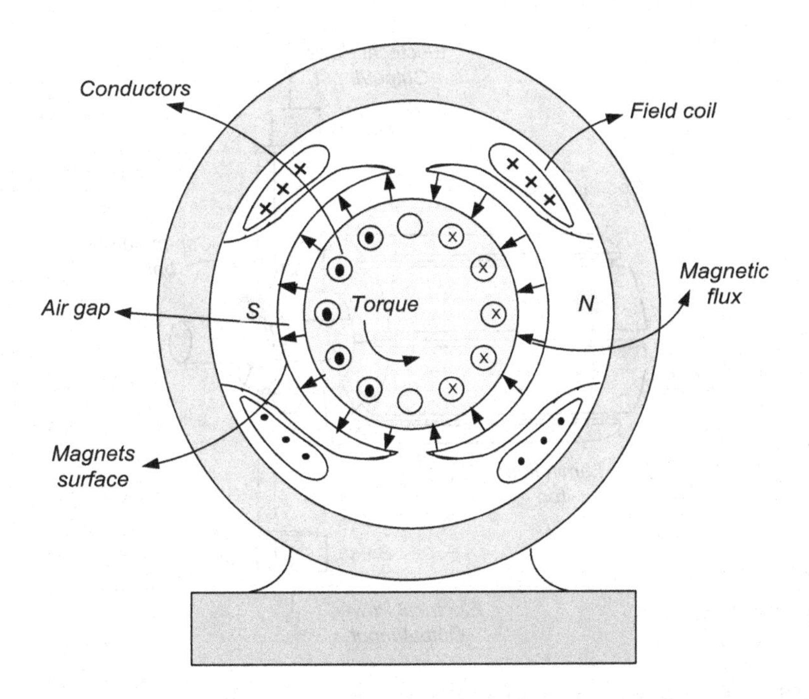

FIGURE 6.16 Practical DC machine.

consumes only a small fraction of the electrical power of a DC machine. Current is supplied to (or received from) the rotating armature windings through commutator segments.

A DC current should pass through the field winding. This current can be supplied either from the armature winding's DC source or from another DC voltage source. Electrical power is supplied through or received from the armature windings in DC form.

Inside the stator is a rotor consisting of a laminated iron cylinder mounted on a shaft that can rotate (Figure 6.16). The stator and rotor cores are made of iron, steel, or another ferromagnetic metal in order to reduce the magnetic reluctance. Slots are cut lengthwise into the surface of the rotor to contain the armature windings. Magnetic flux tends to take the path of least reluctance.

Thus, the flux in the air gap is perpendicular to the surface of the rotor and to the armature conductors, and is nearly constant (Figures 6.17 and 6.18).

6.5 DC MACHINE IN ACTION

Under the pole faces (Figure 6.16), the conductors, the field, and the direction of motion are mutually perpendicular. Thus, a nearly constant voltage is induced (torque developed) in each conductor as it moves under a pole. However, as the conductors move between poles, the field becomes zero and then direction reverses (Figure 6.18). Therefore, the induced voltages (torque) fall to zero and build up with

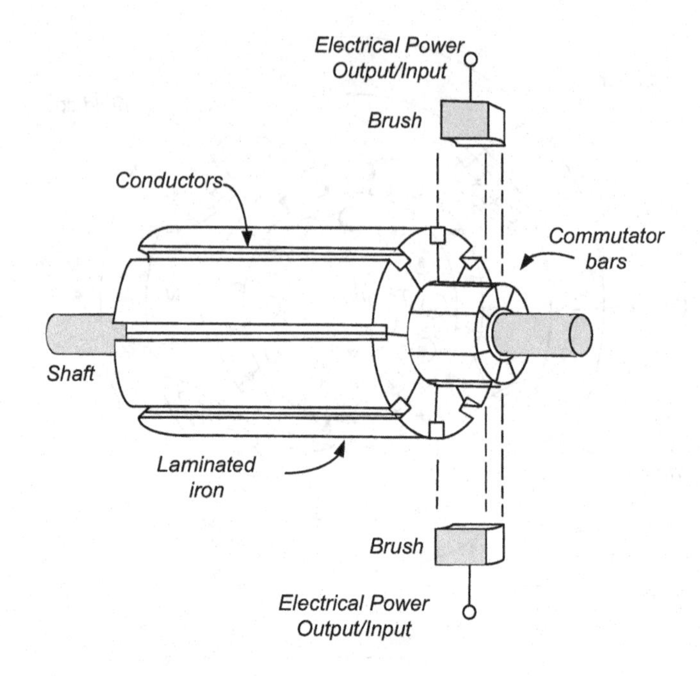

FIGURE 6.17 The rotor of a practical DC machine.

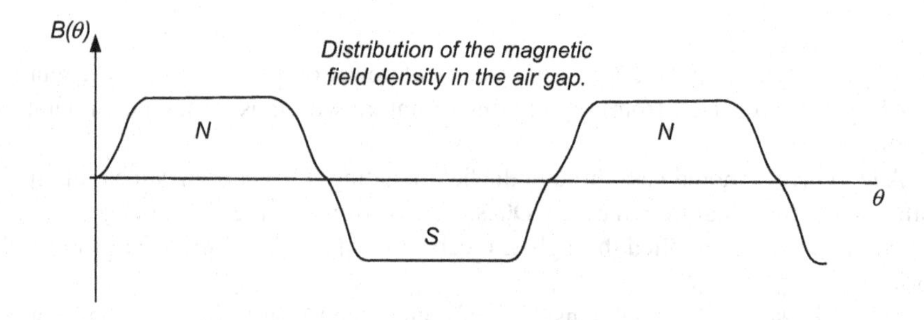

FIGURE 6.18 Magnetic field in the air gap.

the opposite polarity. A commutator (a mechanical switch) reverses the connections to the conductors as they move between poles (Figures 6.17 and 6.19).

This way the polarity of the induced voltage (direction of motion) seen from the external machine terminals stays the same. Commutators in typical machines contain 20–50 segments (Figure 6.19). Because only part of the coils are commutated at a time, the terminal voltage of a real machine shows relatively little fluctuations compared to the two-segment example.

Actual armatures consist of a large number of conductors placed around the circumference of the rotor (Figures 6.17 and 6.20).

To attain high terminal voltages, conductors are placed in series or parallel according to the machine design. A four-pole machine is shown in Figure 6.20. The current directions shown result in a counterclockwise torque. Notice that the directions of

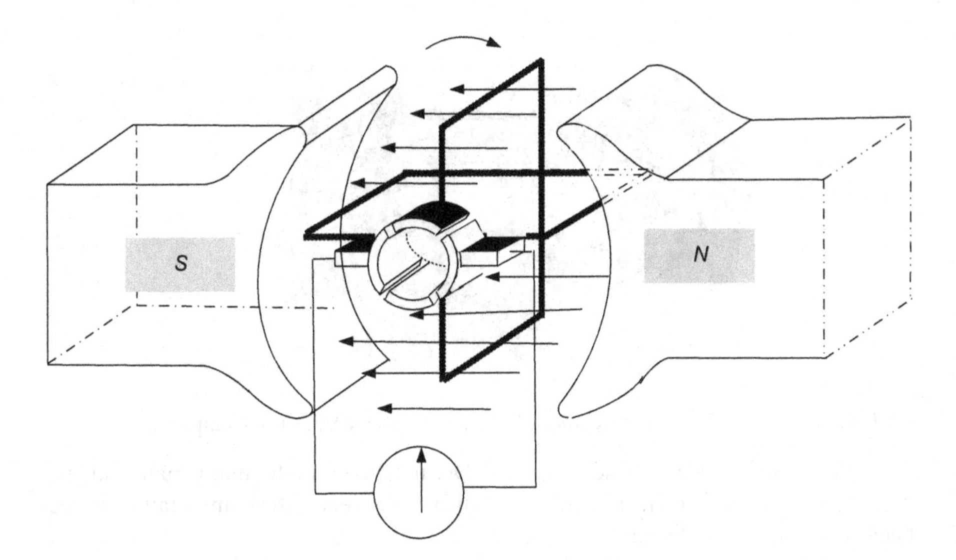

FIGURE 6.19 DC machine with multiple segments.

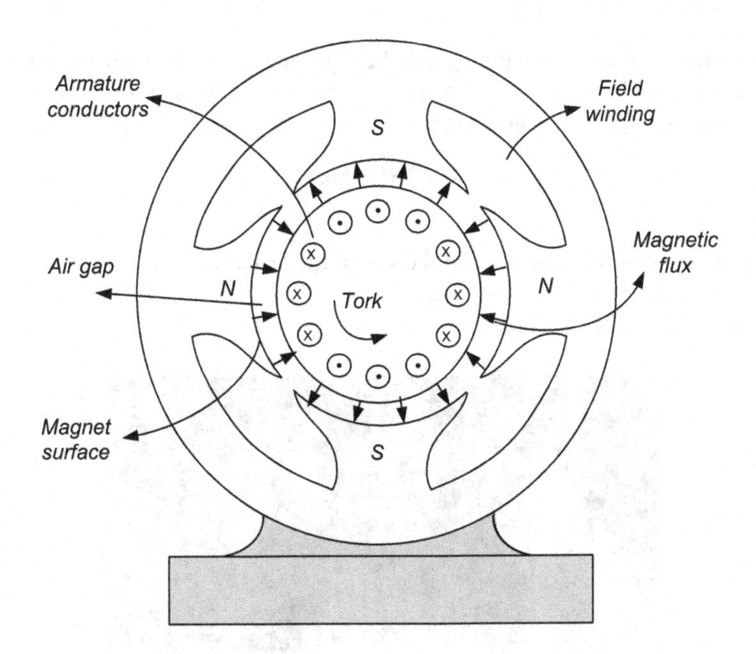

FIGURE 6.20 A four-pole DC machine.

the currents in the armature must be reversed when switching poles to achieve aiding contributions to total torque.

Generally, the commutator segments are copper bars insulated from one another and from the shaft (Figure 6.17). The brushes contain graphite that lubricates the sliding contact (Figures 6.17 and 6.21).

FIGURE 6.21 Practical DC machine. (Courtesy of Reliance Electric Company.)

Even so, a significant disadvantage of DC machines is the maintenance of the moving contacts and the need to replace brushes and repair the commutator surface because of mechanical wear.

6.5.1 Armature Voltage

As the armature rotates in the magnetic field produced by the stator poles (Figures 6.16 and 6.22), voltage is induced in the armature winding. A turn is made of two conductors; therefore, induced voltage in a turn is

$$e_t = 2B(\theta)l\omega_m r$$

where

l = the length of the conductor in the slot of the armature
ω_m = the mechanical angular speed
r = the radius of the armature.

FIGURE 6.22 Constant torque, low-speed rolling mill DC motor. (Courtesy of Westinghouse Motor Company.)

The average value of the induced voltage in the turn is

$$\bar{e}_t = 2\overline{B(\theta)}\, l\omega_m r$$

Let Φ = Magnetic flux of per pole

$$A = \text{Area per pole } = \frac{2\pi rl}{p}$$

where p is the number of poles, then,

$$\overline{B(\theta)} = \frac{\Phi}{A} = \frac{\Phi p}{2\pi rl}; \quad \rightarrow \bar{e}_t = 2\overline{B(\theta)} l\omega_m r \rightarrow \quad \bar{e}_t = 2 \times \frac{\Phi p}{2\pi rl}\, l\omega_m r$$

$$\bar{e}_t = \frac{\Phi p}{\pi}\, \omega_m$$

If E_a = average terminal voltage
N = total number of turns in the armature winding
a = number of parallel paths
p = number of stator poles
then

$$E_a = \frac{N}{a}\bar{e}_t \quad \rightarrow E_a = \frac{Np}{\pi a}\Phi\omega_m \quad \rightarrow E_a = K_a\Phi\omega_m \tag{6.1}$$

$$K_a = \frac{Np}{\pi a}$$

In the case of

- generator operation, this voltage is known as generated voltage
- in motor operation, it is known as back emf.

6.5.2 Developed Torque

The force on a conductor (placed on the periphery of the armature) (Figure 6.16) is

$$f_c = B(\theta) l i_c = B(\theta) l \frac{I_a}{a}$$

where i_c is the current in the conductor of the armature winding and I_a is the armature terminal current.

The torque developed by a conductor is

$$T_c = f_c r$$

The average torque developed by a conductor is

$$\bar{T}_c = \overline{B(\theta)}l\frac{I_a}{a}r$$

$$\overline{B(\theta)} = \frac{\Phi}{A} = \frac{\Phi p}{2\pi rl}$$

The total torque developed is

$$\bar{T}_c = \frac{\Phi p I_a}{2\pi a}$$

$$T = 2N\bar{T}_c = \frac{N\Phi p}{\pi a}I_a = K_a\Phi I_a \tag{6.2}$$

The same result can be obtained from equating electrical power developed to the mechanical power:

$$E_a I_a = K_a \Phi \omega_m I_a = T\omega_m$$

6.6 EQUIVALENT CIRCUIT OF DC MOTORS

The equivalent circuit of the DC motor is shown in Figure 6.23. The field circuit is represented by a resistance R_F and an inductance L_F in series. Since V_F is a DC voltage,

$$V_F = R_F I_F$$

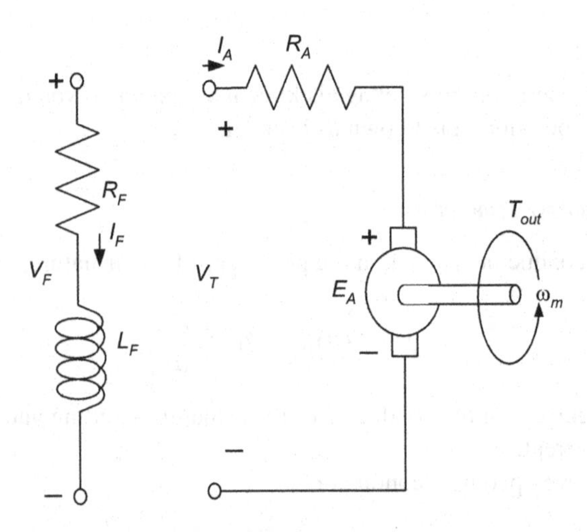

FIGURE 6.23 Equivalent circuit of the DC motor.

E_A = average voltage induced in the armature (it opposes the applied external electrical source V_T).

R_A = the resistance of the armature windings plus the brush

From equations (6.1) and (6.2), we know that

$$E_A = K\Phi\omega_m;\ T_{\text{dev}} = K\Phi I_A$$

The power developed is

$$P_{\text{dev}} = E_A I_A = T_{\text{dev}}.\omega_m$$

Mechanical power delivered is

$$P_{\text{out}} = \omega_m T_{\text{out}}$$

where T_{out} is the mechanical torque driving the load.

The difference

$$P_{\text{losses}} = P_{\text{dev}} - P_{\text{out}}$$

is the rotational losses of the system.

6.6.1 Magnetisation Curve

The magnetisation curve of a DC machine is a plot of E_A versus the field current I_F with the machine being driven at a constant speed (Figure 6.24). E_A can be found by measuring the open-circuit voltage at the armature terminals.

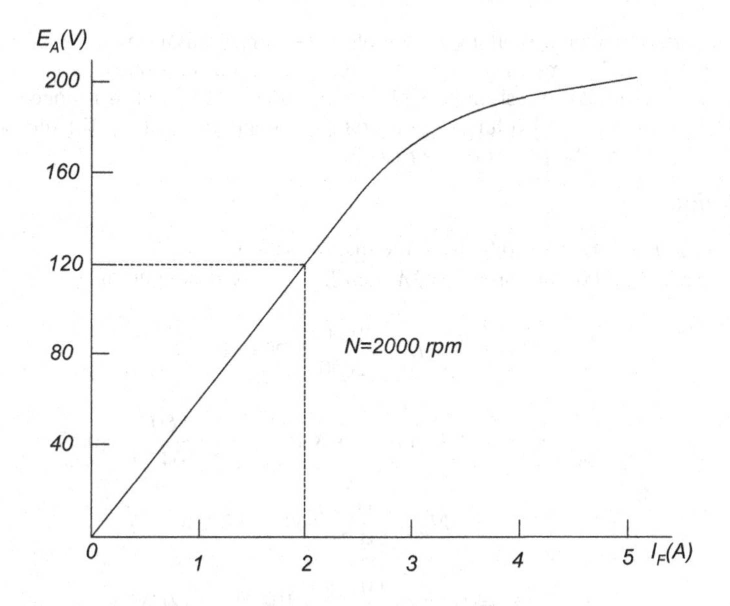

FIGURE 6.24 Magnetisation curve of a DC machine.

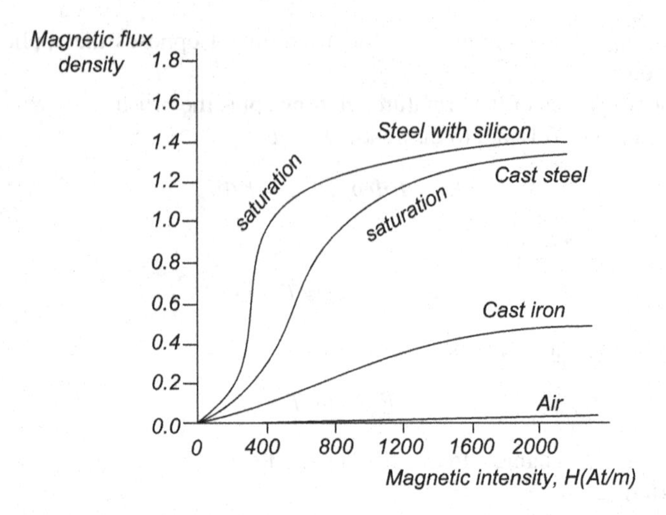

FIGURE 6.25 Magnetisation curves.

Because E_A is proportional to the flux Φ, the magnetisation curve has the same shape as a Φ versus I_F plot. Magnetisation curves of silicon steel, cast steel, and cast iron are illustrated in Figure 6.25.

Hence,

$$E_A = K\Phi\omega_m \qquad \rightarrow \qquad \frac{E_{A1}}{E_{A2}} = \frac{n_1}{n_2} = \frac{\omega_1}{\omega_2}$$

Exercise 6.2:

An electrical machine working as a motor has a magnetisation curve as shown in Figure 6.24. The motor runs at 1,000 rpm with a current intake of $I_A = 60$ A, $I_F = 2$ A. The armature resistance R_A is 0.2 Ω, and the field coil resistance is $R_F = 60$ Ω. Find the field coil voltage V_F, the voltage applied to armature (V_T), the developed torque, and the power of the machine.

Solution:

$V_F = R_F I_F = 60 \times 2 = 120$ V; from the magnetisation curve,
When $n_1 = 2000$ rpm and $I_F = 2A \quad \rightarrow E_{A1} = 120$ V (from curve)

$$E_{A2} = \frac{n_2}{n_1} \times E_{A1} = \frac{1000}{2000} \times 120 = 60 \text{ V}$$

$$\omega_m = \frac{1000}{60} \times 2\pi = 104.71\, r/s \rightarrow K\Phi = \frac{E_A}{\omega_m} = \frac{60}{104.71} = 0.57$$

$$T_{\text{dev}} = K\Phi I_A = \frac{60}{104.71} \times 60 = 34.2 \text{ Nm}$$

$$P_{\text{dev}} = T_{\text{dev}}\omega_m = \frac{60 \times 60}{104.71} \times 104.71 = 3600 \text{ W}$$

Let us check our result for developed power:

$$P_{\text{dev}} = I_A E_A = 60 \times 60 = 3600 \text{ W}$$

Apply Kirchhoff's voltage law to the armature,

$$V_T = I_A R_A + E_A = 0.2 \times 60 + 60 = 72 \text{ V}$$

6.7 SHUNTED DC MOTOR

In Figure 6.26, we see a DC motor with its armature and field coils shunted (paralled). Shunted DC motor is a "self-excited" motor. Let us try to calculate the torque expression that will explain to us how the torque varies with the speed of the motor.

We know that

$$E_A = K\Phi\omega_m \text{ and } T_{\text{dev}} = K\Phi I_A$$

Here, K = machine constant, Φ = magnetic flux created by stator coil, and ω_m = angular velocity of the motor.

$$V_T = I_A R_A + E_A \quad \rightarrow V_T = R_A\left(\frac{T_{\text{dev}}}{K\Phi} + K\Phi\omega_m\right)$$

Therefore, the torque expression in terms of speed would be

$$T_{\text{dev}} = \frac{K\Phi}{R_A}(V_T - K\Phi\omega_m)$$

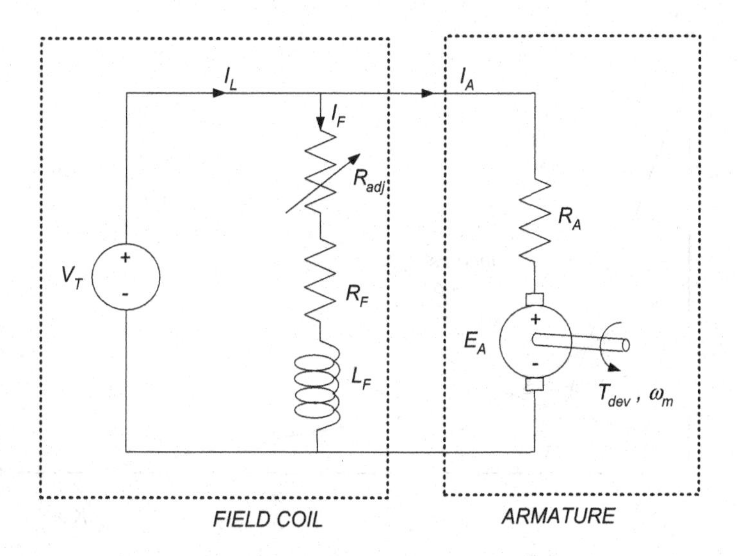

FIGURE 6.26 Shunted DC motor with rheostat changeable field current.

6.7.1 Speed Control and Speed Regulation of a Shunt-Connected DC Motor

There are two ways to achieve the speed control of shunt-connected DC motors:

1. By changing R_{adj} which is in series with the field coil (Figure 6.26), we can change I_f and hence the magnetic field Φ. Torque/speed characteristics will be effected as shown in Figure 6.27.

$$I_f \rightarrow \Phi; \quad \Phi \rightarrow T_{dev} = \frac{K\Phi}{R_A}(V_T - K\Phi\omega_m)$$

2. We can achieve the speed control of the motor by changing the value of the rheostat R_{adj} which is connected in series with the armature winding (Figure 6.28).

   In this case, the torque speed characteristics of the motor would be as shown in Figure 6.29 according to varying values of R_A.

$$R_A \rightarrow \quad T_{dev} = \frac{K\Phi}{R_A}(V_T - K\Phi\omega_m)$$

The speed regulation of DC shunt motor is around 10%. DC shunt motor has a good tendency of keeping its speed regulation low. Therefore, it is very suitable for constant speed operation. Their disadvantage would be their limited starting torque value.

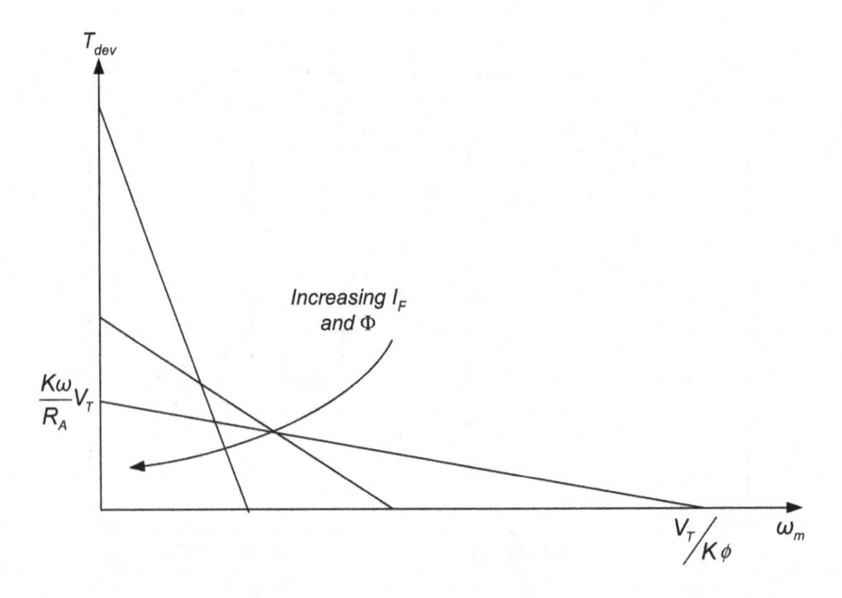

FIGURE 6.27 Effect of field current on torque/speed in shunt connected DC motors.

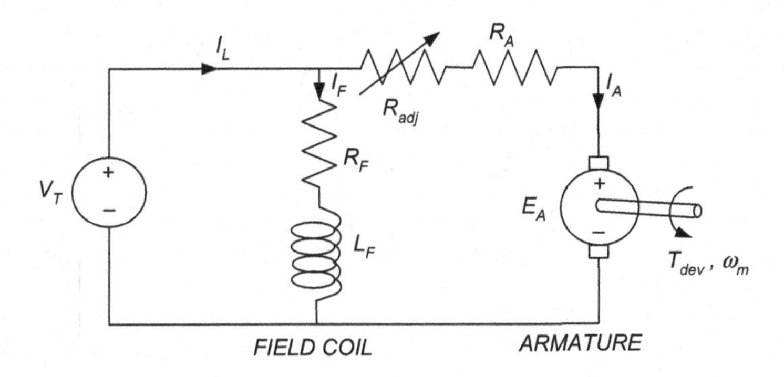

FIGURE 6.28 Shunted DC motor with rheostat changeable armature current.

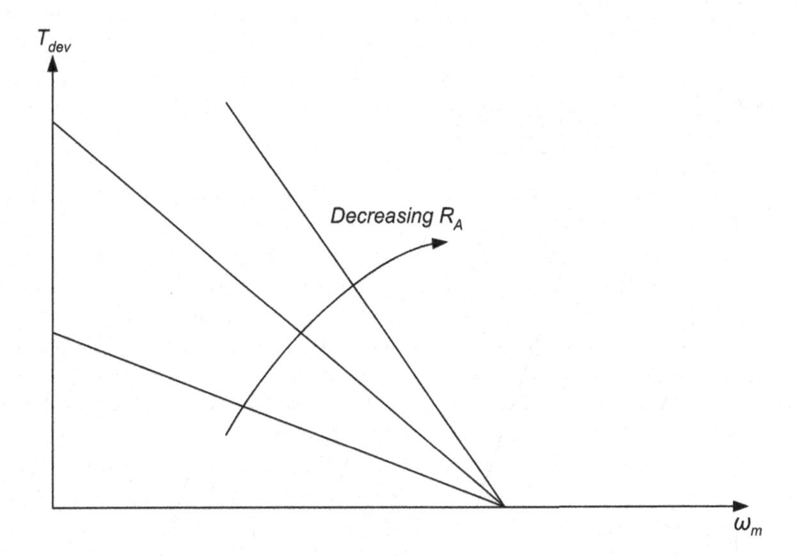

FIGURE 6.29 Effect of armature resistance on torque/speed in shunt connected DC motors.

6.8 SEPARATELY EXCITED DC MOTORS AND SPEED CONTROL

In separately excited DC motors, different voltage sources are used for the armature and the field (Figure 6.30). The torque equation would not be any different than what we found for the shunt-connected DC motors. So the torque expression would be as follows

$$T_{\text{dev}} = \frac{K\Phi}{R_A}(V_T - K\Phi\omega_m)$$

Speed control can be achieved by changing V_T, Φ, or R_A. Figure 6.31 shows how torque changes with V_T. Just like in shunt-connected DC motors, we would be able

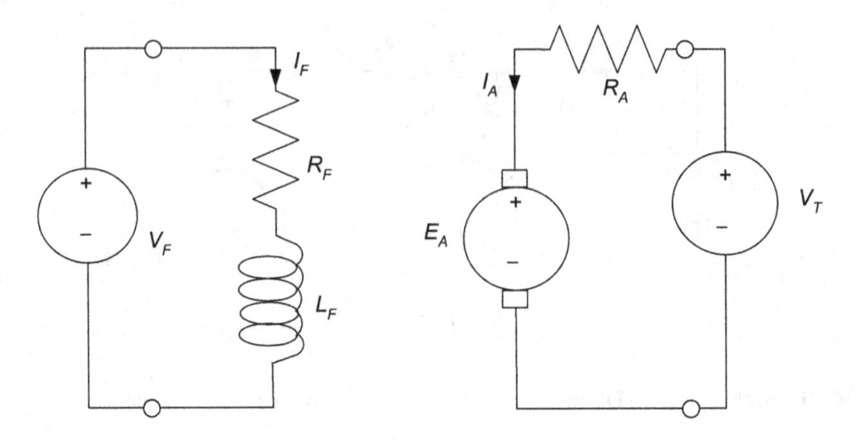

FIGURE 6.30 Separately connected DC motor.

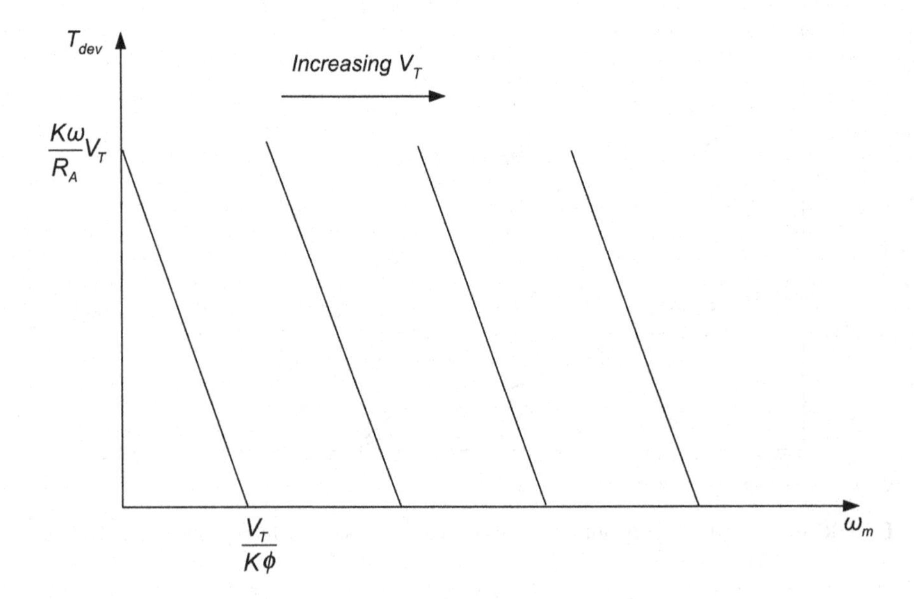

FIGURE 6.31 Effect of V_T on torque/speed in separately excited DC motors.

to obtain the same torque/speed characteristics shown in Figures 6.27 and 6.29 by changing the armature resistance R_A or the field resistance R_F.

However in separately connected DC motors, the torque/speed characteristics obtained by changing R_F can also be obtained by changing V_F. This solution would require a variable DC voltage source which is a rather expensive solution. On the other hand, torque control solution obtained by changing R_F is an inefficient solution due to the losses in the resistor R_F.

Speed regulation characteristics are as expected similar to shunted DC motors.

6.9 SERIES-CONNECTED DC MOTOR

In these types of DC motors, field and armature coils are connected in series (Figure 6.32). Like shunted DC motor, series-connected DC motor is also a "self-excited" motor.

$$\Phi = K_F I_F = K_F I_A \rightarrow E_A = K\Phi\omega_m = K(K_F I_F)\omega_m$$

where ω_m is the angular velocity of the motor. We also have

$$T_{\text{dev}} = K\Phi I_A = KK_F I_A^2$$

It follows that

$$V_T = I_A R_A + I_A R_F + E_A = I_A R_A + I_A R_F + KK_F I_A \omega_m$$

Hence,

$$I_A = \frac{V_T}{R_A + R_F + KK_F\omega_m} \quad \rightarrow \quad T_{\text{dev}} = KK_F \left(\frac{V_T}{R_A + R_F + KK_F\omega_m} \right)^2$$

Torque can be lowered for a fixed value of speed by increasing R_A (Figure 6.33). Speed control achieved by changing R_A is not efficient, and the speed control achieved by changing the source voltage V_T is expensive.

Speed regulation of these motors is poor (more than 35%). However, their starting torque values are their greatest advantage.

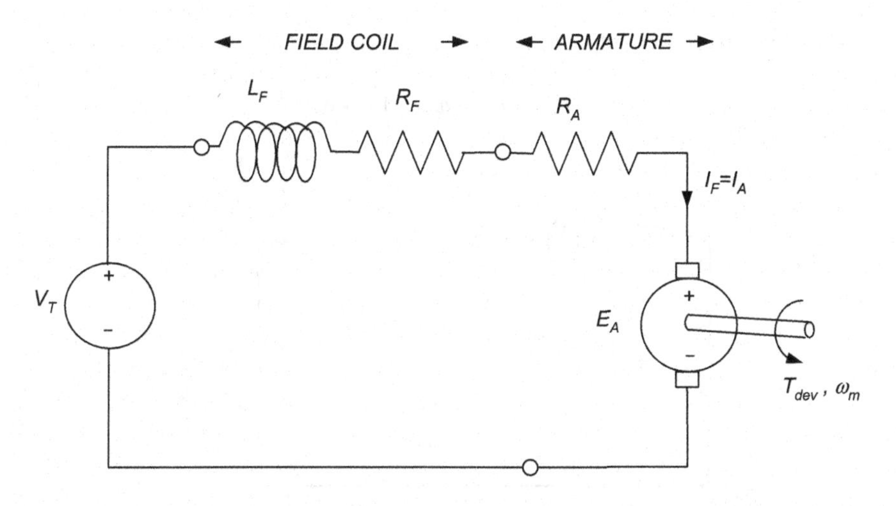

FIGURE 6.32 Series-connected DC motor.

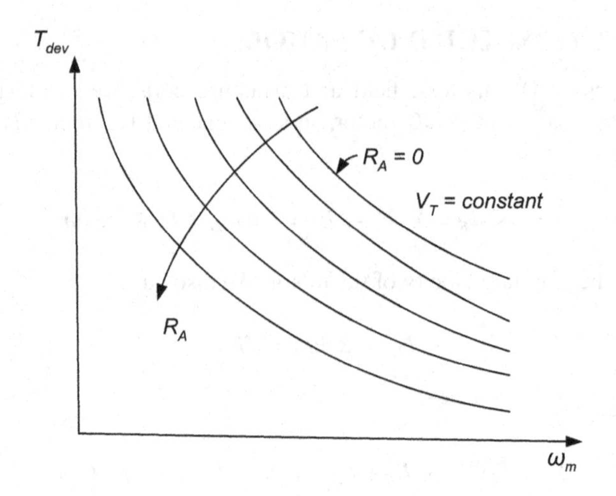

FIGURE 6.33 Effect of R_A on torque/speed in series connected DC motors.

Exercise 6.3:

A series-connected DC motor (Figure 6.34) is driving a load that demands a torque of 10 Nm when running at a speed of $n_{m1} = 1{,}800$ rpm. Neglecting the resistances, rotational loss, and saturation effects, find the power output of the motor. Find the new speed and output power if the load torque increases to 40 Nm.

Solution:

The angular velocity of the motor is

$$\omega_{m1} = n_{m1} \times \frac{2\pi}{60} = 188.50 \text{ rad/s}$$

Therefore, the power output is

$$P_{dev1} = T_{out1} \times \omega_{m1} = 1885 \text{ W}$$

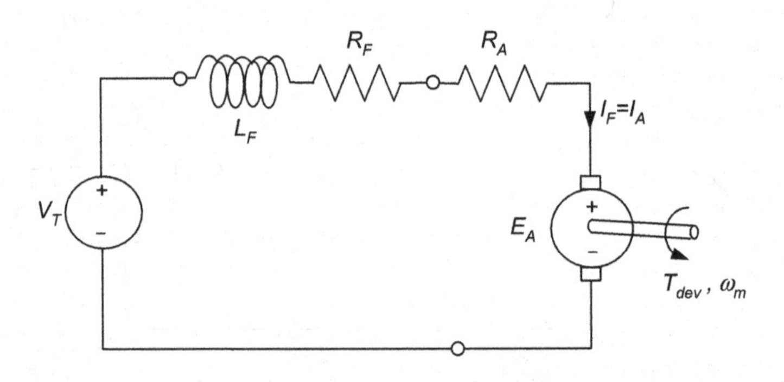

FIGURE 6.34

We know that

$$R_A = R_F = 0$$

Therefore, if we arrange the torque equation according to above, T_{dev} becomes

$$T_{\text{dev}} = KK_F\left(\frac{V_T}{R_A + R_F + KK_F\omega_m}\right)^2 = \left(\frac{V_T}{KK_F\omega_m}\right)^2$$

Therefore, from

$$\frac{T_{\text{dev1}}}{T_{\text{dev2}}} = \frac{\omega_{m2}^2}{\omega_{m1}^2}$$

we derive ω_{m2} as

$$\omega_{m2} = \omega_{m1}\sqrt{\frac{T_{\text{dev1}}}{T_{\text{dev2}}}} = 188.5\sqrt{\frac{10}{40}} = 94.25 \text{ rad/s}$$

$$n_{m2} = \frac{94.25}{2\pi}\times 60 = 900\ d/d$$

$$P_{\text{out2}} = T_{\text{dev2}}\omega_{m2} = 94.25\times 40 = 3770 \text{ W}$$

Exercise 6.4:

A shunt-connected DC motor is driving a load that demands a torque of 10 Nm when running at a speed of n_{m1} = 1,500 rpm. Neglecting the resistances, rotational loss, and saturation effects, find the power output of the motor. Find the new speed and output power if the load torque increases to 20 Nm.

Solution:

With R_A = 0 and V_T constant, E_A and hence the speed must be constant and independent of torque. Thus,

$$n_{m1} = n_{m2} = 1500\ d/d$$

$$P_{\text{out1}} = \omega_{m1}T_{\text{out1}} = \frac{1500}{60}\times 2\pi\times 10 = 1570 \text{ W}$$

$$P_{\text{out2}} = \omega_{m1}T_{\text{out2}} = \frac{1500}{60}\times 2\pi\times 20 = 3140 \text{ W}$$

6.10 UNIVERSAL MOTORS

The torque equation we found in series-connected DC machines was

$$T_{\text{dev}} = KK_F\left(\frac{V_T}{R_A + R_F + KK_F\omega_m}\right)^2$$

This equation reveals that torque is proportional to the square of the source voltage. Thus, the series-connected machine (Figure 6.35) can be operated from a single-phase AC source (provided that the stator is laminated to avoid excessive losses due to eddy currents). The rotor's field coils are connected in series with the stator windings through a commutator. If an AC motor has brushes and commutator, it must be a universal motor. Series DC motors that are intended for use with AC sources are called universal motors because in principle they can operate from either AC or DC. Since the field and armature inductances have nonzero impedances for ACs, the current would not be as large for an AC source as it would be for a DC source of the same average magnitude.

Also a significant part of the input voltage is dropped across the inductive reactances, and therefore $E_A = K\Phi\omega_m$, and hence, speed ω_m is smaller for a given input voltage during AC operation than that during DC operation. The torque remains the same, but the speed appears to be lower (Figure 6.36).

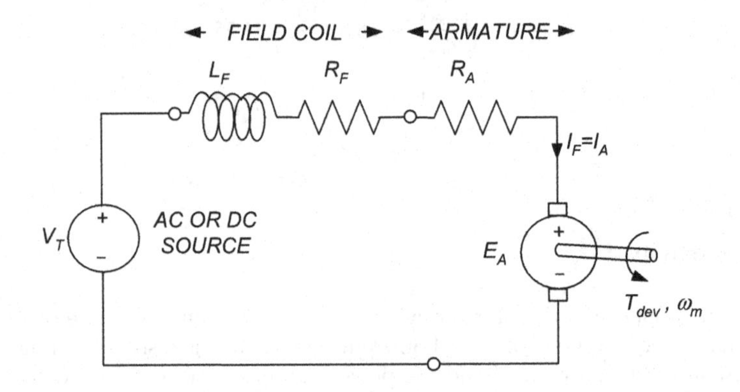

FIGURE 6.35 Equivalent circuit of universal motor.

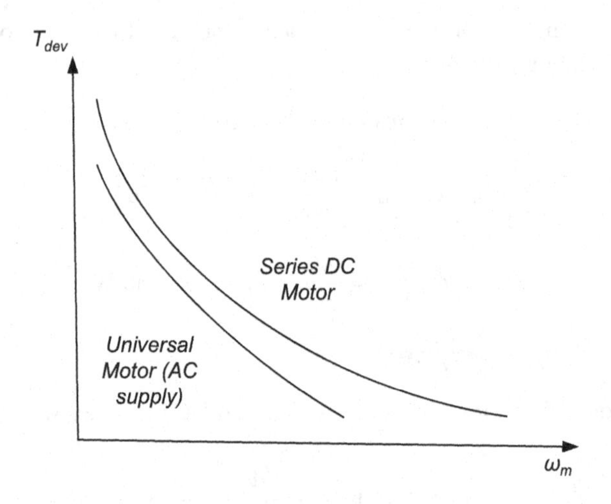

FIGURE 6.36 Universal motor torque–speed characteristics.

Advantages over other types of AC motors:

1. For a given weight, universal motors produce more power than other types. This is a great advantage for handheld tools and small appliances, such as drills, saws, mixers, and blenders.
2. The universal motor produces large starting torque without excessive current.
3. When load (torque) increases, the universal motor slows down more than the series or separately connected DC motors or AC induction motors. Hence, the power produced is relatively constant, and the current magnitude remains within reasonable bounds.* Thus, the universal motor is more suitable for loads that demand a wide range of torque, such as drills and food mixers.†
4. Universal motors can be designed to operate at very high speeds, whereas we will see that other types of AC motors are limited to 3,600 rpm, assuming a 60-Hz source.

As with DC series motors, the speed control of universal motors is best done by varying its rms input voltage. Wearing of the brushes and commutators is a major disadvantage of this type of AC motors.

6.11 COMPOUNDED DC MOTORS

A compound wound DC motor is a self-excited motor and is actually a combination of DC series motor and shunted DC motor. These motors have both shunt and series field windings. They have a combination of the characteristics of the shunt- and series-connected DC motors. There are two types of compounded DC motors. A compound wound DC motor is said to be cumulatively compounded when the shunt field flux produced by the shunt winding enhances the effect of main field flux, produced by the series winding. The cumulatively compounded types have better speed regulation than series-connected DC motors and better starting torque values than shunted motors. Therefore, they are widely used.

Similarly, a compound wound DC motor is said to be differentially compounded when the flux due to the shunt field winding diminishes the effect of the main series winding. The percentage of speed regulation of DC differential compound motor is between 3 and 5 and is superior among all the other motors.

6.12 STARTING OF DC MOTORS

When the motor is starting or the magnetic field is cut off due to any reason, a large current would flow in the armature coil. To prevent such excessive current from flowing in the armature windings, a starter box is used as shown in Figure 6.37.

---

* In contrast, shunt DC motors or AC induction motors tend to run at constant speed and are more prone to drawing excessive currents for high-torque loads.

† For the same reason, series DC motors are used as starter motors in automobiles.

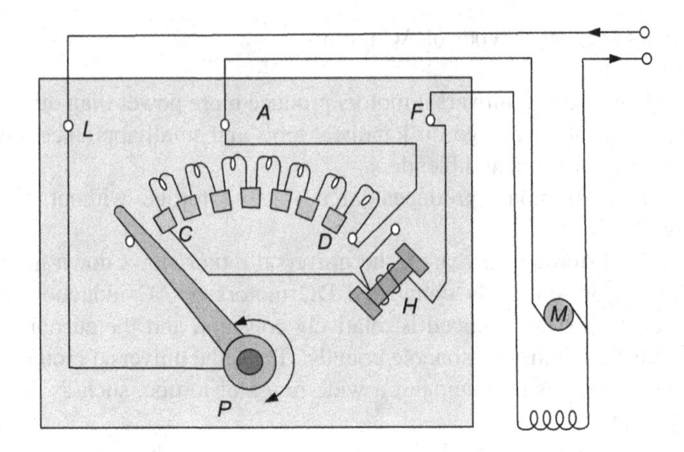

FIGURE 6.37 Starter box.

When starting the motor, the operator moves the lever onto *C*. This closes the circuit to the field coil through electromagnet *H*, and also closes the circuit to the armature windings through the high-resistance coils in the starting box.

As the armature builds up speed, the operator moves the lever to the right to reduce the value of resistance. Finally, the operator moves the lever to the extreme right position and the electromagnet holds it in this position. If the power should fail or the field coils should open for any reason, the electromagnet becomes deenergised and the lever is returned automatically to the off position by spring *P*.

6.13 BRUSHLESS DC MOTORS

Conventional DC motors are useful in many fields, but the commutators and the brushes are a disadvantage. Moving contacts are difficult from servicing and maintenance point of view. The newer idea is the "brushless DC motor" which is essentially a permanent magnet stepping motor with position sensors. Their control units are more enhanced than the conventional ones.

As in the stepper motor, DC power is applied to each stator winding at a time and when the sensor indicates the alignment with the stator winding power is turned off and applied to the next winding. Speed control is achieved by the amplitude and duration of pulses. Performance is similar to the shunt DC motors. They are mostly used in low-power and high-speed applications.

PROBLEMS ON CHAPTER 6

Problem 6.1: A shunt-connected DC motor (Figure 6.38) of 15 kW rated power with armature resistance $R_A = 0.25\ \Omega$ and a shunt field winding resistance of $R_F = 30\ \Omega$ is connected to a DC voltage supply of $V_T = 120$ V. The motor runs at $\omega_m = 800$ rpm at no load and the armature current at no load is $I_A = 16$ A. The magnetisation curve at 800 rpm is shown in Figure 6.39.

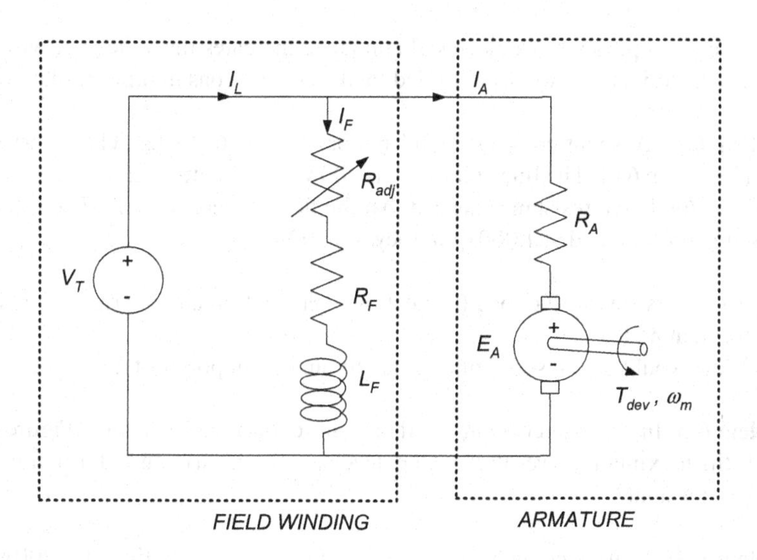

FIGURE 6.38

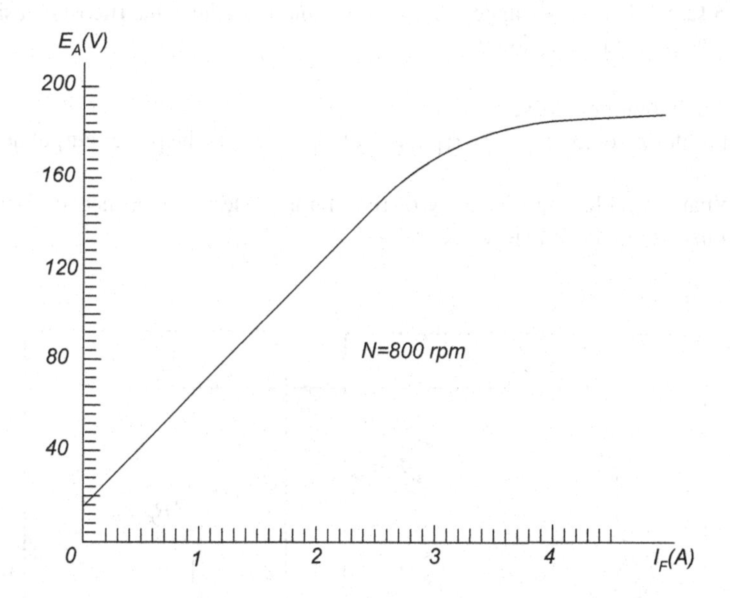

FIGURE 6.39

a. What is the value of the variable rheostat that is in series with the winding resistance R_F?
b. How much are the rotational losses at 800 rpm?
c. When rated current flows in the armature, find the speed (rpm), developed torque T_{dev}, and efficiency of the motor?
d. If rated power is 20 kW instead of 10 kW, find the developed torque of the motor.

Note: The rated power of an electrical appliance indicates the voltage at which the appliance is designed to work multiplied by the current consumption at that voltage.

Problem 6.2: A shunt-connected DC motor (Figure 6.40) has characteristic as shown in Figure 6.41. The line voltage V_T is 100 V, the armature resistance R_A is 0.25 Ω, and the total field resistance is such that the field current is 2.5 A. The rotational losses are constant, and at 1,000 rpm, they are 50 W.

a. Find the speed for no load (neglect the voltage drop on the armature resistance at no load).
b. What would be the speed and torque for an output power of 2 hp?

Problem 6.3: In a separately connected or shunt-connected DC motor (Figure 6.42), what is the maximum power that can be developed in the armature if armature voltage V_T is constant?

Problem 6.4: A shunt-connected DC motor (Figure 6.43) rotating at 2,200 rpm at no load draws $I_L = 11$ A current from the 100 V source. The armature resistance is $R_A = 0.5$ Ω, field coil resistance is $R_F = 20$ Ω, and the adjustable rheostat resistance is $R_{adj} = 80$ Ω. Find the following:

a. Motor rotational losses.
b. If motor is rotating at 1,200 rpm at full load, what is the power output at full load?
c. What would be the efficiency of the motor if the variable rheostat is kept constant at its full load value

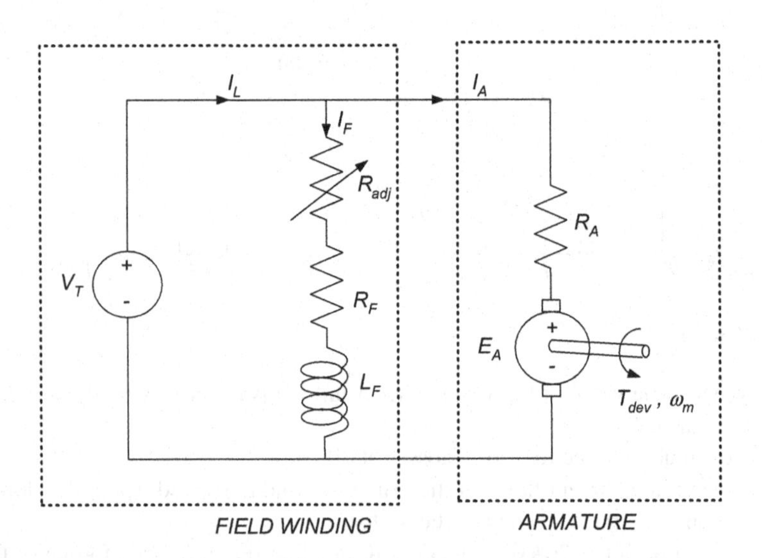

FIGURE 6.40

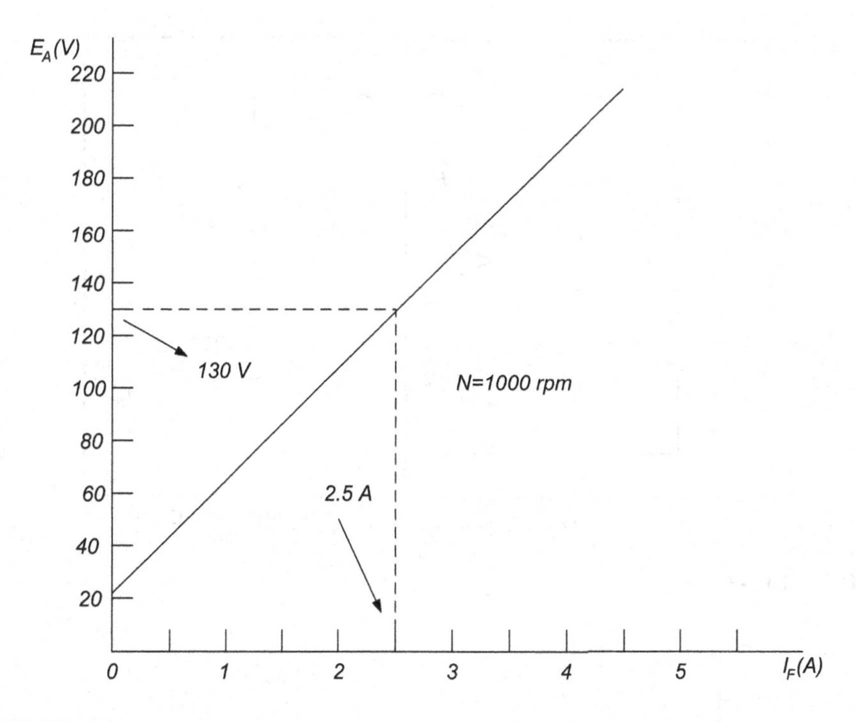

FIGURE 6.41

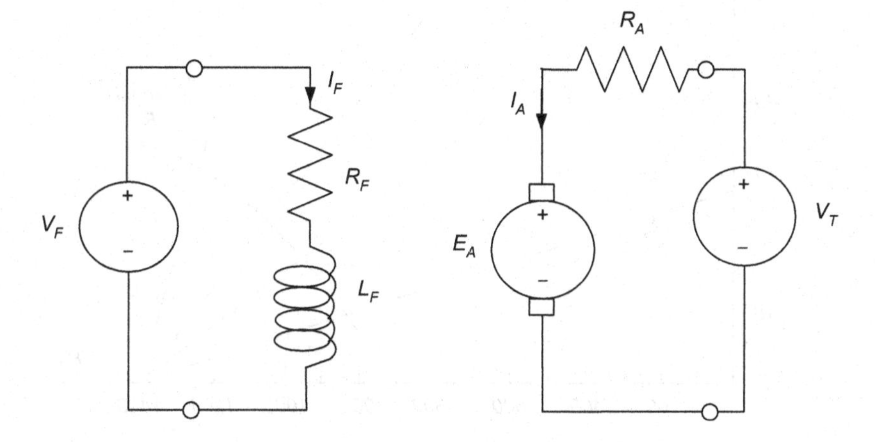

FIGURE 6.42

Problem 6.5: In Figure 6.44, we have the torque–speed characteristics of a shunted DC motor of an electric vehicle. Torque–speed control is achieved by changing the rheostat value R_A.

a. If the vehicle is going up a hill at a speed of 500 rpm with $R_A = R_{A1}$, what would be the power output of the vehicle? (Use values from the graph.)

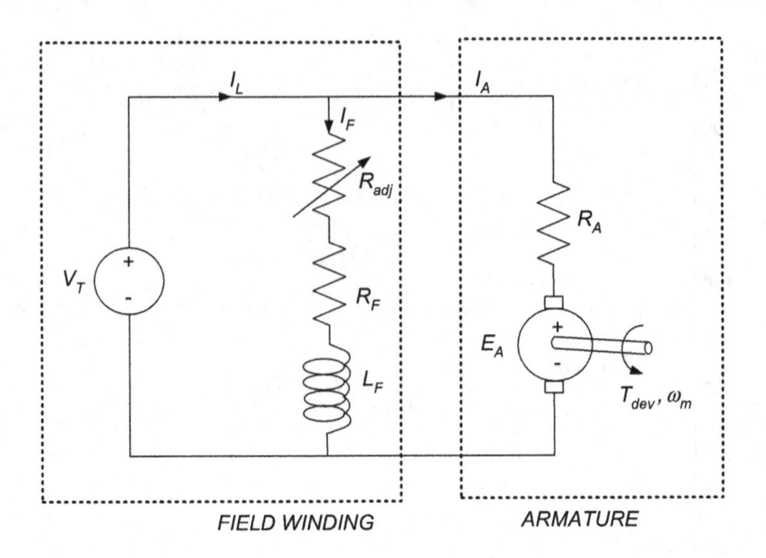

FIGURE 6.43

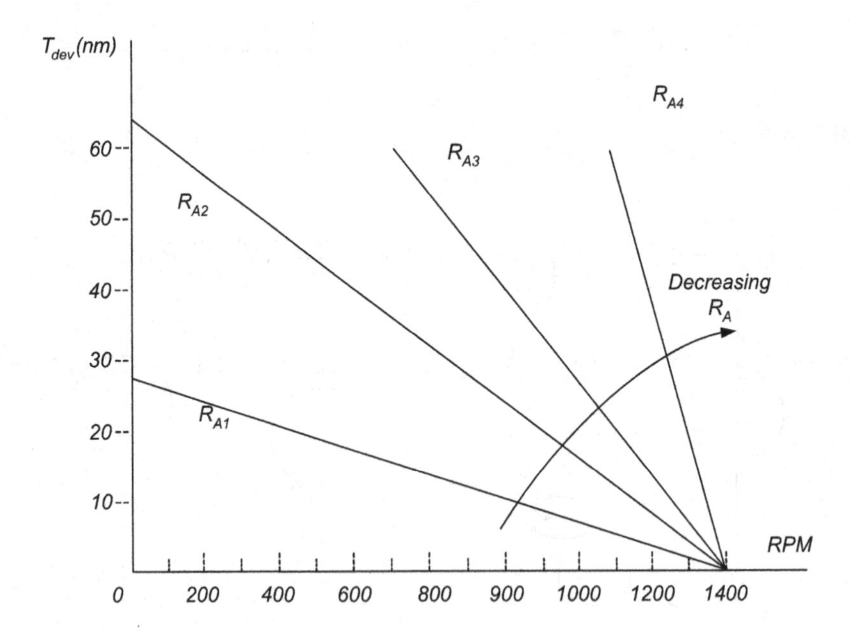

FIGURE 6.44

b. If the vehicle was travelling up the same hill with $R_A = R_{A2}$ or $R_A = R_{A3}$, or $R_A = R_{A4}$, what would be speeds of the vehicle?

Problem 6.6 A separately excited DC motor is supplied from two variable DC sources V_T and V_F as shown in Figure 6.45. When V_T is 500 V, the motor is rotating at 2,000 rpm and providing a torque of 400 nm.

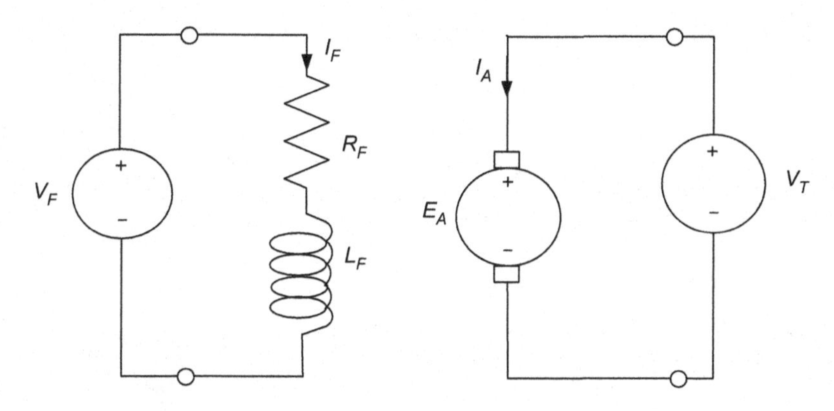

FIGURE 6.45

a. Determine the motor armature current I_A. What is the power developed by the motor (neglect the effect of armature resistance)?
b. While keeping field source V_F constant, V_T is dropped down to 250 V. What would be the new torque and the speed of the motor if the armature current I_A is measured to remain the same? What is the power supplied to the load? Assuming torque is kept constant at this value while V_F remains the same, draw the power/speed characteristic of the motor if we change V_T from 0 to 500 V.
c. When V_T is at 500 V, V_F is changed until the speed is 4,000 rpm while the output power is kept a constant. What is the new value of the torque? Draw on the same graph drawn for (b) the torque speed characteristics of this motor if V_F is brought down to 0 V from the value when V_T is constant at 500 V while keeping the output power constant.

Answers to odd questions:

6.1: 33.16 Ω,

6.3: $\dfrac{V_T^2}{4R_A}$

6.5: $1{,}047\ \text{W}$, for $R_{A2} \rightarrow n = 1{,}100$ rpm, For $R_{A3} \rightarrow n = 1{,}300$ rpm, For $R_{A4} \rightarrow n = 1{,}500$ rpm

7 AC Machines

7.1 INTRODUCTION

AC machines are motors that convert AC electrical energy to mechanical energy and generators that convert mechanical energy to AC electrical energy. They both are important because they mean the power we use at our homes and factories. The fundamental principles of AC machines are simple, but the practical machines used in industry are rather complex in their structures. We shall start explaining their operation using simple examples, and then, consider their complications that occur in actual use. The two major classes of AC machines are synchronous and induction machines.

7.2 PRINCIPLES OF AC MACHINES

The rotating coil shown in Figure 7.1, which consists of a single turn (rotor) that can freely rotate between two magnetic poles (permanent magnet) of opposite polarity (stator), either receives (motor operation) or delivers (generator operation) AC electrical power via the brushes and the slip rings.

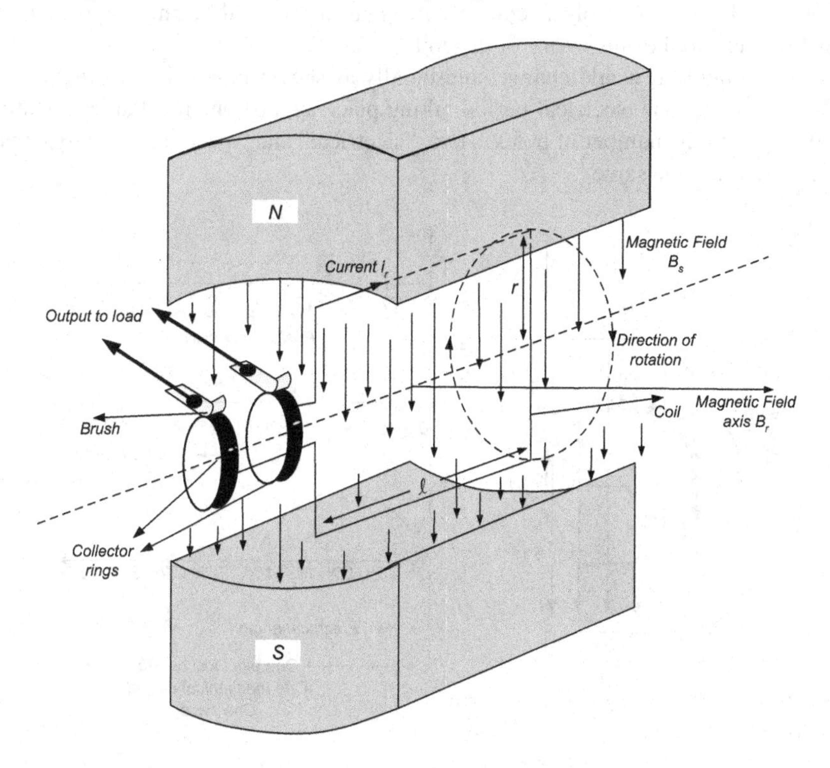

FIGURE 7.1 Simple single-turn permanent magnet AC machine.

Slip rings are conductors that allow the coil to rotate while applying or receiving AC voltage and the brushes are continuously in contact with the rings. This simple single turn machine shown in Figure 7.1 can be operated as a motor or as a generator. If we rotate the coil, we will generate AC voltage from the terminals (brushes) and if we apply an AC voltage through the brushes, we will cause a torque which will rotate the coil. However, this single-loop machine can only be used for laboratory purposes because the amount of energy generated (or consumed) is and have to be quite small. The AC voltage generated by this simple electrical machine depends directly on the number of turns in the armature (the loop), the strength of the magnetic field (a permanent magnet in this case), and the speed of rotation (velocity) of the armature.

7.3 TWO POLE SINGLE-TURN SIMPLE AC GENERATOR

If we apply a mechanical torque and rotate clockwise the coil which consists of a single turn as shown in Figure 7.1, coil sides would cut the magnetic flux and hence induce a voltage in the coil. The induced voltage is taken out by two brushes that are continuously in touch with the two slip rings fixed to two ends of the coil. Because of the opposite directions of movement of the coil sides, at any instant voltage induced in one of the coil sides is equal and opposite to the voltage induced in the other side of the coil. The two emf's are in fact connected in series between the collector rings via the back connection of the coil. For this reason, it would be enough to examine what happens to the emf in one of the coil sides.

Voltage induced would change sinusoidally as shown in the right of Figure 7.2. Here, we notice one electrical period taking place against one mechanical rotation of the coil. As the number of poles is two, "electrical" and "mechanical" frequencies and periods are the same.

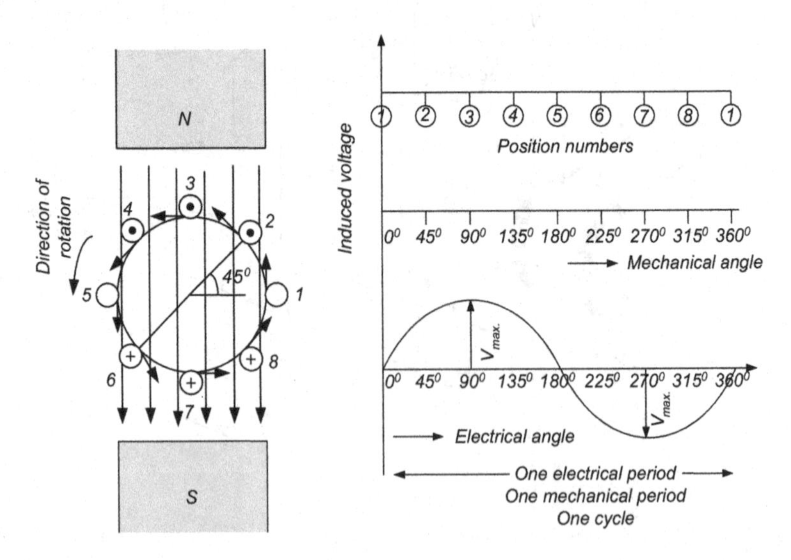

FIGURE 7.2 Voltage induced in one rotation of the coil of Figure 7.1.

In fact, the induced voltage in the coil will cause a current i_r to flow in the coil, which according to Lens's law will create a torque in opposite direction to the mechanical torque applied to the coil. We name this torque as "electromagnetic torque". As the applied mechanical torque accelerates the coil, the developed electromagnetic torque increases to balance the applied mechanical torque and the system reaches the equilibrium point at a speed called the "synchronous speed" n_s where the two torques are equal and opposite. This speed is called synchronous speed because the angular speed of the mechanical rotation is the same as that of the induced electrical voltage. If it hadn't been for this counter torque, the coil in the centre would be flying off with an ever increasing speed. It is because of this electromagnetic torque that we counter great difficulty when trying to feed an electrical heater with an hand generator. We can calculate this developed torque T according to Lorentz's force law as follows (Figures 7.1 and 7.3):

$$T_{\text{dev}} = 2r \times \text{Tangential force} = 2r\left(i_r l B_s \sin\theta\right)$$

where,

$i_r = I_r \cos(\omega_r t) =$ AC current flowing in the coil (rotor)
$B_s =$ magnetic field density caused by the stator (permanent magnet)
$r =$ radius of the loop
$l =$ longitudinal length of the loop
$\theta =$ the angle between rotor and stator magnetic axis $= \omega_m t + \delta_{rs}$
$\omega_m =$ mechanical angular velocity of the loop
$\delta_{rs} =$ starting angle between the rotor and stator magnetic axis.

Notice that the angle between current i_r and stator magnetic field B_s is always 90° and the developed force is always horizontal and in a direction such as to oppose the rotation. To calculate the developed torque, we have to use the torque (tangential) component of this force which is $i_r l B_s \sin\theta$. Since there are two conductors (Figures 7.1 and 7.3),

$$T = 2r i_r l B_s \sin\theta = 2rlB_s I_r \cos(\omega_r t)\sin\left(\omega_m t + \delta_{rs}\right)$$

If the current flowing in the coil is assumed to be DC, there would be no net torque developed as shown in Figure 7.4.

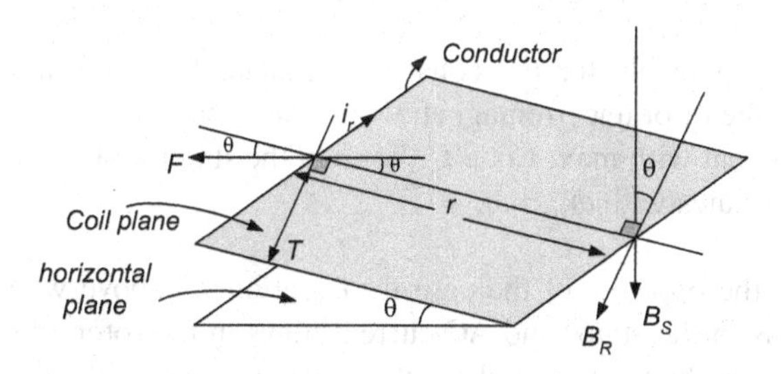

FIGURE 7.3 Torque, $[T \times r = (F \sin\theta) \times r]$, on the left conductor.

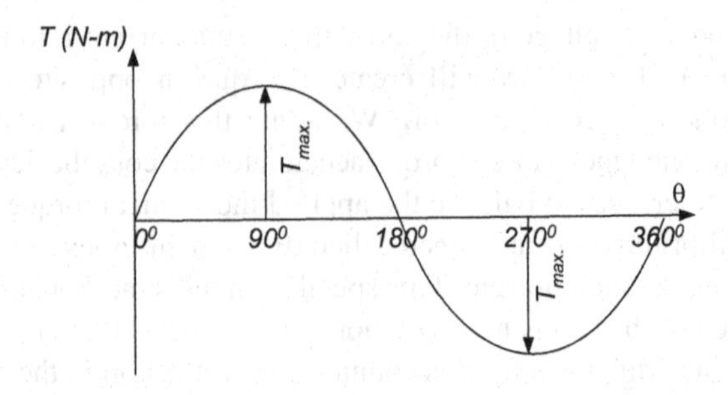

FIGURE 7.4 Induced torque in one electrical cycle if rotor coil current is DC.

If we trigonometrically analyse this torque equation, we will find that it has a mean value only when $\omega_m = \mp\omega_r$ = synchronous speed. From the below given trigonometric identity,

$$\sin x \cos y = \frac{1}{2}\left[\sin(x+y) + \sin(x-y)\right]$$

we can write the developed torque as

$$T = rlB_sI_r\left[\sin(\omega_m t + \delta_{rs} + \omega_r t) + \sin(\omega_m t + \delta_{rs} - \omega_r t)\right]$$

Putting $\omega_m = \omega_r$

$$T = kI_rB_s\left[sin(\delta_{rs}) + \sin(2\omega_r t + \delta_{rs})\right]$$

$$T_{\text{avg}} = kI_rB_s sin(\delta_{rs}) \tag{7.1}$$

Notice the similarities between this equation and equation (4.10) for the average torque in cylindrical synchronous machines when the number of poles (p) is taken as 2:

$$T_{\text{avg}} = -\frac{I_{sm}I_RM}{2}\sin(\delta_{rs}) \tag{4.10}$$

In the derivation of equation (4.10),

- the coil responsible for the constant magnetic field was assumed to be placed in the rotor (electromagnet with DC current I_R)
- an AC current with max. value I_{sm} flows in the stator coil
- the rotor rotates with $\omega_m = \omega_s$

This is exactly the opposite of the case for equation 7.1 above where the constant magnetic field is in the stator and AC current flows in the rotor coil with $\omega_m = \omega_r$. The reason for this change in actual synchronous machines is the difficulties acountered in applying (or receiving) AC power to (or from) a rotating coil.

As we shall see in later chapters, this simple AC machine has a lot of similarities with practical synchronous machines. In fact equation 4.10 and equation 7.1 are of the same form for a single-phase double-pole synchronous machine example. Also we see from the derivation of equation (7.1) that although we develop an average torque at synchronous speed, we cannot avoid the fluctuations at twice the synchronous frequency.

7.4 SIMPLE FOUR-POLE AC GENERATOR

In Figure 7.5 shown is a four-pole AC generator. When the coil rotates one turn, the induced voltage in the coil will cover two periods. Hence,

$$T_{\text{mechanical}} = 2T_{\text{electrical}} \rightarrow \frac{1}{f_{\text{mechanical}}} = \frac{2}{f_{\text{electrical}}}$$

$$\rightarrow f_{\text{electrical}} = 2f_{\text{mechanical}}$$

Therefore, if we define p as the number of machine poles and N as the mechanical (rotor) speed of the machine in revolutions per minute, then

$$f = \text{Frequency of the generated voltage} = \frac{p}{2} \times \frac{N}{60} \text{ Hz}$$

$$\text{Electrical frequency} = \frac{p}{2} \times \text{Mechanical frequency}$$

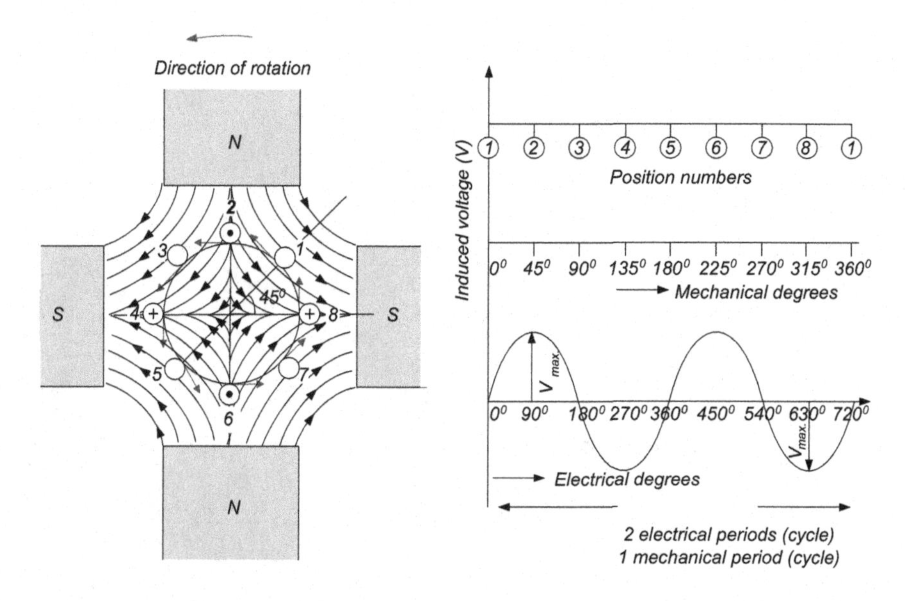

FIGURE 7.5 Four-pole AC generator.

We call this machine a simple "AC generator with four poles." For a p-pole machine, the developed torque equation (7.1) becomes

$$T_{\text{avg}} = \frac{p}{2} k I_r B_s \sin(\delta')$$

where

$$\delta' = \frac{p}{2}\delta_{rs}$$

is the starting electrical angle between the two magnetic axes. There are essentially two AC machines:

1. synchronous machines
2. induction machines.

They both employ essentially the same coil structures in their stators. However, their rotors are different. Synchronous machines use an externally DC-excited electromagnet (or permanent magnet) as their rotor magnetic field whereas induction machines use a simple coil which is not (generally) externally excited. As we shall see in further discussions, none of these motors are "self-starting" except the balanced polyphase induction motors.

8 Synchronous Machines

8.1 INTRODUCTION

As we pointed out earlier in Chapter 7, the simple AC machine described in Sections 7.1 and 7.2 bears a lot of similarities with real synchronous machines. In reality, however delivering power to and from this simple machine through brushes and slip rings would not be very efficient especially in large machines.

Therefore, AC synchronous machines are built with the armature (winding to which power is delivered to/from) on the stator and the field on the rotor. We know that electrical machines can work as both generators and motors. As generators, synchronous machines are the main source of AC electrical power for utility networks. However, as motors, they are not as widely used as synchronous generators. Their applications as motors are specific, and they are used in a narrow range of applications.

Synchronous machines are AC machines in which for a two-pole machine, magnetic field created by the AC current flowing in the armature windings (in the stator) rotates at the same speed as the rotor. This speed is known as the synchronous speed, and for a p-pole machine, the synchronous machine speed in revolutions per minute is

$$N_s = \frac{f \times 60}{p/2}$$

where f is the frequency of the electrical output (in the case of generator) or electrical input (in the case of motor).

Throughout this book, we will assume the machines are always in the steady-state condition. In synchronous machines, alternating currents flow in the armature windings, and DC excitation is supplied to the field windings. The armature winding is nearly always in the stator and is very often a three-phase winding, whereas the field winding is on the rotor.

The salient rotor pole construction is best used in hydroelectric generators and in most motor applications. Hydraulic turbines rotate at a relatively slow speed; hence, a large number of poles are required to obtain the required frequencies of 50–60 Hz. The salient pole construction is a better mechanical solution for the large number of poles required in hydraulic turbines (Figure 8.1A).

The cylindrical rotor construction is generally used for two–four-pole turbine generators. Steam turbines and gas turbines operate at higher speeds, and cylindrical rotor construction is best suited for these kinds of applications.

The DC power for the field excitation is usually supplied through the slip rings, and the DC source is often mounted on the same shaft as the machine. For large turbine generators, AC exciters and solid-state rectifier diodes are used to obtain the necessary DC source.

FIGURE 8.1A Brushless exciter rotor of a low-speed synchronous motor. (Courtesy of TECO-Westinghouse co.)

In Chapter 4, we saw that in a p-pole cylindrical synchronous machine, average torque is (equation 4.10)

$$T_{avg} = -\frac{p}{2}\frac{I_{sm}I_R M}{2}\sin\left(\frac{p}{2}\delta_{rs}\right) \tag{4.10}$$

where

p = number of poles
I_R = DC rotor current to create the rotor magnetic field
I_{sm} = maximum value of the stator AC current
M = maximum value of the mutual inductance between stator and rotor coils
δ_{rs} = is the starting angle between the rotor and stator magnetic axis.

Notice the similarities between the torque equation (7.1) and the torque equation (4.10) which is given above once again.

8.2 PRACTICAL SYNCHRONOUS MACHINES

The rotor of a synchronous machine is usually a p-pole electromagnet with field windings that carry DC currents (Figure 8.1B). In smaller machines, the rotor can be a permanent magnet.

An external DC source can supply the field current through stationary brushes to slip rings mounted on the shaft. A small AC generator known as an exciter and diodes mounted on the same shaft can also be used to rectify the AC to provide the DC field current. This avoids the maintenance associated with brushes and slip rings. Some machines are of round-rotor type, which have slots in the side to hold the field coils (Figures 8.2a). The synchronous machine in Figure 8.2b is called a salient pole machine because the rotor is not symmetric. In either case, the rotor only sees a DC flux, so the rotor does not have to be laminated.

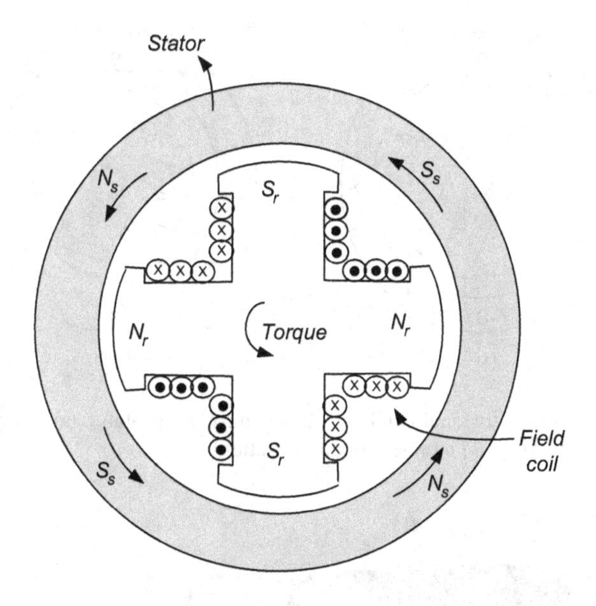

FIGURE 8.1B A four-pole "salient rotor" synchronous motor.

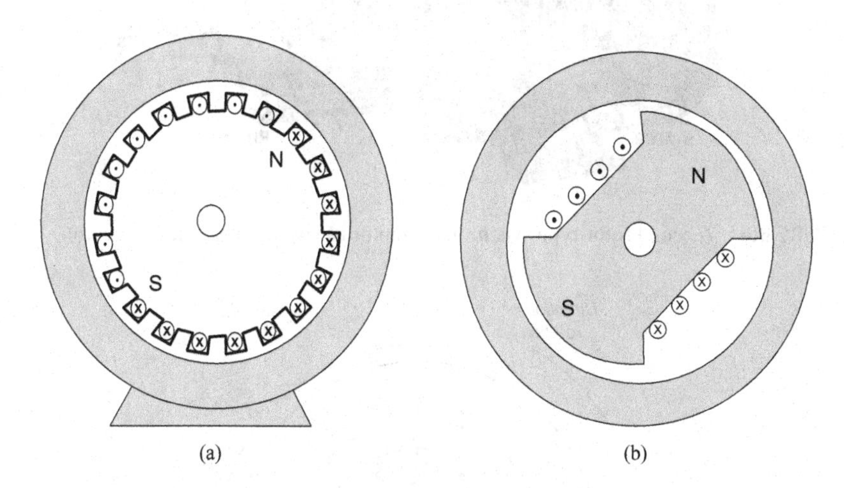

FIGURE 8.2 Synchronous machines of different rotor shapes. (a) Cylindrical rotor; (b) salient rotor.

To develop torque, the same number of poles must be used on stator and rotor. This can be seen with the help of Figure 8.3. With unequal numbers, there will be no rotation because of opposite forces applied to the two ends of the rotor (Figure 8.3a).

Generally, salient-pole construction is less costly but is limited to low-speed machines. High-speed machines usually have cylindrical rotors (Figures 8.4 and 8.5).

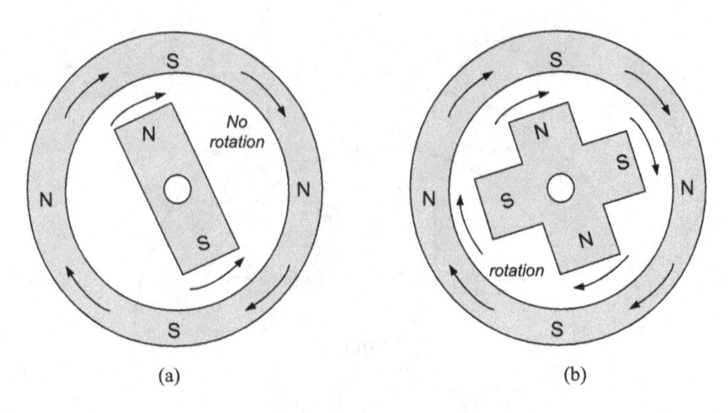

FIGURE 8.3 Effect of unequal number of poles in rotor and stator. (a) Unequal number of poles (no rotation). (b) Equal number of poles (rotation).

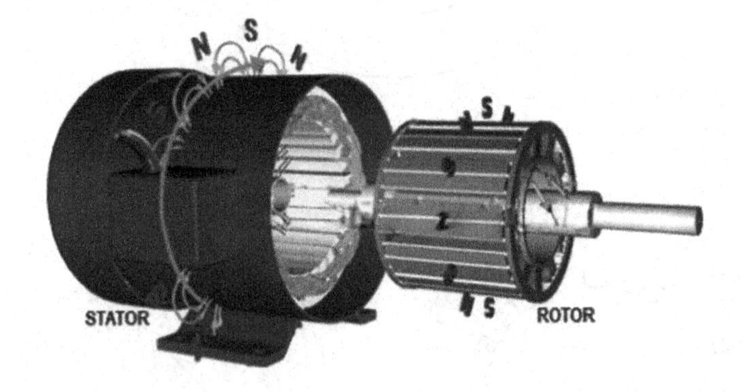

FIGURE 8.4 A round rotor-type synchronous motor (www.learn engineering.com).

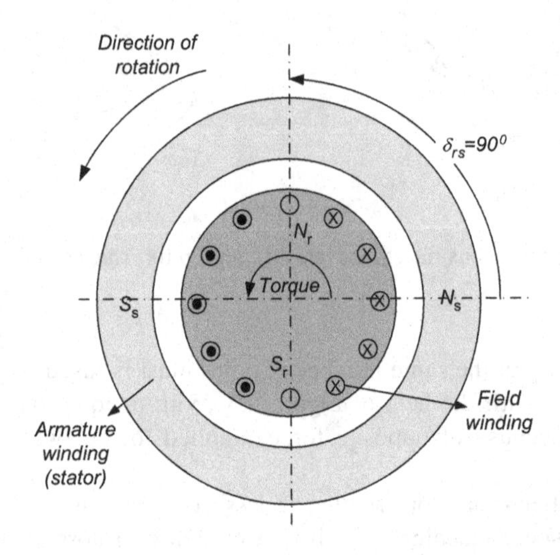

FIGURE 8.5 A cylindrical synchronous machine.

A synchronous machine is a doubly excited machine:

- its rotor poles are excited by a DC source.
- its stator windings are connected to the AC supply.

When working as a motor, depending on the rotor excitation current, it can draw either lagging or leading reactive current from the AC supply. Induction AC motors on the other hand are singly excited machines and always operate at a lagging power factor. Therefore, synchronous machines are sometimes used for compensation of the reactive power in the systems. Because AC currents will flow in the stator windings, the stator must be constructed of laminations to reduce the eddy current losses.

Since

- the output of a synchronous generator is AC
- and DC is required for the field windings,

the generated voltage cannot be directly used for field excitation as with DC machines.

Therefore, all synchronous machines are separately excited. Large synchronous machines are used in hydro, nuclear, or thermal power stations for generating electrical power. In these power generating stations, they can be of power ratings of several hundred MVAs (Figure 8.6).

FIGURE 8.6 A high-power synchronous generator. (Courtesy of Dravske elektrarne Maribor d.o.o).

The alternators found in most automobiles are basically synchronous machines (Figure 8.7). The AC armature voltages are rectified, and the DC obtained is used in the car to power the headlights, charge the battery, and so on.

Synchronous machines run at synchronous speed ω_s, which is given by the following equation:

$$\omega_s = \frac{\omega}{p/2}$$

where ω is the angular frequency of the AC source applied to the armature (stator), and "p" is the number of magnetic poles of the stator or rotor. The torque/speed characteristic of the synchronous machine is shown in Figure 8.8.

FIGURE 8.7 A car alternator.

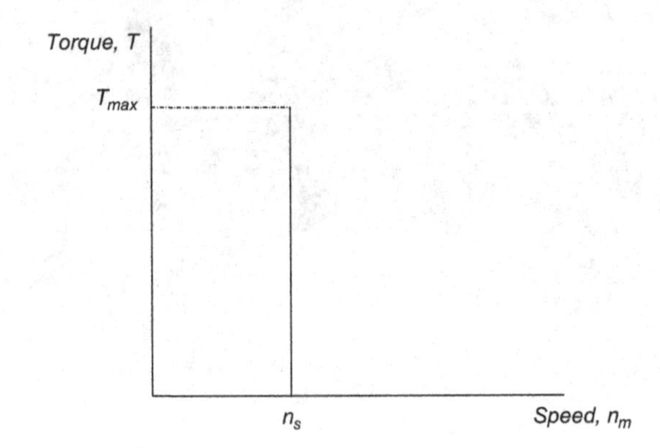

FIGURE 8.8 Torque–speed characteristics of synchronous machines.

During generation, if the prime mover torque is increased beyond the maximum value, the rotor will over speed and fall out of synchronism, and if not disconnected, it can be dangerous.

Problem: What is the synchronous speed of a 50Hz, two-pole, synchronous generator?

Answer:

$$\omega_s = \frac{\omega}{p/2} = 2\pi \times 50 \ r/s$$

$$\rightarrow n_s = \text{Synchronous speed in rpm} = \frac{\omega_s}{2\pi} \times 60 = 2\pi \times 50 \times \frac{60}{2\pi} = 3000 \text{ rpm}$$

8.3 SYNCHRONOUS MOTOR MADE SIMPLE

In Section 7.2, we saw that for the simple two-pole cylindrical machine in Figure 7.1, the instantaneous electromagnetic torque developed on the rotor (loop) is (equation 7.1)

$$T = 2ri_r lB_s sin\theta = 2ri_r lB_s sin\left(\omega_m t + \delta_{rs}\right)$$

This was the developed torque in the generator action. We can see that for a motor, this equation should bear a minus sign as follows:

$$T = -2ri_r lB_s sin\theta = -2ri_r lB_s sin\left(\omega_m t + \delta_{rs}\right)$$

Let us assume that rotor is also an electromagnet or a permanent magnet. Then, we can write this as

$$T = -KB_s B_r sin\left(\omega_m t + \delta_{rs}\right)$$

Now, let us assume that the stator permanent magnet B_s and rotor permanent magnet B_r are at standstill with $\omega_m = 0$. Then, the instantaneous torque becomes

$$T = -KB_{\text{stator}} \times B_{\text{rotor}} \times \sin\delta_{rs} \tag{8.1}$$

This is equivalent to the magnetic circuit shown in Figure 8.9 with the starting angle between rotor and stator magnetic axes $\delta_{rs} = 0°$. According to the above torque equation, with no rotation in the magnetic system, the N and S poles of the rotor will align with those of the permanent magnet A-B to minimise the stored energy in the air gap.

In this stable state, there will be no torque on the rotor. Let us suppose that the permanent magnet A-B starts rotating slowly anticlockwise until finally it reaches the speed s as shown in Figure 8.9. The rotor with no load applied will align its N and S poles with those of the magnet A-B and will start to rotate with the same speed *s* of the rotating magnetic field.

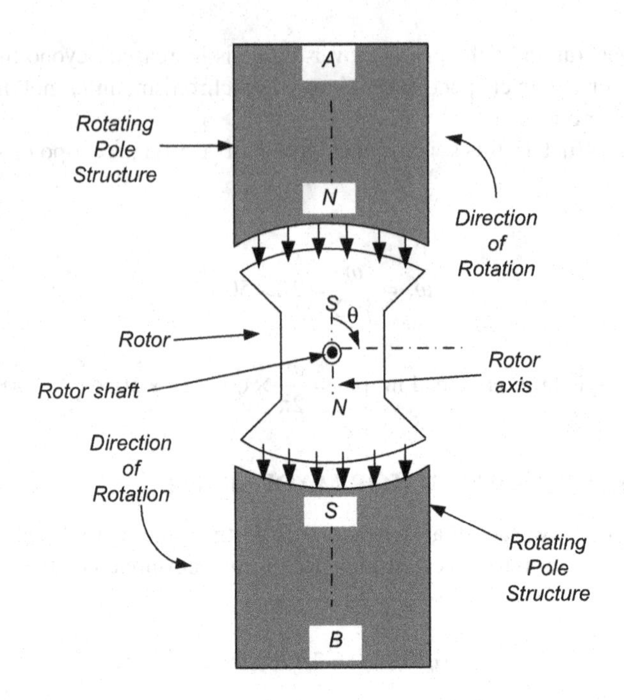

FIGURE 8.9 Illustration of the basic working principles of synchronous motors.

When a load is slowly applied to the rotor, it will slow down and start lagging (by angle δ_{rs}) the rotating field. Hence, a torque of the value

$$T = -KB_{\text{stator}}B_{\text{rotor}} \times \sin\delta_{rs} \tag{8.2}$$

will develop. Torque on the rotor will increase until it is equal to the load. Then, rotor will start rotating at the same speed s in synchronism with the rotating magnetic field again.

This simple example shows that if we can achieve a rotating magnetic field, a permanent magnet placed in this rotating field can rotate in synchronism with the magnetic field. What is more it can carry loads at this synchronous speed (equation 8.2).

In the following section, we explain how to obtain this rotating magnetic field. It is an outstanding design achieved during the 1880s. The practical application of the rotating magnetic field was claimed to be demonstrated by two electrical engineers: the Italian Galileo Ferraris and Serbian-American Nikola Tesla (Figure 8.10).

8.4 ROTATING MAGNETIC FIELD IN BALANCED THREE-PHASE AC MACHINES

In Figure 8.11, the three field windings a, b, and c are embedded in slots on the stator, and their axes are 120° apart from each other. A balanced three-phase AC source (angular velocity = ω) is applied to the windings, and the currents on these windings are shown in Figure 8.12. Now let us evaluate the total flux as time elapses from $t = 0$ seconds:

FIGURE 8.10 Tesla's first rotating field machine (https://teslauniverse.com/nikola-tesla).

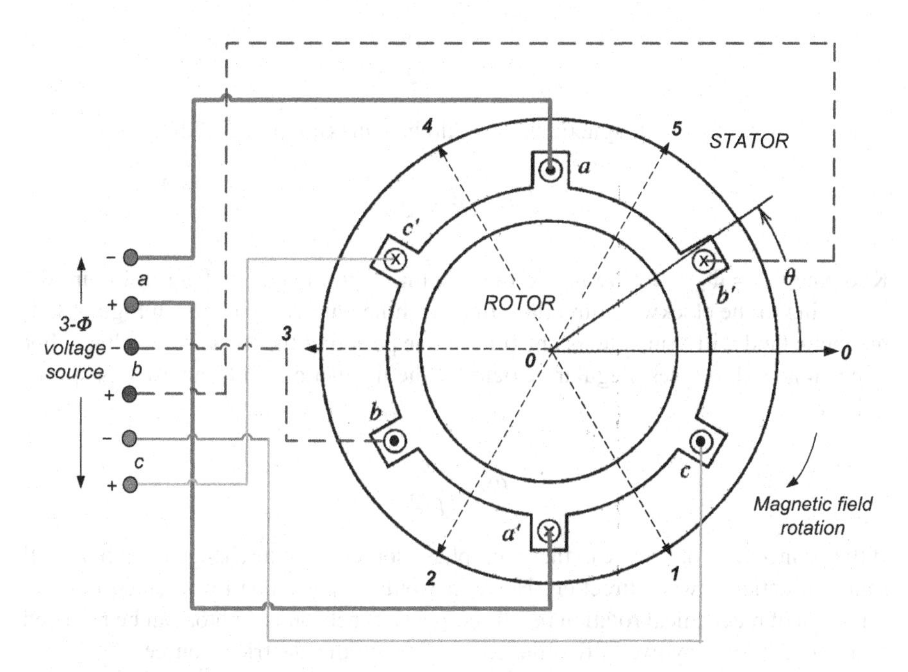

FIGURE 8.11 Stator windings of two-pole three-phase cylindrical synchronous machine.

- At t = 0, the total magnetic field caused by the applied 3 phases of the AC voltage source a, b and c on the windings would be a maximum in 0-0 direction as shown in Figure 8.11:

$$\Phi_{max} = \Phi_a + 2\cos 60^\circ \times \cos 60^\circ \times \Phi_a = 3\frac{\Phi_a}{2}$$

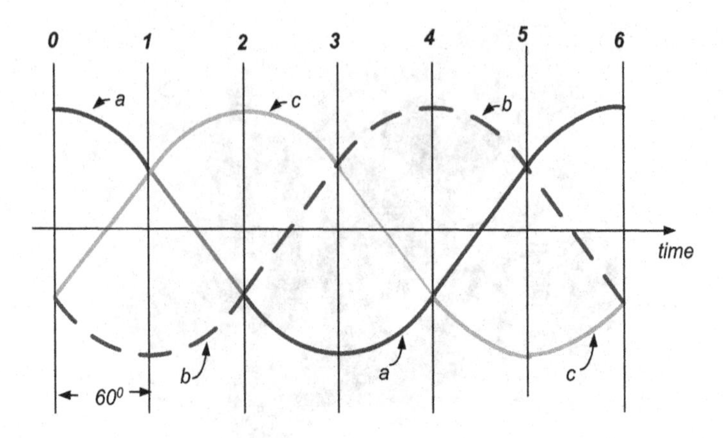

FIGURE 8.12 Three-phase voltage applied to the armature windings.

- At $t = 1$, the total magnetic field would be a maximum in 0-1 direction:

$$\Phi_{\max} = 3\frac{\Phi_b}{2}$$

- At $t = 2$, the total magnetic field would be a maximum in 0-2 direction:

$$\Phi_{\max} = 3\frac{\Phi_c}{2}$$

Knowing $\Phi_a = \Phi_b = \Phi_c$, we notice that the maximum magnetic field in Figure 8.1 is rotating in the clockwise direction. In a machine with two poles as in Figure 8.11, magnetic field will complete its one turn in one period of the AC source voltage. For a machine with p poles, angular velocity of the magnetic field (synchronous speed) will be

$$\omega_s = \frac{\omega}{p/2}$$

If the connections of b and c to the three-phase source are interchanged, the field will rotate in anticlockwise direction. Hence it wouldn't be wrong to conclude that the direction of mechanical rotation in a three-phase synchronous motor can be reversed by interchanging any two of the line connections to the electrical source.

8.5 MATHEMATICAL ANALYSIS OF THE ROTATING FIELD IN BALANCED THREE-PHASE AC MACHINES

We see that in Figure 8.13, the magnetic field density in the air gap due to current $i_a(t)$ in winding aa' would be,

$$B_a = Ki_a(t)\cos\theta$$

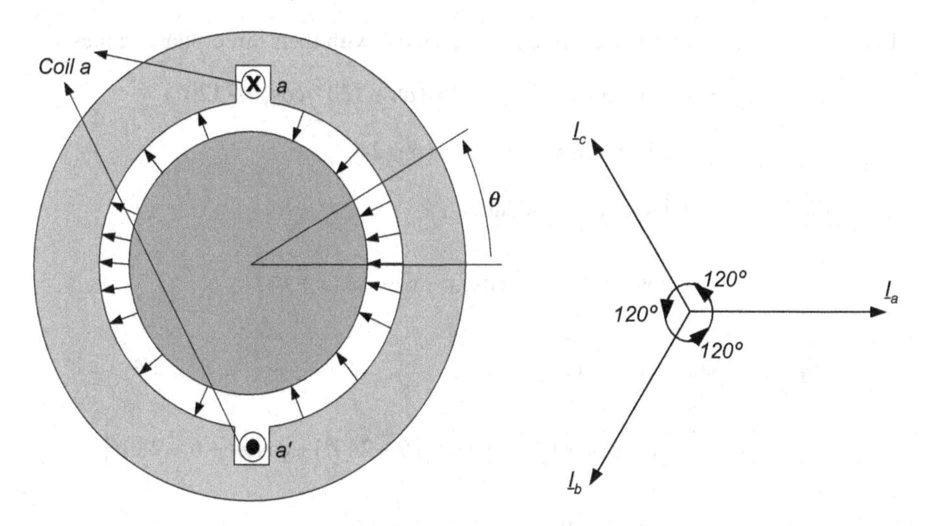

FIGURE 8.13 (a) Magnetic field density in winding aa'. (b) Three phase currents in stator windings.

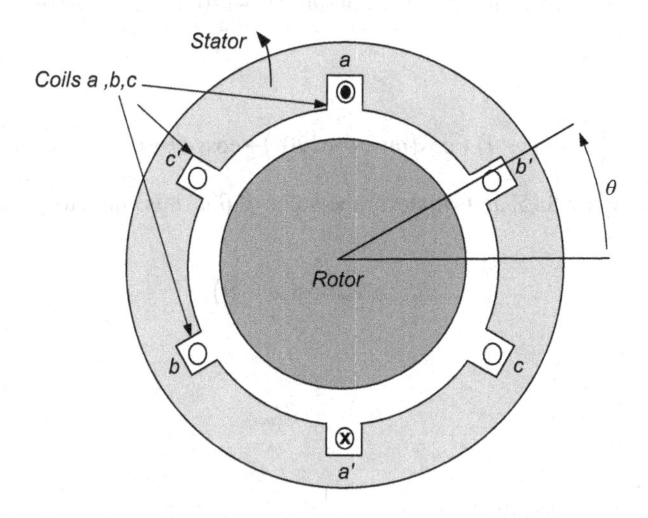

FIGURE 8.14 Positions of the three stator windings.

The other two windings (b and c) are identical to winding a, except that they are rotated in space by 120° and 240°, respectively (Figure 8.14). Thus, the fields in the air gap due to windings b and c are given by

$$B_b = Ki_b(t)\cos(\theta - 120°)\ ; B_c = Ki_c(t)\cos(\theta - 240°)$$

The total field is

$$B_{\text{gap}} = B_a + B_b + B_c = Ki_a(t)\cos\theta + Ki_b(t)\cos(\theta - 120°) + Ki_c(t)\cos(\theta - 240°)$$

In this equation, if we replace currents i_a, i_b, and i_c with their three-phase values,

$$B_{\text{gap}} = Ki_m \cos\omega t \cos\theta + Ki_m \cos(\omega t - 120°)\cos(\theta - 120°)$$
$$+ Ki_m \cos(\omega t - 240°)\cos(\theta - 240°)$$

Using the following trigonometric identity,

$$\cos x \cos y = \frac{1}{2}\left[\cos(x - y) + \cos(x + y)\right]$$

$$B_{\text{gap}} = \frac{3}{2} Ki_m \cos(\omega t - \theta)$$
$$+ \frac{1}{2} Ki_m \left[\cos(\omega t + \theta) + \cos(\omega t + \theta - 240°) + \cos\omega t + \theta - 480°\right]$$
$$= \frac{3}{2} Ki_m \cos(\omega t - \theta)$$
$$+ \frac{1}{2} Ki_m \left[\cos(\omega t + \theta) + \cos(\omega t + \theta + 120°) + \cos\omega t + \theta - 120°\right]$$

The term

$$\left[\cos(\omega t + \theta) + \cos(\omega t + \theta + 120°) + \cos(\omega t + \theta - 120°)\right]$$

is shown in Figure 8.15, and whatever the value of θ is, it is equal to "zero".

Hence,

$$B_{\text{gap}} = B_m \cos(\omega t - \theta)$$

where

$$B_m = \frac{3}{2} KI_m$$

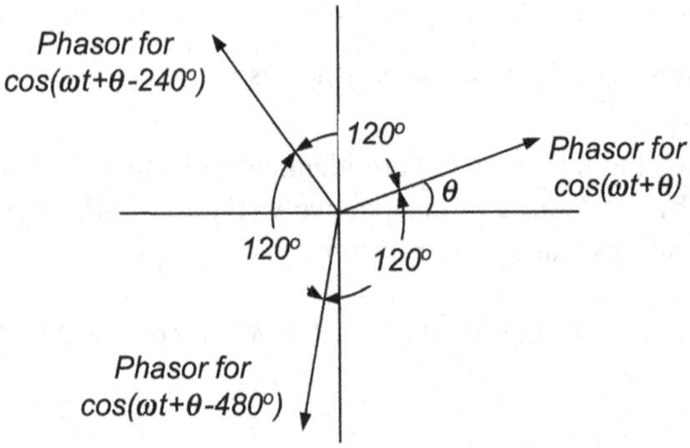

FIGURE 8.15 Addition of phasors in the equation for total magnetic field.

and

$$I_m = \text{The amplitude of the three-phase current}$$

Notice that the maximum flux density occurs for

$$\theta = \omega t$$

This is only possible if

$$\frac{d\theta}{dt} = \omega$$

which is equivalent to saying that in a two-pole three-phase stator winding structure of the above type, the point of maximum flux rotates counterclockwise or clockwise depending on how the stator windings are connected, with an angular velocity of ω. In an AC machine, this speed is called the synchronous speed ω_s.

For a machine with p poles,

$$\text{Synchronous speed} = \frac{\omega}{p/2}$$

Let us remember that the direction of rotation of the field in a three-phase AC machine can be reversed by interchanging any two of the line connections to the electrical source.

8.6 ROTATING MAGNETIC FIELD IN BALANCED TWO-PHASE AC MACHINES

Similar analysis would reveal that a rotating magnetic field can also be obtained if a two-phase balanced AC power source is applied to two stator windings separated by 90° instead of three windings placed at 120° separation.

$$B_{\text{gap}} = B_a + B_b = Ki_a(t)\cos\theta + Ki_b(t)\sin\theta$$

$$B_{\text{gap}} = KI_m\cos(\omega t)\cos\theta + KI_m\sin(\omega t)\sin\theta$$

$$B_{\text{gap}} = KI_m[\cos\omega t\cos\theta + \sin\omega t\sin\theta] = KI_m\cos(\omega t - \theta)$$

Here, we can see that the flux density rotates in space just like the balanced three-phase machines except for the smaller magnitude. In fact we can generalise this by saying that any appropriately placed stator field windings excited by balanced poly-phase currents produce a constant amplitude rotating magnetic field. Notice that the amplitude of the rotating field with balanced three-phase windings is 3/2 times more than with the balanced two-phase windings. Two-phase machines exist, but it is rare.

8.7 ROTATING MAGNETIC FIELD IN SINGLE-PHASE AC MACHINES

If a single-phase voltage source is connected to a single winding stator coil (aa') which is actually distributed in slots to produce a sinusoidal space distribution of mmf,

$$B_{\text{gap}} = Ki_m \cos \omega t \cos \theta$$

Using the following trigonometric identity,

$$\cos x \cos y = \frac{1}{2}\cos(x-y) + \frac{1}{2}\cos(x+y)$$

$$B_{\text{gap}} = \frac{1}{2}Ki_m \cos(\omega t - \theta) + \frac{1}{2}Ki_m \cos(\omega t + \theta) \tag{8.3}$$

In a single-phase single winding stator coil, there are two rotating fields revolving at the same speed ω but in opposite directions. As we will see in later sections, the second rotating field creates a number of disadvantages to single-phase AC machinery.

8.8 ROTATING MAGNETIC FIELD IN UNBALANCED POLYPHASE AC MACHINES

Similar analysis for unbalanced polyphase AC machines would reveal that although a forward component dominates and produces a forward rotating magnetic field for these kinds of machines, there is a backward component of the magnetic field which is responsible for the fluctuations in the torque as well as the extra losses. This emphasizes the importance of the discovery of balanced polyphase rotating field we see in all balanced polyphase AC machines which have no magnetic field fluctuations and extra losses since there is no backward component of rotating field.

8.9 AC MACHINES IN GENERAL

At this stage, it is worth mentioning that essentially there are two types of AC machines, "induction machines" and "synchronous machines," and they both use the same stator winding structures to create the rotating magnetic fields described above.

In synchronous machines, rotor of the machine is a DC electromagnet which mechanically rotates at synchronous speed (rotating magnetic field speed), whereas in induction machines, rotor has a different speed than the synchronous speed and the rotor windings have induced AC currents circulating in them. There is no external source applied to most of rotor windings of the induction machines.

In single-phase AC machinery, according to equation (8.3), there are two rotating fields: one in the positive direction and the other in the negative direction. Note that since there is no revolving magnetic field at start-up, a single-phase AC motor (induction or synchronous) will produce no starting torque at standstill.

For single-phase induction motors which do not have to run at synchronous speed, there are ways to get around this problem like placing a starter winding physically separated from the main winding to give an initial impetus to the rotor. This in fact is no different than making the machine look like an unbalanced two-phase machine at start which does have a dominating positive rotating magnetic field. Consequently, it causes a positive initial torque. After reaching approximately 75% of the synchronous speed, auxiliary winding is disconnected by means of a centrifugal switch and then the motor takes over. This method of starting single-phase induction machines is called "split-phase" method. This and other techniques will be described in Chapter 9.

Single-phase induction machines are not torque-wise inferior to polyphase induction machines. However as explained in Chapter 9 in detail, the torque would be pulsating because of the backward rotating magnetic field. These effects inherent in single-phase induction machines can be totally eliminated in balanced polyphase induction machinery. Balanced three-phase induction motors do have a net starting torque and a uniform operational torque which are great advantages. However, precautions must be taken for excessive current intake of these machines during start-up (see Chapter 9).

Balanced polyphase synchronous machines which do have a uniform torque once brought to synchronous speed are not self-starting either. They must somehow be brought to the synchronous speed. This is achieved mostly by constructing the rotor of these machines like the polyphase induction motors at start-up. Problems for the single-phase synchronous machines are twofold:

1. They must be brought to the synchronous speed which can only be achieved by induction motor techniques.
2. The torque at synchronous speed would be pulsating which cannot be avoided.

With the discovery of constant amplitude rotating field in balanced polyphase machines, it has become possible to operate AC machines at constant torque (or constant power in the case of generators). This way the double-frequency time-varying torque in single-phase and unbalanced polyphase machinery is eliminated. We can imagine the scale of the difficulties that would arise if single-phase generators were used in high-power AC power stations. That is why the discovery of constant amplitude rotating field in balanced polyphase machinery is important. Another outstanding outcome of this discovery was the self-starting balanced polyphase induction motor which also leads up to the self-starting balanced polyphase synchronous motor.

8.10 DIFFERENCES BETWEEN SINGLE- AND THREE-PHASE SYNCHRONOUS MACHINES

Synchronous motors (three phase or single phase) are not self-starting because they cannot produce torque unless they run at their synchronous speed. In equation (8.3), if we insert

$$\theta = \omega t \rightarrow \frac{d\theta}{dt} = \omega$$

$$B_{\text{gap}} = \frac{1}{2} K i_m + \frac{1}{2} K i_m \cos(2\omega t) \tag{8.4}$$

From equation (8.4), we can see that for single-phase synchronous machines rotating at synchronous speed, magnetic field has two components. The second term is a double-frequency component whose average value is zero and the first term is responsible for the average magnetic field (torque). The total of these is a pulsating field but nevertheless produces a net average value responsible for an average torque (or generated voltage). We can eliminate this double-frequency time-varying torque inherently built into single-phase synchronous motors by using balanced three-phase (or polyphase) currents to produce a constant air-gap magnetic field rotating at synchronous speed, which is described in Section 8.3.5. This is a very important outcome implying that it is possible to operate synchronous machines under conditions of constant torque (or constant electrical power). We can imagine the difficulties associated with power stations in the order of megawatts having pulsating power requirements of double the supply frequency. In this respect, the discovery of rotating field in balanced three-phase AC machines was a major breakthrough in supplying electrical power to utilities as well as providing simple, rugged, reliable, self-starting polyphase induction motors. Balanced polyphase AC machines also reduce the rotor losses to a great scale.

However for synchronous motors whether they are single or three phase, the problem of bringing the rotor to the synchronous speed remains to be solved. This is a major defect for synchronous motors (see the next section).

In synchronous motors, a three (or two or single)-phase AC power is applied to the stator windings (armature) and a rotating field of angular speed

$$\omega_s = \text{Synchronous speed} = \frac{\omega}{p/2}$$

is created.

$$\text{Synchronous speed in rpm} = \frac{60f}{p/2}$$

A DC source is applied to the rotor windings through slip rings (Figure 8.16) to create an electromagnet. Rotor will try to catch up with the magnetic field and hence rotate with synchronous speed.

8.11 STARTING OF SYNCHRONOUS MOTORS

Since there is no initial torque available at standstill and an average torque is only available at the synchronous speed, synchronous motors (polyphase or single phase) are not self-starting and some special measures have to be taken for their starting. What is more their rotors are heavy. Instead, they will vibrate when starting from standstill. As the magnetic field rotates, the rotor tries to align itself with the magnetic field. But because of its moment of inertia before it can align itself with

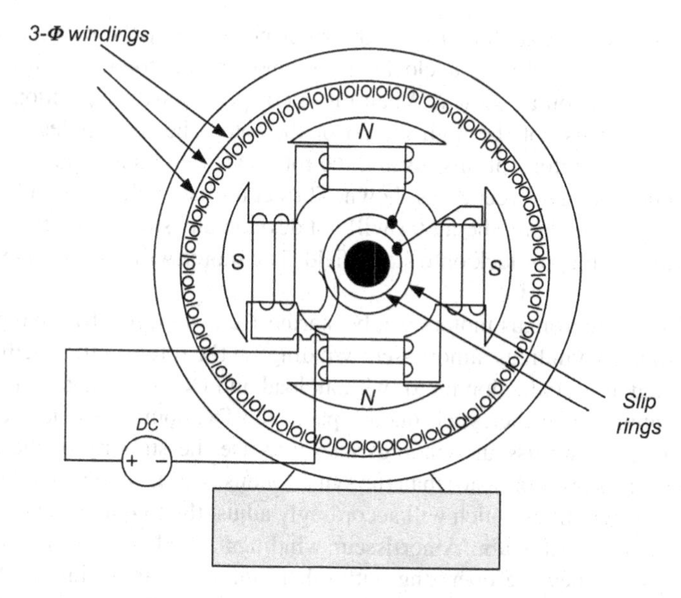

FIGURE 8.16 DC source applied to the rotor windings.

the magnetic field, magnetic field reverses its direction and hence the rotor will not receive a net torque; instead, it will start to vibrate with big noise.

We can visualize this with the help of Figure 8.17 in which a two-pole synchronous machine's rotating field is represented with a pair of N-S poles rotating with a synchronous speed of n (e.g., 2,000 rpm).

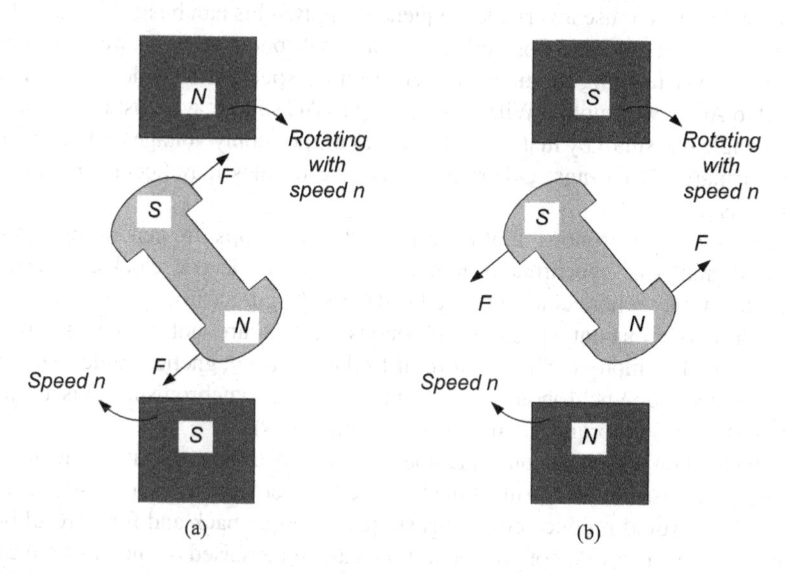

FIGURE 8.17 Starting of synchronous motors. (a) At $t = 0$ (b) short while after $t = 0$.

Let us assume that at start ($t = 0$), the rotor poles are at the position shown in Figure 8.17a. There will be a clockwise torque on the rotor, making it rotate in clockwise direction as shown. When the stator poles move to position shown in Figure 8.17b because of the high inertia of the rotor, the rotor poles would have hardly moved. Therefore, at this instant, the rotor experiences a torque making it to rotate in anticlockwise direction as shown. The net torque on the rotor in one revolution is zero, and therefore, the motor will not develop any starting torque. The rotor poles cannot catch up or lock onto stator field. The motor will not speed up but will vibrate.

Polyphase synchronous motors can be started from standstill by having squirrel cage-like shorted windings (amortisseur winding) in the rotor. In this method motor can be started as an induction motor without load and DC field excitation. After the motor has approached the synchronous speed, the DC source is connected to the rotor windings. Amortisseur windings also increase the stability of the machine. If the rotor runs slower or faster than the synchronous speed, voltage will be induced in amortisseur windings which will accordingly adjust the torque to bring the speed to synchronous speed again. Amortisseur windings are also used in synchronous generators when they are operating with other generators in an infinite bus. This way, variations on the shaft torque will be compensated by the opposing torque generated by these windings. That is why the name "amortisseur" is given to these windings. This method cannot readily be used for single-phase synchronous motors as single-phase induction motors are not self-starting either.

Extra measures have to be taken for single-phase induction motors to start from standstill. One way of doing this, i.e., "split-phase" method, has already been discussed previously. Other methods will be covered in the next chapter. Single-phase synchronous motors are not as widely used as single-phase induction motors.

Another way that can be used for starting both single and polyphase synchronous machines is to use a variable frequency supply. This can be accomplished with power electronic circuits. For both types of synchronous motors, we can also use a prime mover to bring the motor to synchronous speed after which motor is connected to AC power supply. When connecting to AC source, we must make sure that power angle δ is small by matching the phases of the supply voltage to those induced in the armature windings. Otherwise, excessive torques may occur which can be dangerous.

Large-size synchronous motors are used for pumps in generating stations, whereas small-size synchronous motors are used in electric clocks, timers, and record turntables where constant speed is desired (Figure 8.18).

Because of constant speed, synchronous motors are not as widely used as induction or DC motors. There is a limit to the electromagnetic torque that can be produced by the synchronous motor without losing synchronism. This torque is named as "pull-out torque" (equation 8.2) (Figure 8.19).

If the load on a synchronous machine was to exceed the pull-out torque, it would no longer be possible for the machine to drive the load at synchronous speed. Then, the machine would produce enormous surges in torque back and forth, resulting in great vibration. The value of pull-out torque can be increased by increasing the field current.

FIGURE 8.18 Synchronous motor.

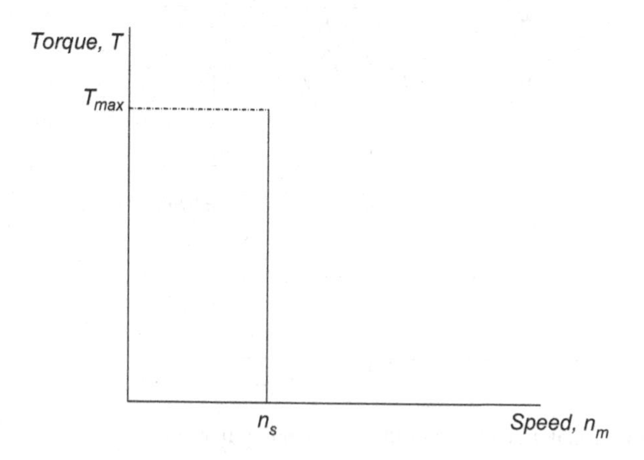

FIGURE 8.19 Torque-speed characteristics of synchronous motors.

8.12 EQUIVALENT CIRCUIT OF SYNCHRONOUS MOTORS

Equivalent circuit of the one phase of the armature winding of a three-phase cylindrical synchronous motor is shown in Figure 8.20. We shall mention about the effect of salient poles in the following sections. Stator winding resistance is generally (especially in large machines) small, so it is neglected. Phase voltage $\underline{V}_a$ of the three-phase source supplies current to the armature (stator) winding. The AC voltage induced in the armature by the rotor field is represented by the voltage source $\underline{E}_r$:

$$\underline{E}_r = k\underline{B}_r$$

Here, B_r is the magnetic field caused by the rotor on the stator.

The DC voltage source V_f supplies the DC field current I_f to the rotor (Figure 8.21). An adjustable resistance R_{adj} is included in the field circuit so that the field current

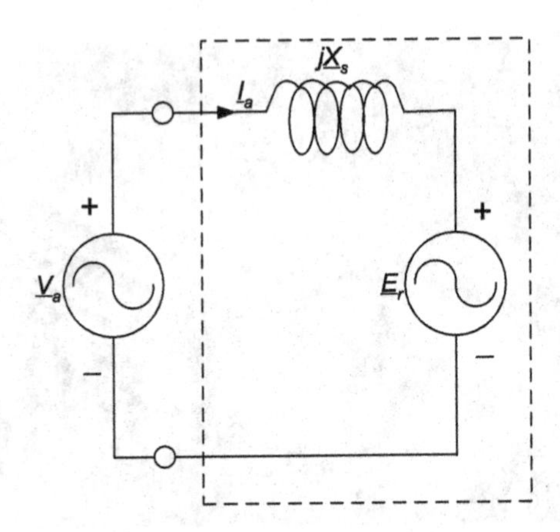

FIGURE 8.20 Equivalent circuit of one phase of armature (stator) winding.

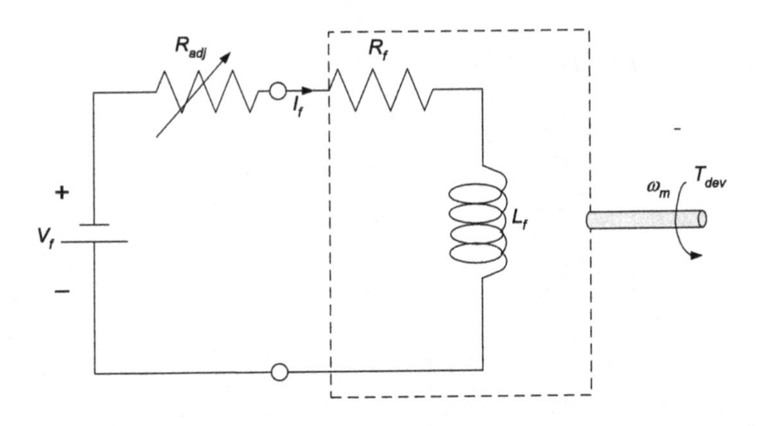

FIGURE 8.21 Excitation circuit of the field winding in rotor.

can be varied. This in turn adjusts the magnitudes of the rotor field $\underline{B}_r$ and the resulting induced voltage $\underline{E}_r$. If

$$\underline{V}_a = |\underline{V}_a| \angle 0° \text{ then} \rightarrow \underline{E}_r = |\underline{E}_r| \angle(-\delta°)$$

Here, δ is named as "power angle" (also known as "torque angle") and it is the angle between the rotor magnetic field B_r (hence $\underline{E}_r$) and the total magnetic field B_{total} (hence V_a). This angle is positive for generators and negative for motors. Rotating stator field B_s induces another voltage component in each stator (armature) winding which we denote as E_s. This voltage component is given by

$$\underline{E}_s = k\underline{B}_s$$

Neglecting the resistance of the stator winding,

$$\underline{E}_s = j\underline{X}_s\underline{I}_a$$

where $\underline{X}_s$ is known as the armature synchronous reactance, and $\underline{I}_a$ is the phasor for the armature current. The voltage observed at the terminals of the armature winding is the sum of these two components. Thus, we can write

$$\underline{V}_a = \underline{E}_r + \underline{E}_s$$

where $\underline{V}_a$ is the phasor for the terminal voltage for the a-phase armature winding. Hence,

$$\underline{V}_a = \underline{E}_r + j\underline{X}_s\underline{I}_a$$

Also we can write

$$\underline{V}_a = k'\underline{B}_{\text{total}}$$

Therefore, the equivalent circuit's phasor diagram would be as shown in Figure 8.22.

The torque equation for synchronous motors was found to be (equation 8.2),

$$T = -k|\underline{B}_S||\underline{B}_R|\sin\delta_{RS}$$

where δ_{RS} is the angle between rotor poles ($\underline{B}_R$) and the stator poles ($\underline{B}_S$). We can see from the vector diagram that

$$|\underline{B}_s|\sin\delta_{RS} = |\underline{B}_{\text{total}}|\sin\delta$$

Here, δ is the torque angle between $\underline{B}_{\text{total}}$ and $\underline{B}_R$. Therefore, it is clear that

$$T = -k|\underline{B}_{\text{total}}||\underline{B}_R|\sin\delta \tag{8.5}$$

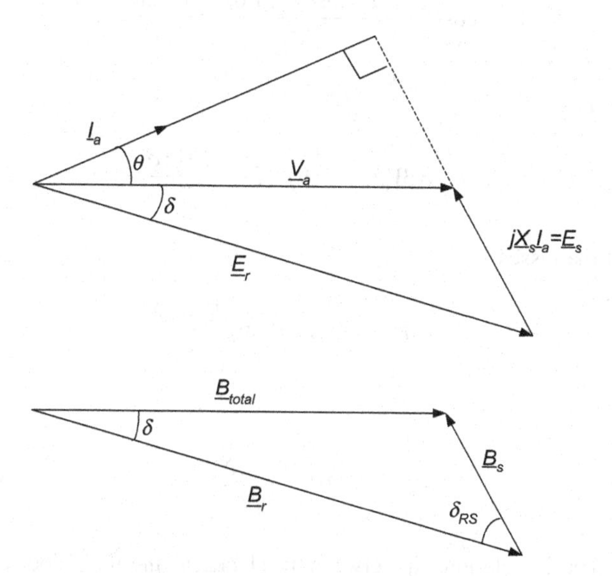

FIGURE 8.22 Phasor diagram.

The armature windings can be connected either in

- a wye
- or in a delta configuration.

For either connection (regardless of the manner in which the machine is connected),

- $\underline{V}_a$ is the voltage across a phase winding.
- $\underline{I}_a$ is the current through a phase winding.

In a wye connection, $\underline{V}_a$ corresponds to the line-to-neutral voltage, whereas in a delta connection, $\underline{V}_a$ corresponds to the line-to-line voltage. Similarly, $\underline{I}_a$ is the current through the a winding, which corresponds to the line current in a wye connection but $\frac{\underline{I}_{a\text{line}}}{\sqrt{3}}$ in a delta connection. The input power taken from the three-phase AC source is given by

$$P_{\text{in}} = 3V_a I_a \cos(\theta) \tag{8.6}$$

in which θ is defined to be the angle by which the phase current $\underline{I}_a$ lags or leads the phase voltage $\underline{V}_a$.

Here, we can write

$$\underline{I}_a = \frac{\underline{V}_a - \underline{E}_r}{jX_s} = \frac{V_a\angle 0^\circ - E_r\angle(-\delta^\circ)}{jX_s} = \frac{V_a\angle(-90^\circ)}{X_s} - \frac{E_r\angle(-90^\circ - \delta)}{X_s}$$

$$I_a \cos\theta = -\frac{E_r \cos(90^\circ + \delta)}{X_s} = \frac{E_r \sin\delta}{X_s}$$

From equation (8.6),

$$P_{\text{in}} = 3V_a I_a \cos(\theta) = 3V_a \frac{E_r \sin\delta}{X_s}$$

If we neglect the losses,

$$P_{\text{in}} = P_{\text{out}} = \omega_m T_m = 3V_a \frac{E_r \sin\delta}{X_s}$$

Hence,

$$T_m = 3V_a \frac{E_r \sin\delta}{\omega_m X_s}$$

Let us remember the relationship between the electrical angular velocity ω_e, mechanical angular velocity ω_m, and synchronous angular velocity ω_s which is as follows:

$$\omega_s = \text{Synchronous angular speed} = \omega_m = \frac{\omega_e}{p/2}$$

Therefore, the torque equation becomes

$$T_m = 3V_a \frac{E_r \sin\delta}{\omega_s X_s} = 3V_a \frac{p}{2} \frac{E_r \sin\delta}{\omega_e X_s} \tag{8.7}$$

Above approach can also be used for salient pole machines with relatively minor errors. However to be more precise the reluctance torque caused by the variation of the self-inductance L_{ss} of stator of the salient pole rotor may also be taken into account (see Section 4.2). As we have shown in Chapter 4, this reluctance torque is proportional to $\sin(2\delta_{rs})$. The full expression for the torque developed in a p pole salient synchronous motor is

$$T = -3\frac{V_a E_r}{\omega_s X_d}\sin\delta - 3\frac{V_a^2}{2\omega_s}\left(\frac{X_d - X_q}{X_d X_q}\right)\sin 2\delta$$

The first term in the above torque expression is the torque developed in cylindrical machines whereas the second term is the contribution coming from the reluctance torque. Here, X_d is the direct axis reactance and X_q is the quadrature axis reactance. Throughout the examples in this book, we will ignore the effect of saliency in the rotor structures.

Exercise 8.1:

A 520-V-rms, four-pole delta-connected synchronous motor has a synchronous reactance of $X_s = 1.6\ \Omega$. Motor runs at a speed of 1,800 rpm and operates with a developed torque of 300 Nm at a power factor of 0.85 leading.

a. find the frequency of the supply voltage and the developed power (assume no losses).
b. determine the values of I_a, E_r, and the torque angle (neglect stator losses).
c. suppose that the excitation remains constant and the torque angle increases until the developed power is 100 hp. Determine the new values of I_a, E_r, the torque angle, and the power factor.

Solution:

a. Since it is a synchronous motor and it has no losses, it runs at the same speed under all loads. Hence,

$$n_s = 1800 = \frac{f \times 60}{p/2} \rightarrow f = 60 \text{ Hz}$$

$$P = T\omega = 300 \times \frac{1800}{60} \times 2\pi = 56548.6 \text{ W}$$

b. The windings are delta-connected, so we have $V_a = V_{line} = 520\,\text{V}$ (rms). From the power equation, we can solve for the phase current I_a:

$$I_{a1} = \frac{P_{dev}}{3V_a \cos\theta_1} = \frac{56548.6}{3 \times 520 \times 0.85} = 42.6 \text{ A rms}$$

$$\cos\theta_1 = 0.85 \rightarrow \theta_1 = 31.79^\circ$$

Since the power factor is given as leading, we know that the phase of I_{a1} is positive. Thus, we have

$$\underline{I}_{a1} = 42.6\angle 31.79^\circ \text{ A rms}$$

$$\underline{E}_{r1} = \underline{V}_{a1} - jX_s\underline{I}_a = 520 - j(1.6)(42.6\angle 31.79^\circ)$$

$$= 520 - j68.16\cos 31.79^\circ - j68.16(j\sin 31.79^\circ)$$

$$= 520 - j57.93 + 35.9 = 555.9 - j57.93 = 558.9\ \angle[-5.95^\circ] \text{ V rms}$$

Hence, torque angle is $\delta_1 = -5.95^\circ$

c.

$$\frac{P_2}{P_1} = \frac{\sin\delta_2}{\sin\delta_1} \rightarrow \sin\delta_2 = \frac{P_2}{P_1}\sin\delta_1 = \frac{74600}{56548.6} \times \sin 5.95^\circ = 0.136$$

$$\delta_2 = 7.82^\circ$$

$$E_{r2} = 558.9\ \angle(-7.82^\circ)$$

We know that E_{r2} lags $V_a = 520\angle 0^\circ$ because the machine is working as motor. Now we can find the new current:

$$\underline{I}_{a2} = \frac{\underline{V}_a - \underline{E}_{r2}}{jX_s} = \frac{520 - 558.9\ \angle[-7.82^\circ]}{j1.6}$$

$$= \frac{520 - 558.9\cos(-7.82^\circ) - j558.9\sin(-7.82^\circ)}{j1.6} = \frac{520 - 553.7 + j76}{j1.6}$$

$$= \frac{-33.7 + j76}{j1.6} = \frac{83.1\angle 114^\circ}{1.6\angle 90^\circ} = 51.93\angle 24^\circ \text{ A rms}$$

$$PF = \cos 24^\circ = 0.91$$

8.13 SYNCHRONOUS GENERATORS

If we apply external force to rotate the rotor of a synchronous machine with three stator windings placed at 120° intervals, and if we apply through a pair of brushes a DC voltage to the rotor windings, we can obtain a three-phase voltage from the stator windings (Figure 8.23).

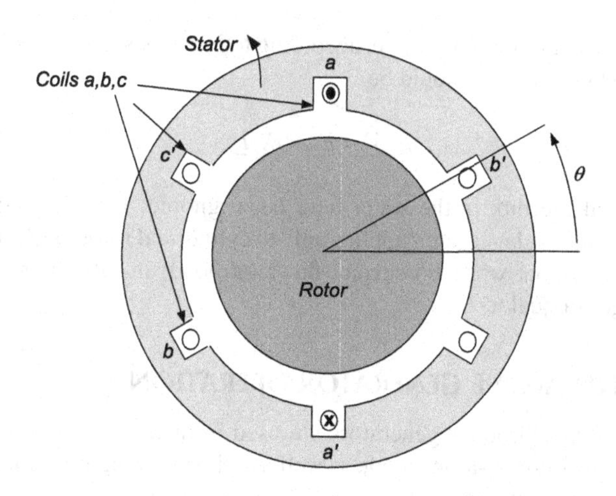

FIGURE 8.23 Cylindrical synchronous generator.

Big AC generators are of this type. The DC voltage applied to the rotor windings will control the level of voltage at the stator windings. Depending on the speed of the turbine used, they can be a "salient rotor type" (Figure 8.16) or "nonsalient rotor type" of machine (Figure 8.23).

8.14 EQUIVALENT CIRCUIT OF THE SYNCHRONOUS GENERATOR

Rotor of the synchronous generator is a DC electromagnet. The equivalent circuit of the one phase of a cylindrical-type synchronous generator would be as shown in Figure 8.24.

The effect of the sinusoidally changing rotor flux on the stator coils is represented by the excitation voltage $\underline{E}_r$. This voltage is controlled by the DC field (rotor) current $\underline{I}_r$. It is also controlled by the torque applied to the rotor. The effect of the current $\underline{I}$ flowing in the stator windings would be a voltage produced by the product of $jX_s\underline{I}$

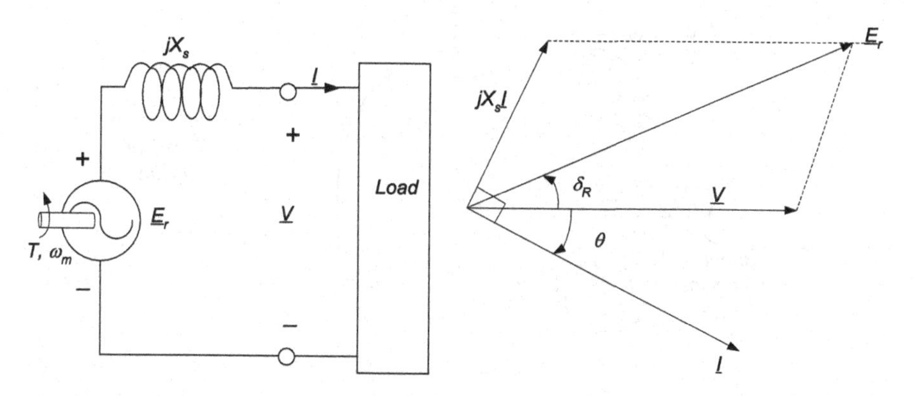

FIGURE 8.24 Equivalent circuit of the one phase of a cylindrical synchronous generator.

due to the right-hand rule. Therefore, according to Faraday's law, the voltage induced per phase in stator coils $\underline{V}$ should be

$$\underline{V} = \underline{E}_r - jX_s\underline{I}$$

Losses and leakage flux in the stator have been ignored. However, because of the symmetry, the equivalent circuit applies only to cylindrical rotor machines. It is quite clear from above that when no current flows (no load) in the stator windings, the output voltage is equal to E_r.

8.15 STAND-ALONE GENERATOR OPERATION

Although most synchronous generators are used in interconnection networks, they find applications in stand-alone mode as well. As shown in Figure 8.25, there are two mechanisms (inputs) to control the output.

1. Rotor field DC voltage (current $\underline{I}_r$)
2. Mechanical torque applied to the rotor shaft.

System frequency and voltage output of the system are dependent on the load. Thus, for constant frequency, speed control is required. Field current partly controls the output voltage. To maintain constant terminal voltage, automatic field current regulation is required (see next exercise for a better understanding).

Exercise 8.2:

A four-pole, three-phase Y-connected stand-alone generator is supplying 30 A to a load at a PF of 0.85 lagging at a terminal voltage of 420 V. The synchronous reactance of the generator is 2 Ω. Assume there are no losses and leakage reactance is negligible.

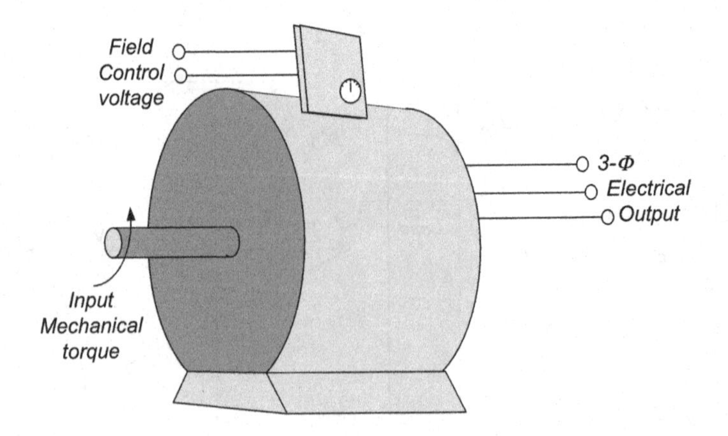

FIGURE 8.25 Stand-alone synchronous generator.

a. what is the power supplied to the load? If the rotor is driven at a speed of 1,500 rpm, what would be the frequency of the generated voltage?
b. what would be the reading of a voltmeter put across the open-circuited terminals?
c. if the terminal voltage is maintained at 420 V, what would be the new value of the open-circuit voltage if load is dropped down by half at the same PF of 0.85?

Solution:

a. We can find the power from

$$P = 3\times 30\times \frac{420}{\sqrt{3}}\times 0.85 = 18550.2\ \text{W}$$

The rotor is rotating at 1,500 rpm, so

$$\omega_s = \frac{\omega_e}{P/2} \rightarrow \omega_e = \frac{1500}{60}\times 2\pi\times 2 = 100\pi \rightarrow f = \frac{100\pi}{2\pi} = 50\ \text{Hz}$$

b. We can find the open-circuit output voltage from

$$\underline{E}_r = \underline{V} + jX_s\underline{I}_a = \frac{420}{\sqrt{3}}\angle 0^\circ + j\times 2\times 30\angle\left(-\cos^{-1}0.85\right)$$

$$= 242.5\angle 0^\circ + j60\angle(-31.79^\circ)$$

$$= 242.5 + j60\times 0.85 + j60\times j\sin(-31.79^\circ)$$

$$= 242.5 + j51 - 60\times\sin(-31.79^\circ)$$

$$= 242.5 + j51 + 31.61$$

$$= 274.11 + j51 = 278.8\angle 10.5^\circ\ \text{V rms}$$

Voltage reading would be 278.8 V

c. If terminal voltage and power factor are the same and power is halved, then current I_a must also be halved. Hence,

$$I_a = 15\ \text{A}$$

The new $\underline{E}_r$ can be calculated from

$$\underline{E}_r = \underline{V} + jX_s\underline{I}_a = \frac{420}{\sqrt{3}}\angle 0^\circ + j\times 2\times 15\angle(-31.79^\circ)$$

$$= 242.5 + j25.5 + 15.81 = 258.31 + j25.5 = 259.56\angle 5.64^\circ\ \text{V}$$

The new open-circuit voltmeter reading would be 259.56 V.

8.16 THE INFINITE BUS

Synchronous generators are very seldom used in stand-alone mode. In general, they are used in infinite bus or grid systems in which many synchronous generators are connected in parallel to each other by varying lengths of transmission lines and they work in synchronism with each other. Figure 8.26 shows the schematic display of such a grid system. Individual generators connected to the grid cannot significantly affect the system voltage or frequency. Therefore, we can think of the grid as a constant-voltage constant-frequency source which explains why it is called an "infinite bus system."

The operation of connecting a synchronous generator to the infinite bus is known as paralleling with the infinite bus. Before the generator can be connected to the infinite bus, we must make sure that they both have the same

1. voltage
2. frequency
3. phase sequence
4. phase.

In practice, these conditions are checked by an instrument known as a synchroscope, shown in Figure 8.27. Synchroscope is a device that indicates the degree to which generators or power networks are synchronised with each other.

8.17 SYNCHRONOUS GENERATORS IN INFINITE BUS

Because many large-sized generators are connected together, the voltage and frequency of the infinite bus can be assumed to be constant. Thus, if we increase the mechanical drive to an individual generator, we do not increase frequency as in stand-alone operation; rather, we contribute more real power to the grid.

8.17.1 Effect of Increasing Mechanical Drive to an Infinite Bus-Connected Synchronous Generator

The constant voltage and frequency of the grid system will impose on every machine an air gap magnetic field which rotates at synchronous speed as determined by the grid-system frequency. This way we can visualise that all generators in the grid must run at the same speed (assuming the same number of poles).

Therefore, if we increase the mechanical drive to one of the generators its speed power curve dynamically lifts up from AA_1 to BB_1 to increase its load share (active power) from P_1 to P_1' and its speed (frequency) from A to B by Δf (Figure 8.28). In the meantime, another generator (generators) in the grid decreases its power and frequency from P_2 to P_2' and its speed (frequency) from A to C by equal amounts.

This way the other generators in the grid will automatically drop off some of their loads to reach a balance at the same grid frequency.

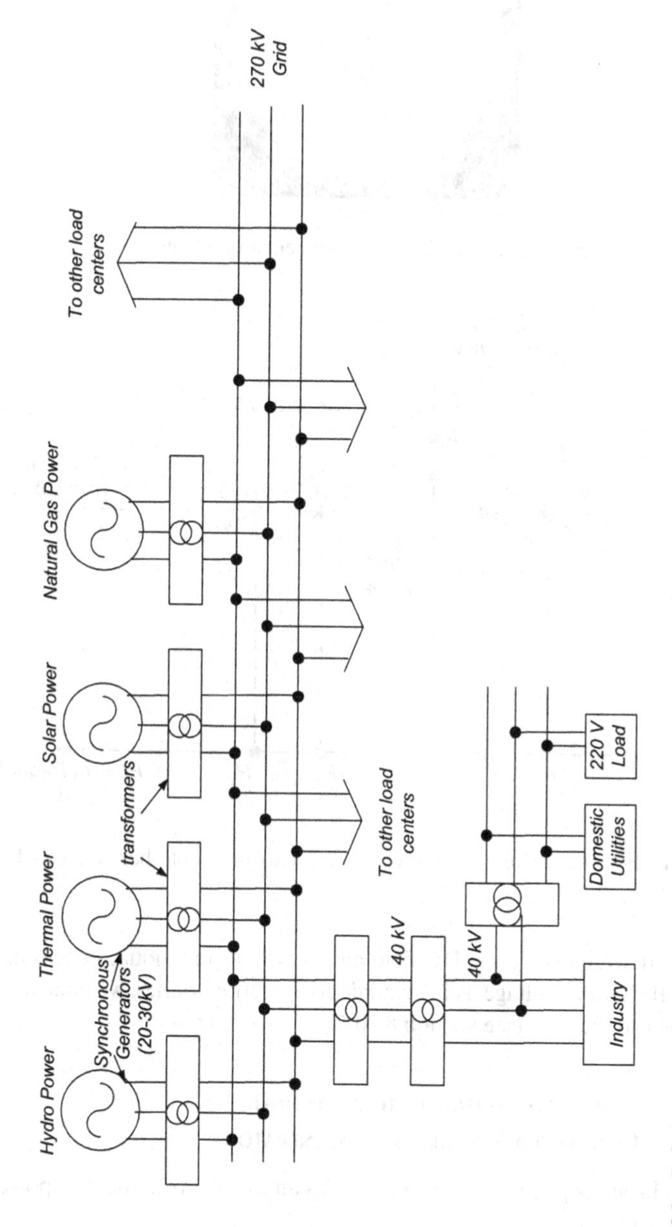

FIGURE 8.26 Infinite bus grid system.

FIGURE 8.27 A synchroscope https://www.rammeter.com/yokogawa.

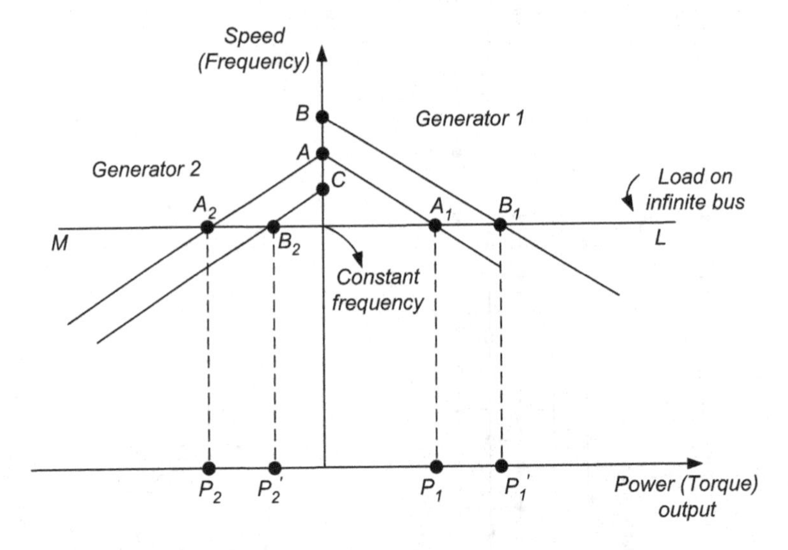

FIGURE 8.28 Effect of increasing mechanical drive to an infinite-bus connected synchronous generator.

Likewise, if we increase the DC field current on an individual generator, we do not increase the output voltage as in standalone operation; rather, we contribute more reactive power to the grid (see section 8.10.2).

8.17.2 Effect of Field Excitation to an Infinite Bus-Connected Synchronous Generator

When a synchronous generator is working into an infinite bus, the real power P out of the generator is

$$P_{\text{out}} = 3V_{\text{rms}}I_{\text{rms}}\cos\theta$$

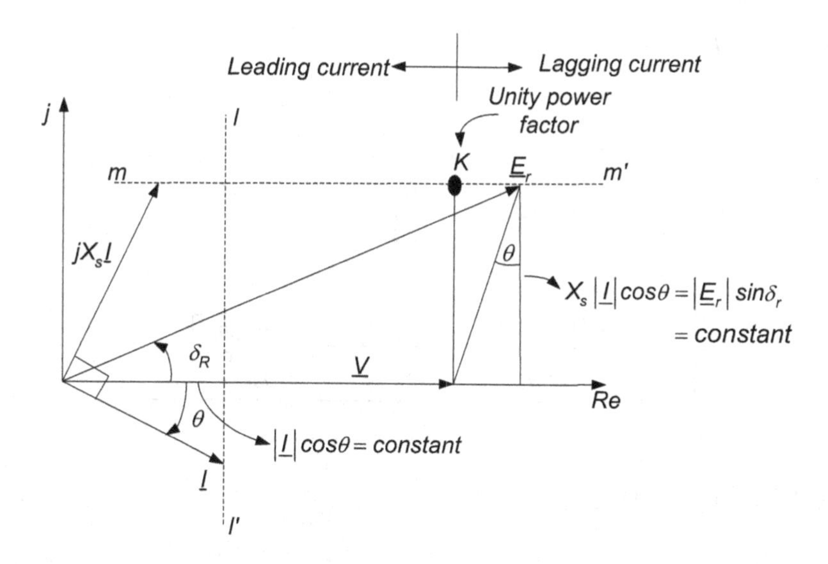

FIGURE 8.29 Effect of field excitation to an infinite-bus connected synchronous generator.

where θ is the angle between output phase voltage $\underline{V}$ and phase current $\underline{I}$. The phasor diagram is shown in Figure 8.29.

Since the generator is working into an infinite bus grid system, we can assume that the mechanical drive and the frequency (speed) are kept constant. When the field excitation is changed, we keep the in-phase component of $\underline{I}$ (i.e., $|\underline{I}|\cos\theta$) a constant too. However, the out-of-phase component varies in magnitude and sign.

Therefore, the tip of current $\underline{I}$ should be on the dashed line ll' shown in Figure 8.29. Also we can see from the diagram that

$$X_s|\underline{I}|\cos\theta = |E_r|\sin\delta_r$$

Therefore, the tip of E_r should be on the other dashed line mm' shown in the figure. We can visualise from the figure that as we decrease the field excitation, we would be increasing the power factor to unity and if we further decrease the field excitation current, $\underline{I}$ would start leading the voltage $\underline{V}$.

The angle of the current relative to the output voltage is controlled, therefore, by the magnitude of the excitation voltage, which is controlled in turn by the DC field current. Hence, the reactive power exchanged between the generator and power system is controlled by the DC field current.

Figure 8.30 shows the effect of the DC field current on the current magnitude and power factor of a synchronous generator with constant output power. For small field currents, the current $\underline{I}$ leads the voltage $\underline{V}$ (negative reactive power) (under excited generator), and for large field currents, the current $\underline{I}$ lags the voltage $\underline{V}$ (positive reactive power) (overexcited generator). Current reaches a minimum current magnitude for unity power factor at Q.

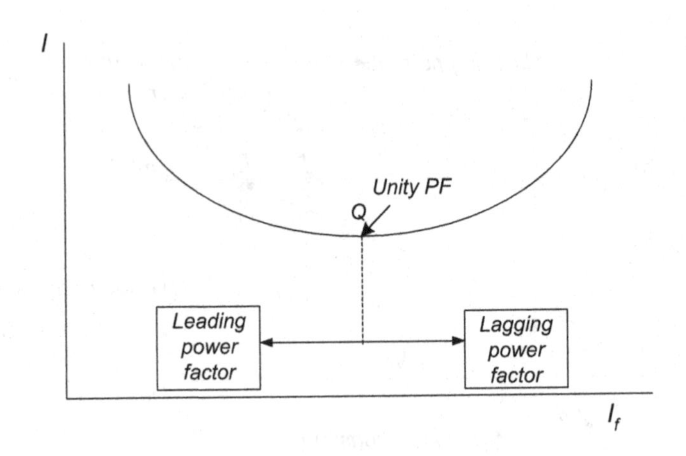

FIGURE 8.30 Field current against output current.

PROBLEMS ON CHAPTER 8

Problem 8.1: A 450-V-rms, 50-Hz eight-pole delta-connected synchronous motor (Figure 8.31) operates with a developed power (including losses) of 150 kW at a power factor of 0.9 lagging. The synchronous reactance is $X_s = 1.6\ \Omega$. How much must the field current be increased to produce 100% power factor? Assume that B_r is proportional to rotor field current I_f with no magnetic saturation.

Problem 8.2: An eight-pole 220-V-rms 60-Hz delta-connected synchronous motor operates with a constant developed power of 50 hp, unity power factor, and a torque angle of 30°. Then, the field current is increased such that rotor field B_r doubles in magnitude. Find the new torque angle and power factor. Is the new power factor leading or lagging? How much is the torque?

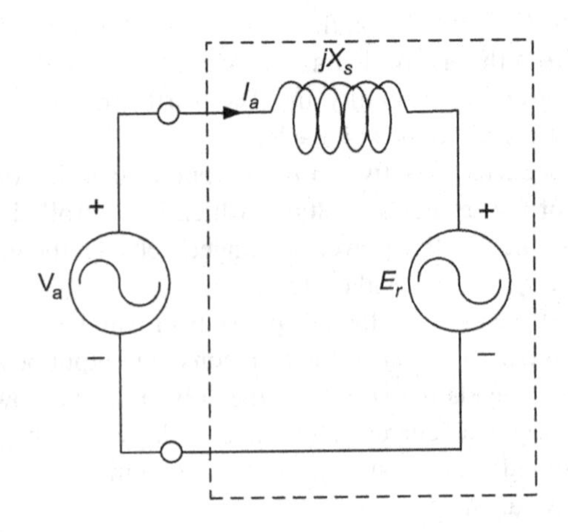

FIGURE 8.31

Problem 8.3: A four-pole 50-Hz synchronous motor is operating with a developed power of 10 kW and a torque angle of 6°. Find the speed and developed torque. What would be the torque angle if the developed torque needs to be doubled? How much is the pull-out torque and the maximum developed power of this motor?

Problem 8.4: A 420 V delta-connected synchronous motor is operating with zero developed power (assuming no losses) and draws a positive reactive power of 31.5 kVAR. The synchronous reactance is 6 Ω. If the rotor field magnitude is proportional to field current, find what percentage the field current must be increased to reduce the armature current to zero.

Problem 8.5: A three-phase, 40-hp, six-pole, 50-Hz, 500-V synchronous motor is providing a 50 nm of output torque with a power factor of 0.85. Motor total losses are 500 W. Calculate the input current supplied to the motor.

Problem 8.6: A two-pole synchronous motor is driving a load which requires a torque of 40 Nm at a speed 1,500 rpm. Assuming no losses, what is the electrical frequency of the supply voltage? What is the power the motor is delivering to the load? If the maximum torque this motor can provide is 80 Nm, find the initial value of the power angle.

Problem 8.7: A cylindrical rotor, three-phase, 50-Hz, 2,000 V, six-pole Y-connected synchronous motor is driving a load which requires a power of 400 hp. The AC voltage induced in the armature (E_r) by the rotor field excitation is 1,200 V, and the motor operates with a power factor of 1. Find the current intake of the motor. How much is the synchronous reactance and what would be the torque and the rotor power angle?

Answers to odd-numbered questions:

8.1: 1.19
8.3: 1500 rpm, 63.6 Nm, 12.12°, 608.4 Nm, 95567 W;
8.5: 7.79 A
8.7: $86.14\angle 0°$ A rms, 3.8 Ω, 2849.5 Nm, −15.8°.

9 Induction Machines

9.1 INTRODUCTION TO INDUCTION MACHINES

Induction machine (asynchronous machine) is another type of AC machine which we will cover in this book (Figure 9.1).

In these types of machines, AC currents flow in both the stator and the rotor windings. In practice, induction machines mostly appear in the form of induction motors. In induction motors, torque (force) is obtained as a result of the interaction between the stator and the rotor magnetic fields (force of interaction). Operation of induction machines as generators is very rare as its performance characteristics as a generator are poor. As generators, they are actually best suited for wind-power applications.

Induction motors are very widely used in the industry as a driver for pumps, fans, compressors, and grinders. This is due to its simplicity and structure. The geniously designed three-phase induction motor was patented by Nikola Tesla in 1888, and it accounts for a very large percentage of the motors used in the industry.

Nikola Tesla (1856–1943)

Therefore, in this chapter, we shall mainly concentrate on the induction motors and include only a short section on induction generators.

The stator windings of induction motors (Figure 9.2) are essentially the same as those of synchronous motors.

However, their rotor windings are different and can be of two types. The first type resembles a squirrel cage, and that is why they are named as "squirrel-cage rotors." It consists of bars of aluminium which are shorted at their ends by conducting rings (Figure 9.3). The squirrel cage is cast as molten aluminium bars into the slots cut in the laminated rotor. Since there are no external connections to the rotor of a squirrel-cage induction motor, they are simple, inexpensive, economical, and reliable. The electrical contactless (brushless) structure of the rotor is a great advantage

FIGURE 9.1 Induction motor. (With the courtesy of ABB.)

FIGURE 9.2 Cutaway view of an induction motor. (With the courtesy of TECO Westinghouse Comp.)

to these types of machines. AC currents are induced in the rotor windings by induction (transformer action). That is why they are named as "induction machines."

In synchronous machines, the rotating flux produced by the windings in the stator (armature) rotates synchronously with the rotor. However, in induction motors, the short-circuited rotor structure travels at a different speed than the rotating field, and this lag depends on the amount of load on the machine.

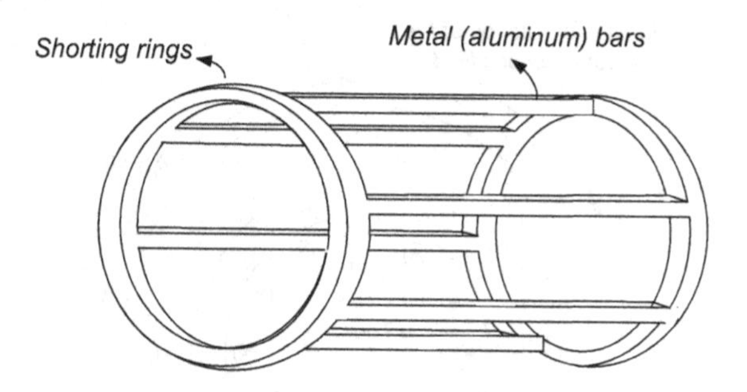

FIGURE 9.3 Rotor conductors of squirrel-cage induction motors.

The other type of rotor construction is the "wound rotor" type. In this type of induction motors, the rotor contains a set of coils placed in slots. The winding terminals are brought out to external terminals through slip rings and brushes. It is because of this that although the starting torque is better, the machine is less rugged and more expensive. They are not as commonly used as the squirrel-cage motors.

Hence, in this chapter, we will concentrate generally on "squirrel-cage" type of induction motors. Where necessary we will indicate the advantages of the wound rotor type of induction motors. Cage type of induction motors are by far the most commonly used motors in daily life.

9.2 THREE-PHASE INDUCTION MOTORS

The simplicity and the ruggedness of squirrel-cage type induction motors played an important role in choosing between DC and AC powers in the late nineteenth century. These types of motors can be fed from a single-, two-, or three-phase source. We will start understanding the concept of this motor with balanced three phase types which are particularly preferred for high power.

In the squirrel-cage induction machine, there are no external electrical connections to the rotor.

Let us assume that rotor conductors are purely resistive and have no inductive component. As the stator magnetic field B_s rotates anticlockwise (Figure 9.4) and moves past a conductor, a voltage will be induced essentially only on the conductor that is directly under the magnetic field B_s. Since the magnetic field, the length of the conductor, and the direction of relative motion are mutually perpendicular, according to Faraday's law, a voltage

$$v = B_s l u \tag{9.1}$$

will be induced on this conductor. Here, u is the relative speed $(n_m - n_s)$ between the conductor of length l and the stator magnetic field B_s. For simplicity, we can neglect the small voltages induced in other conductors that are not directly under the rotating field B_s. The voltage induced on these conductors will be smaller due to smaller B_s and smaller angle between the magnetic field B_s and the direction of relative motion.

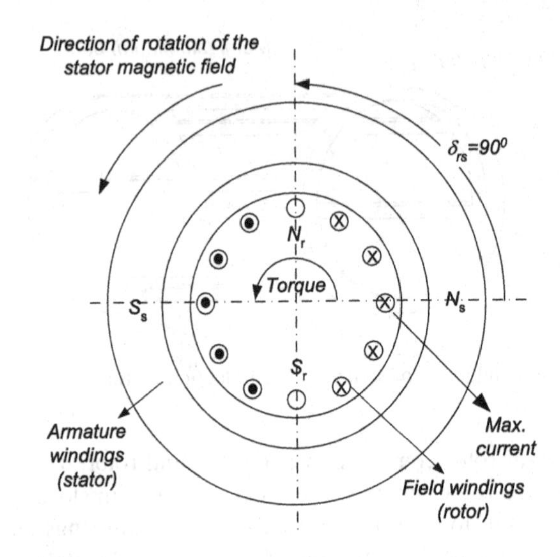

FIGURE 9.4 Squirrel-cage rotor and stator of induction motors (front view).

Although we will consider them negligible, currents flowing in these conductors will only improve the torque produced by the conductors directly under the magnetic field B_s.

The voltage v of equation (9.1) causes currents to flow in the rotor conductors. The direction of the current flow can be found by Flemings' right-hand rule. Notice that from the rotor point of view, the direction of rotation of the rotor conductors would be in the clockwise direction (when stator field is assumed not to rotate) as shown in Figure 9.5. Hence, the direction of current flowing in the conductors would be as shown in Figures 9.4 and 9.5. This current will cause a torque on the conductors according to

$$T = B_s l i_r r \tag{9.2}$$

Hence, according to the right-hand rule, the rotor will start to rotate in anticlockwise direction.

Actually, the rotor currents establish magnetic poles on the rotor (E_r). North stator poles are where the magnetic flux lines leave the stator, and south stator poles are where the magnetic flux lines enter the stator (see Figures 9.4 and 9.5). The same applies to rotor magnetic field B_r. Notice that B_r shown in Figures 9.4 and 9.5 is inside the rotor and B_s is outside the stator. It is the interaction of the rotor poles with the stator poles that produces torque. The north rotor pole attempts to align itself with the south stator pole.

If the inductance of the rotor conductors is neglected, the rotor magnetic field B_r (hence rotor poles) will be $\delta_{rs} = 90°$ away from stator magnetic field B_s (hence rotor poles). Notice that when there is no rotor conductor inductance, no matter what the speed of the machine is, δ_{rs} is always 90°. The torque on the rotor will speed the rotor up in the anticlockwise direction. Hence, the induced voltages on the rotor

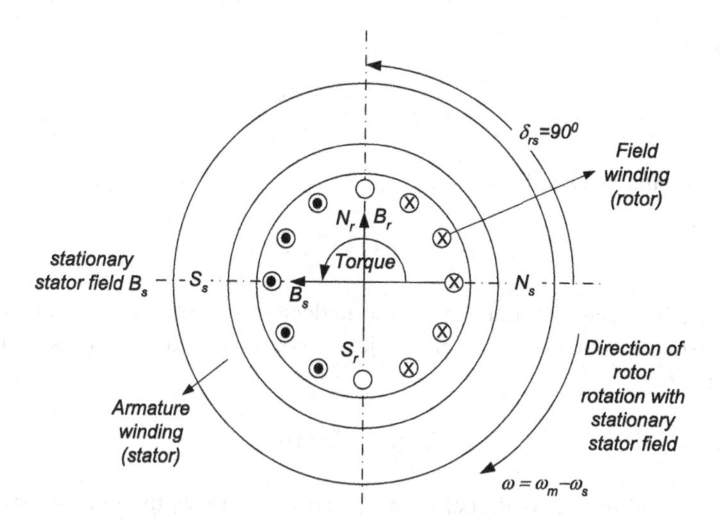

FIGURE 9.5 Squirrel-cage induction motors (front view) with stator field forced to stand still.

conductors will decrease due to decrease in relative speed u. Consequently, the torque will drop down linearly until the rotor speed reaches the synchronous speed n_s where the torque will be zero.

The current and voltage in the rotor will be induced because of the different speeds of the rotor and the rotating stator magnetic field. We can imagine this with the help of Figure 9.5 in which the anticlockwise rotating stator field (speed n_s) is forced to be at standstill. As a result of this assumption, the rotor would relatively appear to be rotating in the negative (clockwise) direction with relative speed $(n_m - n_s)$ and induced rotor currents in the rotor conductors on the right-hand side would be flowing into the page and on the left-hand side, they would be flowing out of the page (Figures 9.4 and 9.5). With this assumption and no rotor inductance, it is easier to visualise how $\delta_{rs} = 90°$ remains the same throughout the operation of the machine.

Hence, if

ω = angular velocity of the current in the stator windings
ω_s = angular velocity of the rotating stator magnetic field (synchronous speed)

then we can write the following relationship:

$$\omega_s = \frac{\omega}{P/2}$$

Suppose now ω_m is the angular (shaft) velocity of the rotor. We can now write the ω_{slip} angular velocity of the rotor current or voltage as

$$\omega_{slip} = (\omega_s - \omega_m) \times P/2$$

If we define

$$s = \text{slip} = \frac{\omega_s - \omega_m}{\omega_s}$$

then ω_{slip} would be as follows:

$$\omega_{\text{slip}} = s\omega_s \times p/2 = s\omega$$

This is an important relationship in the understanding of the behaviour of these motors. Now since the stator magnetic field is constant, torque on the rotor conductors would be

$$\text{Torque} \propto i_r \sin \delta_{rs} \tag{9.3}$$

Here, δ_{rs} is the torque angle between B_r and B_s, and i_r is the induced sinusoidal current on the rotor conductors. Note that with no rotor conductor inductance, $\delta_{rs} = 90°$.

9.2.1 Torque/Speed Characteristics of Balanced Three-Phase Induction Motors

The torque equation 9.2 given in Section 9.2 holds only with the assumption of no rotor conductor inductance. Because the conductors are embedded in iron, there is significant series inductance associated with each conductor in the rotor. Because of this inductive reactance, the rotor conductor current lags the induced voltage. This lag is large (nearly 90°) if the motor is at standstill and gets smaller as the motor speeds up towards synchronous speed and power factor approaches to unity. Consequently, the peak rotor current occurs somewhat after the stator pole is passed by. The equivalent circuit of a rotor conductor would be as shown in Figure 9.6. Hence,

$$\text{Rotor conductor impedance per phase} = \underline{Z}_r = R_r + js\omega L_r$$

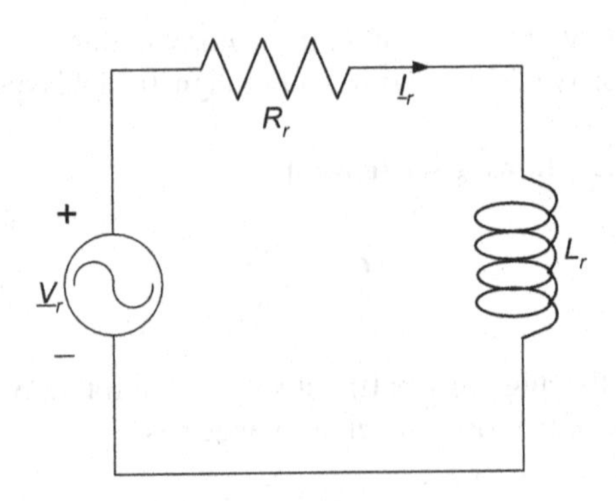

FIGURE 9.6 Squirrel-cage rotor conductor equivalent circuit for one phase.

where

R_r = resistance per phase of the rotor conductor
L_r = inductance per phase of the rotor conductor
s = slip
ω = angular velocity of the current in the stator windings.

Hence, rotor current would be

$$\underline{I}_r = \frac{\underline{V}_r}{R_r + js\omega L_r} \tag{9.4}$$

As the rotor speed starts increasing from standstill, slip s will get smaller and induced voltage on the conductors ($\underline{V}_r$) will proportionally decrease (Faraday's law) (equation 9.1). However, because of the dominant effect of the decrease in inductive reactance $js\omega L_r$ in equation (9.4), the magnitude of the current in the conductors will increase and its power factor will improve (angle δ_{rs} approaches 90°). The overall effect on the torque (equation 9.3) will be a steady increase in magnitude as the rotor speeds up from zero. As we get closer to the synchronous speed n_s, the effects of decreasing inductive reactance and improving power factor will level with the effect of the decrease in speed (hence, voltage v) and torque will reach a point of maximum after which the decrease in induced voltage will begin to be dominant and torque will drop almost linearly with slip s.

We can assume that for small slips, the inductive reactance of the rotor conductors, given by $s\omega L_r$, is negligible and hence torque decreases linearly with s (Figure 9.7).

In fact, we can further go on with this argument for the speeds in the opposite direction (negative speed) as well. If the speed increases from zero to negative values, the slip s will further increase and as a result the reactance X_r and hence the impedance $|\underline{Z}_r|$ will increase and become more and more inductive. This will not

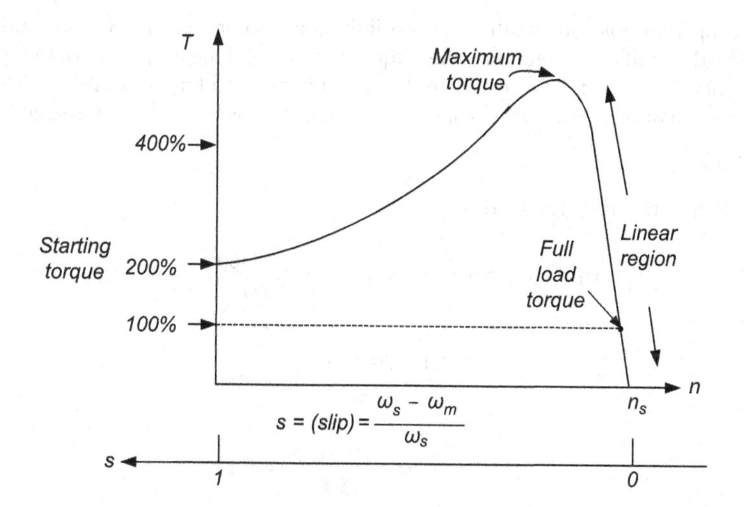

FIGURE 9.7 Typical torque–speed characteristics of three-phase induction motors.

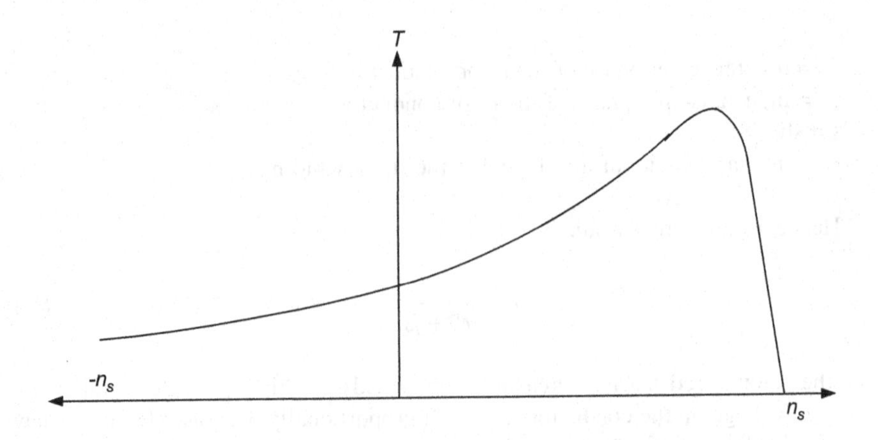

FIGURE 9.8 Torque-speed characteristics of three phase-induction motors with negative speeds.

only decrease the current but also decrease the power factor and increase the torque angle to near 180° (Figure 9.8). As a result torque will get smaller as speed increases in the opposite direction. This argument will be useful in understanding the behaviour of the single-phase induction motors where there are two rotating stator fields in opposite directions.

By changing the structure of the rotor (e.g., the cross section and depth of the rotor conductors), it is possible to change the shape of the torque–speed curve to suit certain applications. We also notice that by driving the motor to beyond the synchronous speed, we change the motor action to generator action.

Exercise 9.1:

A 10-hp four-pole 50-Hz three-phase induction motor runs at 1300 rpm under full-load conditions. Determine the torque, the slip, and the frequency of the rotor currents at full load. Also, estimate the speed if the load torque drops to 40% of full load (assume linear torque–speed characteristics near synchronous speed).

Solution:

The output power at full load is

$$p_{\text{out}} = 10 \times 746 = 7460 \text{ W} = T\omega \rightarrow T = \frac{7460}{\frac{1300}{60} \times 2\pi} = 54.8 \text{ Nm}$$

$$n_s = \frac{f \times 60}{P/2} = \frac{50 \times 60}{4/2} = 1500 \text{ rpm}$$

$$s = \frac{n_s - n_m}{n_s} = \frac{1500 - 1300}{1500} = 0.13$$

$$\omega_{\text{slip}} = s\omega \rightarrow f_s = 0.13 \times 50 = 6.5 \text{ Hz}$$

$$\frac{T_1}{T_2} = \frac{1}{0.4} = \frac{s_1}{s_2} \rightarrow s_2 = 0.13 \times 0.4 = 0.052$$

$$0.052 = \frac{1500 - n_m}{1500} \rightarrow n_m = 1500 - 1500 \times 0.052 = 1422 \text{ rpm}$$

9.2.2 Rotor Equivalent Circuit per Phase

We know that the angular velocity of the rotor currents is $s\omega$. If

E_r = the induced voltage in phase a of the rotor when it is not rotating ($s = 1$)
R_r = rotor resistance per phase
L_r = rotor inductance per phase
$X_r = \omega L_r$ = reactance under locked rotor ($s = 1$) condition,

then the equivalent circuit when the rotor is not rotating would be as shown in Figure 9.9b.

The equivalent circuit when the rotor rotates with slip s would be as shown in Figure 9.9a.

Here, we see that as s gets smaller (in other words, as speed approaches synchronous speed), the rotor resistance R/s starts dominating the rotor circuit. As a result, the rotor current (hence torque) decreases linearly with slip s.

9.2.3 Complete Induction Motor Equivalent Circuit per Phase

As in a transformer, the induced rotor voltage E_r under locked rotor conditions is related to the stator voltage by the turn ratio. Thus, we can reflect the rotor impedances of Figure 9.9b to the primary (stator) side of the equivalent circuit. We denote

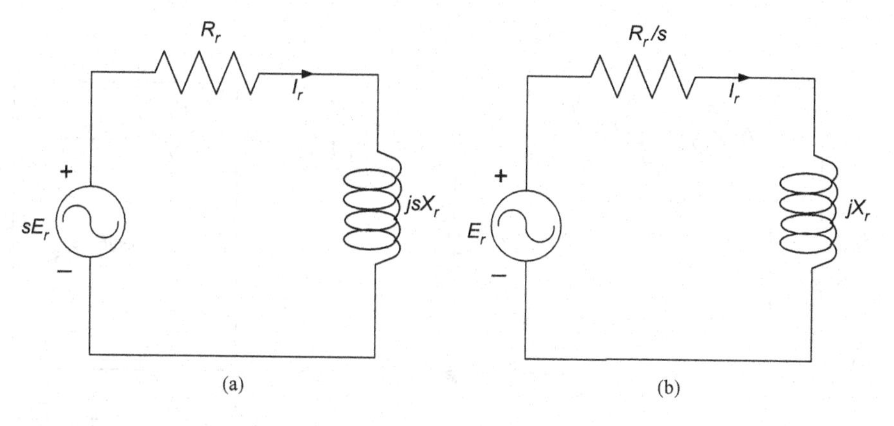

FIGURE 9.9 The equivalent circuit of one phase of the rotor windings. (a) Rotating rotor and (b) Locked rotor.

the reflected values of X_r and R_r/s by X'_r and R'_r/s, respectively, (Figure 9.10). Here, the resistive component R_r/s is transferred to the stator side as

$$\frac{R'_r}{s} = R'_r + \frac{1-s}{s}R'_r$$

In the equivalent circuit,

R_s = the resistance per phase of the stator winding.
X_s = the stator leakage reactance per phase.
X_m = the magnetising reactance per phase that accounts for the current needed to set up the rotating stator field.

The equivalent circuit of one phase of the stator winding with the reflected rotor values and the related power flow is illustrated in Figure 9.11.

The developed torque is

$$T_{dev} = \frac{P_{dev}}{\omega_m}$$

The power P_{ag} that crosses the air gap into the rotor is delivered to the rotor resistances. Thus,

$$P_{ag} = P_r + P_{dev}$$

$$P_{ag} = 3R'_r(I'_r)^2 + 3\times\frac{1-s}{s}R'_r(I'_r)^2$$

$$P_{ag} = 3\times\frac{1}{s}R'_r(I'_r)^2 \tag{9.5}$$

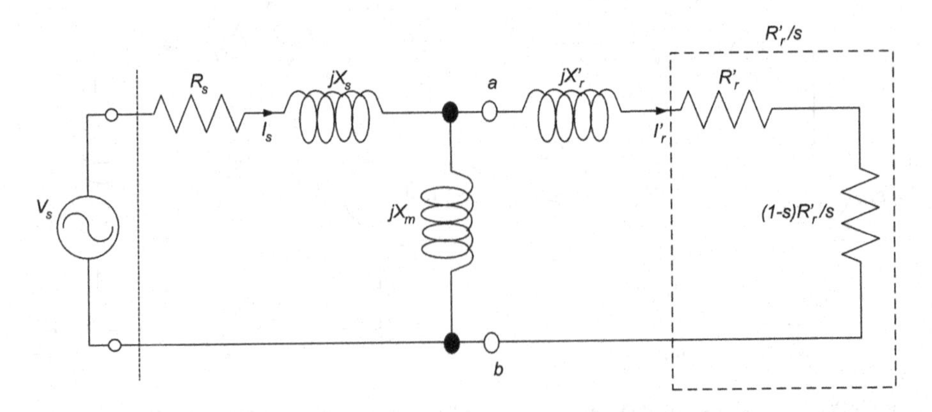

FIGURE 9.10 The equivalent circuit of one phase of the stator windings.

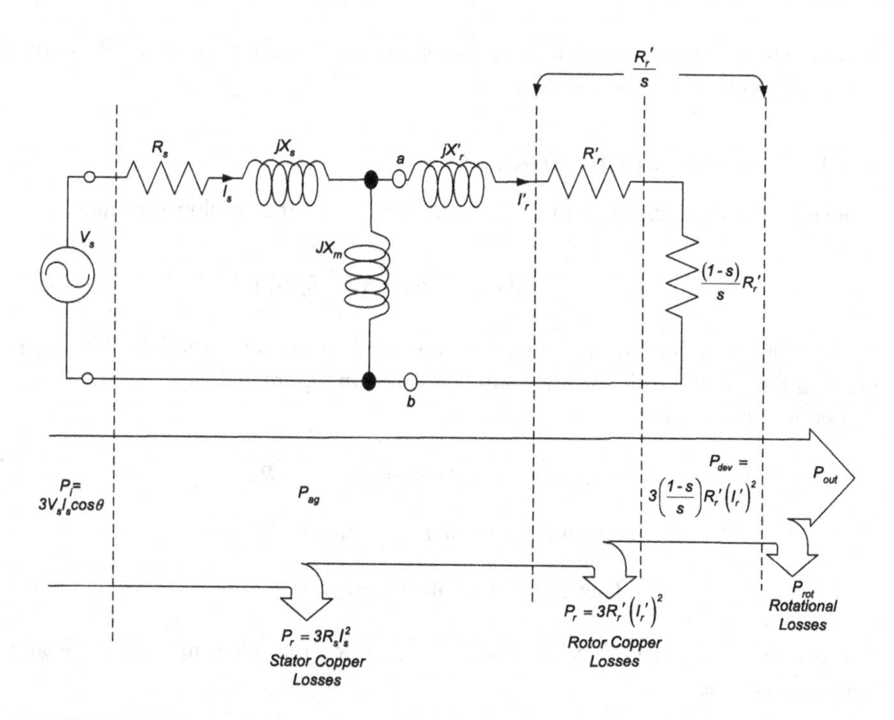

FIGURE 9.11 The equivalent circuit of one phase of the stator winding and the related power flow diagram.

Hence,

$$P_r = sP_{ag}$$

$$P_{dev} = (1-s)P_{ag}$$

Thus,

$$T_{dev} = \frac{(1-s)P_{ag}}{\omega_m}; \ \omega_m = (1-s)\omega_s$$

$$T_{dev} = \frac{P_{ag}}{\omega_s} \tag{9.6}$$

To speed up from standstill, the starting torque must be larger than the torque required by the load. Under starting conditions (i.e., $\omega_m = 0$), we have $s = 1$ and

$$P_{ag} = 3R_r'(I_r')^2 = P_r$$

Then, the starting torque can be computed by using the following equation:

$$T_{dev} = \frac{P_{ag}}{\omega_s} \rightarrow T_{dev(start)} = \frac{P_r}{\omega_s}$$

Rotor circuit parameters of an induction motor can be varied by changing the diameter or the depth of the rotor conductors.

9.2.4 Calculation of the Torque Expression

From above paragraphs, we can write the expression for the developed torque as

$$T_{\text{dev}} = \frac{P_{\text{ag}}}{\omega_s} \text{ where} \rightarrow P_{\text{ag}} = 3 \times \frac{1}{s} R_r' (I_r')^2 \tag{9.7}$$

To calculate the rotor current I_r', we can represent the circuit to the left of points a and b in Figure 9.11 with the Thevenin equivalent in Figure 9.12.

Let us assume that

$$\text{Thevenin's equivalent resistance} = R_{\text{th}}$$

$$\text{Thevenin's equivalent reactance} = X_{\text{th}}$$

$$\text{Thevenin's equivalent voltage} = V_{\text{th}}$$

Because $X_m \gg X_s$ and $X_m \gg R_s$ and $X_m + X_s \gg R_s$, the Thevenin values would approximately be

$$\underline{V}_{\text{th}} = \underline{V}_a \frac{jX_m}{R_s + j(X_s + X_m)} \cong \underline{V}_a \frac{X_m}{X_s + X_m}$$

$$Z_{\text{th}} = \frac{jX_m(R_s + jX_s)}{R_s + j(X_s + X_m)} = \frac{jX_m(R_s + jX_s)[R_s - j(X_s + X_m)]}{R_s^2 + (X_s + X_m)^2}$$

$$= \frac{jX_m\left[R_s^2 - jX_sR_s - jX_mR_s + jX_sR_s + X_s^2 + X_sX_m\right]}{R_s^2 + (X_s + X_m)^2}$$

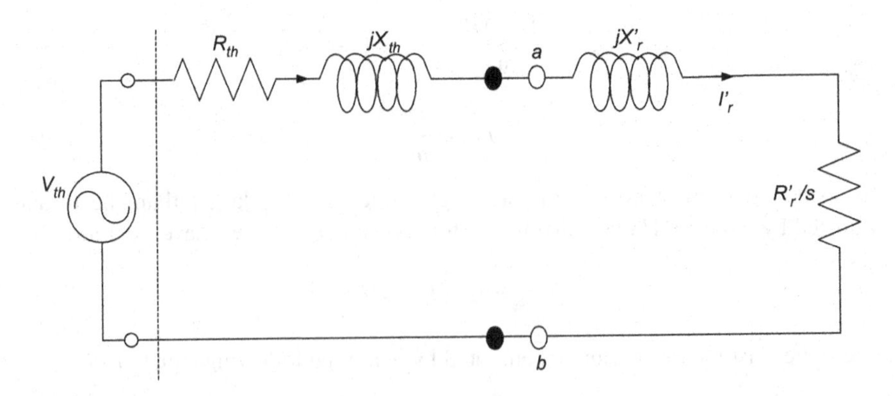

FIGURE 9.12 The equivalent circuit of one phase of the stator with Thevenin's theorem applied at points a and b.

$$= \frac{jX_m\left[R_s^2 - jX_mR_s + X_s^2 + X_sX_m\right]}{R_s^2 + (X_s + X_m)^2}$$

$$= \frac{R_sX_m^2}{R_s^2 + (X_s + X_m)^2} + \frac{jX_m\left[R_s^2 + X_s^2 + X_sX_m\right]}{R_s^2 + (X_s + X_m)^2}$$

$$\cong \frac{R_sX_m^2}{(X_s + X_m)^2} + \frac{jX_m\left[X_s^2 + X_sX_m\right]}{(X_s + X_m)^2}$$

$$= \frac{R_sX_m^2}{(X_s + X_m)^2} + \frac{jX_mX_s}{(X_s + X_m)} \cong \frac{R_sX_m^2}{(X_s + X_m)^2} + \frac{jX_mX_s}{X_m}$$

$$= R_s\left(\frac{X_m}{X_s + X_m}\right)^2 + jX_s = R_{\text{th}} + jX_{\text{th}}$$

Then, the magnitude of the rotor current I_r' would be

$$I_r' = \frac{V_{\text{th}}}{\sqrt{\left(R_{\text{th}} + \frac{R_r'}{s}\right)^2 + (X_{\text{th}} + X_r')^2}}$$

So from equation (9.7), torque would be equal to

$$T_{\text{dev}} = \frac{3V_{\text{th}}^2\frac{R_r'}{s}}{\omega_s\left[\left(R_{\text{th}} + \frac{R_r'}{s}\right)^2 + (X_{\text{th}} + X_r')^2\right]} \tag{9.8}$$

Figures 9.7, 9.8, and 9.13 are plots of torque against speed derived from equation 9.8. Maximum torque will occur when resistor R_r'/s draws maximum power. According to the maximum power transfer theorem, when the angle of the load impedance is fixed, maximum power transfer to the load will occur when the magnitude of the source impedance is equal to the magnitude of the load impedance. Here, Z_{source} is given by the following equation:

$$Z_{\text{source}} = R_{\text{th}} + jX_{\text{th}} + jX_r'$$

Hence, according to maximum power transfer theorem,

$$\frac{R_r'}{s} = \sqrt{R_{th}^2 + (X_{th} + X_r')^2}$$

Solving this for s,

$$s_{max} = \frac{R'_r}{\sqrt{R_{th}^2 + (X_{th} + X'_r)^2}} \tag{9.9}$$

We notice that slip for maximum torque is directly proportional to the referred rotor resistance R'_r (Figure 9.13).

If we replace s in the equation for torque by the value of s_{max} above, we find the equation for the maximum torque as

$$T_{max} = \frac{3V_{th}^2}{2\omega_s\left[R_{th} + \sqrt{R_{th}^2 + (X_{th} + X'_r)^2}\right]} \tag{9.10}$$

Maximum torque is proportional to the square of the voltage supplied to the stator windings. It is also inversely proportional to the stator impedance and rotor reactance. Smaller rotor or stator reactances will improve the maximum torque value. Slip at which maximum torque occurs is directly dependent on the value of the rotor resistance (Figure 9.13), but the value of the maximum torque is independent of the rotor resistance. With a squirrel-cage motor, it is not possible to change the rotor resistance as there is no terminal attached to the rotor. But if the motor is a wound rotor type, then it would be possible to move the maximum torque value to different speeds by adding a resistance to the rotor circuit. When using wound rotor-type induction motors for very heavy loads, we move the maximum torque point to the start of the machine by changing the rotor resistance. After start off, we take out the added resistance so that maximum torque occurs at a normal speed of the motor.

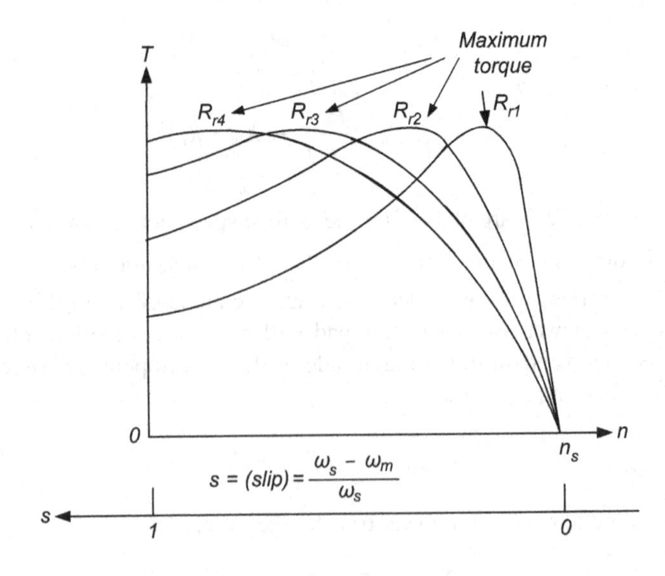

FIGURE 9.13 Change of maximum torque with rotor resistance of a wound rotor induction motor.

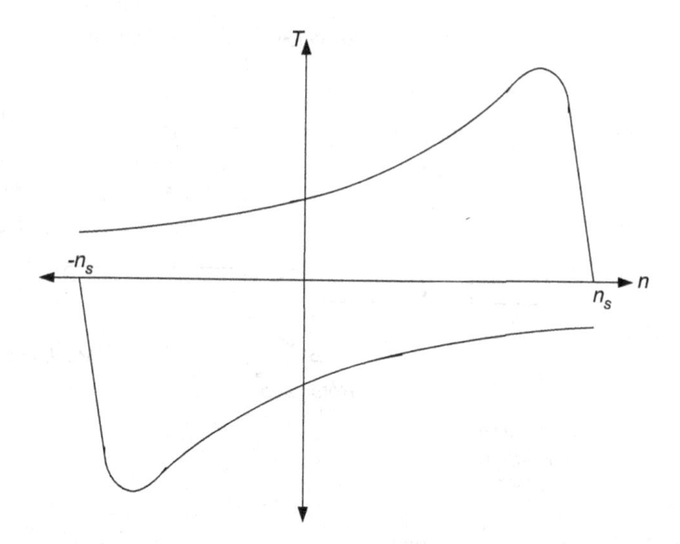

FIGURE 9.14 The individual torque/speed curves produced by the sole forward and sole backward rotating fields (not existing at the same time).

When the motor is rotating at synchronous speed, if suddenly the direction of the rotating magnetic field is changed, the rotor currents ought to be very high, but since the rotor frequency will be very high, the reactance of the rotor will be very much higher than its resistance. Therefore, the rotor current will lag the rotor voltage by 90° which will produce a magnetic field that is almost 180° from the stator magnetic field. At this point, the induced torque will reverse but will be very small due to this torque angle. Hence, the induced torque will change its direction (Figure 9.14) and will first stop the machine and then speed it up in the opposite direction. In fact, this way by switching any two of the phase voltages, we can stop the machine very rapidly.

9.2.5 Power Relationships

We can summarise the power relationships in induction motors as shown in Figure 9.15.

In Figure 9.10, voltage V_s is the voltage across each stator winding and current I_s is the current through each of the stator windings. Therefore, they are called phase voltage and phase current of each of the stator windings. The voltage and current ratings of a machine are invariably given as line-to-line, whereas in the equivalent circuit diagrams given, they are shown as phase voltage and currents. Therefore, we have to convert line-to-line voltage and current values to phase values in these diagrams. For example, for a delta-connected motor, line voltage must be taken as equal to the phase voltage, but line current must be divided by $\sqrt{3}$ for phase current.

On the left-hand side of the power flow diagram in Figure 9.15, we see the stator iron and copper losses. As the name implies, we can examine these losses under two separate headings. The first one is the I_s^2R copper losses encountered in the stator

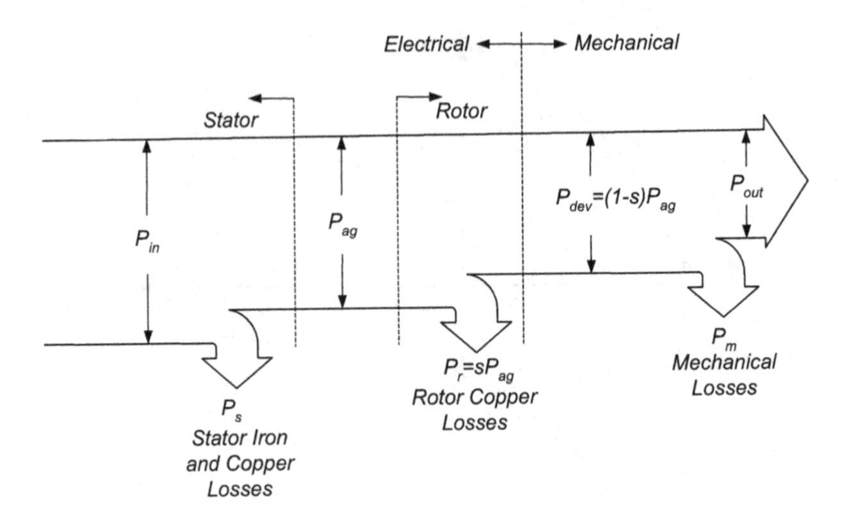

FIGURE 9.15 The power flow diagram.

windings. The second one is in fact the core losses encountered as hysteresis and eddy current losses. Core losses occur mainly in the stator windings. As the motor runs generally near the synchronous speed, *s* is very small and therefore the core losses in the rotor are negligible compared to the core losses in the stator windings. However, the higher the speed of an induction motor, the higher will be the mechanical losses like friction, windage, and stray losses. Therefore, rotor core losses and mechanical losses are often lumped together as rotor rotational losses. Rotational losses are considered as constant.

The power remaining after stator iron and copper losses is the power (P_{AG}) transferred to the rotor of the machine through the air gap.

Exercise 9.2:

A three-phase, six-pole, 50-Hz induction motor has a full-load output power of 6 hp at 900 rpm. The motor efficiency is 80%, and the mechanical losses are 50 W. Determine all the power quantities.

Solution:

The synchronous speed is

$$n_s = \frac{50 \times 60}{6/2} = 1000 \text{ rpm}$$

The input power is

$$\frac{6 \times 746}{0.8} = 5595 \text{ W}$$

and hence the total losses are

$$5595 - 6 \times 746 = 1119 \text{ W}$$

The mechanical losses are 50 W, and electrical losses are therefore

$$1119 - 50 = 1069 \text{ W}$$

The developed power is

$$P_{\text{dev}} = P_{\text{out}} + P_m = 6 \times 746 + 50 = 4526 \text{ W}$$

$$s = \frac{n_s - n}{n_s} = \frac{1000 - 900}{1000} = 0.1$$

$$P_{\text{ag}} = \frac{P_{\text{dev}}}{1 - s} = \frac{4526}{1 - 0.1} = 5029 \text{ W}$$

$$P_R = sP_{\text{ag}} = 0.1 \times 5029 = 502.9 \text{ W}$$

Hence, the stator loss

$$P_s = 1069 - 502.9 = 566.1 \text{ W}$$

Exercise 9.3:

A three-phase six-pole delta-connected induction motor (Figure 9.16) has 25 hp, 420 V rms, 50 Hz, and

$$R_s = 1.1\ \Omega;\ X_s = 2.1\ \Omega;\ X_m = 60\ \Omega;\ R_r' = 0.6\ \Omega;\ X_r' = 0.7\ \Omega$$

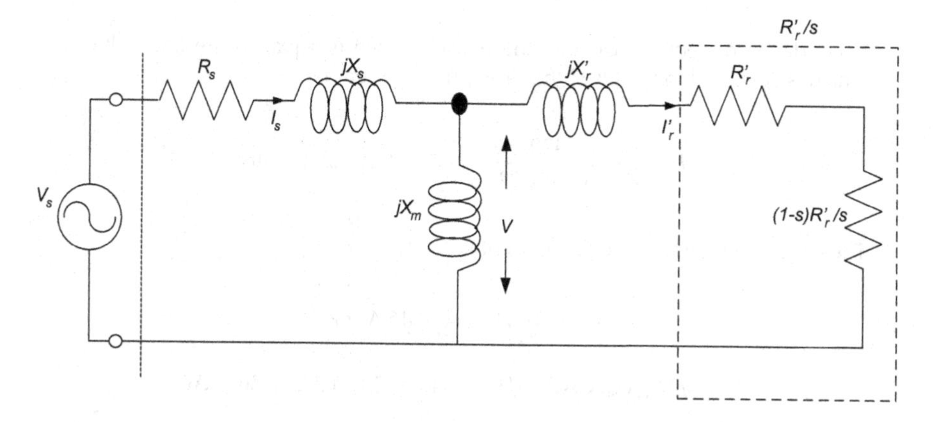

FIGURE 9.16 The equivalent circuit of induction motor.

Under load, the machine operates at 950 rpm and has rotational losses of 500 W. Find the power factor, the line current, the output power, copper losses, output torque, and efficiency.

Solution:

$$n_s = \frac{f \times 60}{P/2} = \frac{50 \times 60}{6/2} = 1000 \text{ rpm}$$

$$\text{slip} = s = \frac{n_s - n_m}{n_s} = \frac{1000 - 950}{1000} = 0.05$$

$$Z_s = 1.1 + j2.1 + \frac{j60(12 + j0.7)}{j60 + 12 + j0.7}$$

$$= 1.1 + j2.1 + \frac{j720 - 42}{j60.7 + 12}$$

$$= 1.1 + j2.1 + \frac{(j720 - 42)(-j60.7 + 12)}{(j60.7 + 12)(-j60.7 + 12)} = 1.1 + j2.1 + \frac{43704 + j8640 + j2549.4 - 504}{3684.49 + 144}$$

$$= 1.1 + j2.1 + \frac{j11189.4 + 43200}{3828.49} = 1.1 + j2.1 + j2.92 + 11.28$$

$$= j5.02 + 12.38 = 13.36\angle 22^\circ \ \Omega$$

Because of the inductive impedance, we know that the power factor is lagging. Hence,

$$pf = \cos 22^\circ = \%92.7^\circ \text{ lagging}$$

For a delta-connected machine, the phase voltage is equal to the line voltage, which is specified to be 420 V rms. The phase current is

$$I_s = \frac{V_s}{Z_s} = \frac{420\angle 0^\circ}{13.36\angle 22^\circ \ \Omega} = 31.44\angle(-22^\circ) \text{ A rms}$$

Thus, the magnitude of the line current is

$$I_{\text{line}} = 31.44\sqrt{3} = 54.46 \text{ A rms}$$

$$P_{\text{in}} = \sqrt{3} I_{\text{line}} V_{\text{line}} \cos\theta = \sqrt{3} \times 54.46 \times 420 \times 0.927 = 36.7 \text{ kW}$$

$$\underline{V} = \underline{I}_s \frac{j60(j0.7+12)}{j60+j0.7+12} = (31.44\angle-22°)\frac{-42+j720}{j60.7+12}$$

$$= (31.44\angle-22°)\frac{(-42+j720)(-j60.7+12)}{(j60.7+12)(-j60.7+12)}$$

$$= (31.44\angle-22°)\frac{j2549.4-504+43704+j8640}{3684.49+144}$$

$$= (31.44\angle-22°)\frac{j11189.4+43200}{3828.49}$$

$$= (31.44\angle-22°)(j2.92+11.28)$$

$$= (31.44\angle-22°)(11.65\angle14.51°)$$

$$= 366.27\angle-7.49° \text{ V rms}$$

Now,

$$I_r' = \frac{V}{j0.7+12} = \frac{366.27\angle-7.49°}{12.02\angle3.34°} = 30.47\angle-10.83° \text{ A rms}$$

The copper losses in the stator and rotor are

$$P_s = 3R_sI_s^2 = 3\times1.1\times31.44^2 = 3262 \text{ W}$$

$$P_r = 3R_r'I_r'^2 = 3\times0.6\times30.47^2 = 1671 \text{ W}$$

Finally, the developed power is

$$P_{\text{dev}} = 3\times\frac{1-s}{s}\times R_r'I_r'^2 = 3\times\frac{1-0.05}{0.05}\times0.6\times30.47^2 = 31752$$

As a check, we note that

$$P_{\text{in}} = P_{\text{dev}} + P_s + P_r$$

9.2.6 Speed Control of Induction Motors

In induction motors, synchronous speed varies linearly with voltage source frequency and inversely with the number of poles. We can change speed as follows:

1. by changing the frequency of the voltage applied to the stator windings
2. by changing the slip of the rotor for a given load
3. by changing the number of poles.

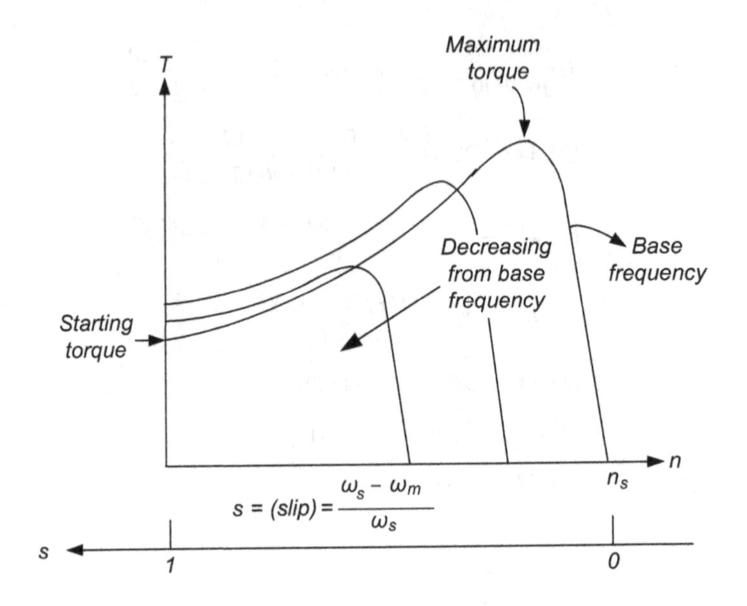

FIGURE 9.17 The effect of decreasing the base frequency.

In the early days when semiconductor technology was not as well advanced as today, it was hard to control the speed of induction motors by changing the frequency of the stator voltage source. Nowadays, it is easier to change the frequency of a voltage supply by using converters and inverters. When base frequency is lowered, the no-load point in the torque–speed characteristic curve will move to left (Figure 9.17). Similarly when base frequency is increased, the no-load point in the torque-speed characteristic curve will move to right (not shown in Figure 9.17).

However as the stator voltage frequency is changed, some safety precautions have to be taken related to the voltage level applied to the stator windings to prevent saturation and excessive magnetisation currents. Changing the number of poles involves simple changing of coil connections. Speed can also be controlled over a limited range by changing the line voltage.

9.2.7 Starting of Three-Phase Induction Motors

Unlike synchronous motors, three-phase induction motors can be started by just connecting them to the AC source. However, they draw high currents when starting. Therefore, precautions must be taken when starting induction motors. Especially for high-power motors, the high starting current will produce a severe voltage drop and will affect the operation of other equipment. Normally with motors beyond 5 hp, starters are provided.

Star--delta starters are the easiest and cheapest of these starters. The main idea in this kind of starters is to apply low starting voltages to the stator windings during start-up. For low start up currents,

1. We can use star/delta combinations. When starting, we prefer the star connection (Figure 9.18) to make sure lower voltage is applied to the windings. This way, we decrease the excess current that would otherwise flow in the windings. But in this method of starting induction motors, we have the disadvantage of decreased starting torque by square of the voltage reduction.

   After the motor starts up, we change to delta connection (Figure 9.19) to increase the applied voltage to the windings for rated output. We can also use a transformer to reduce the starting voltage.

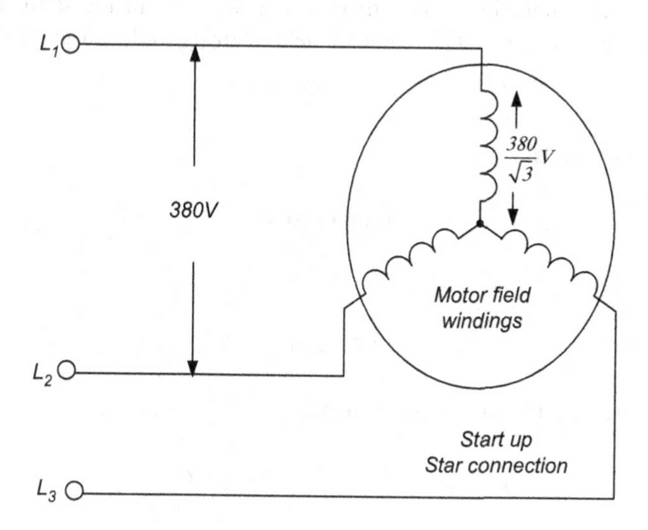

FIGURE 9.18 Star/delta starter at start up.

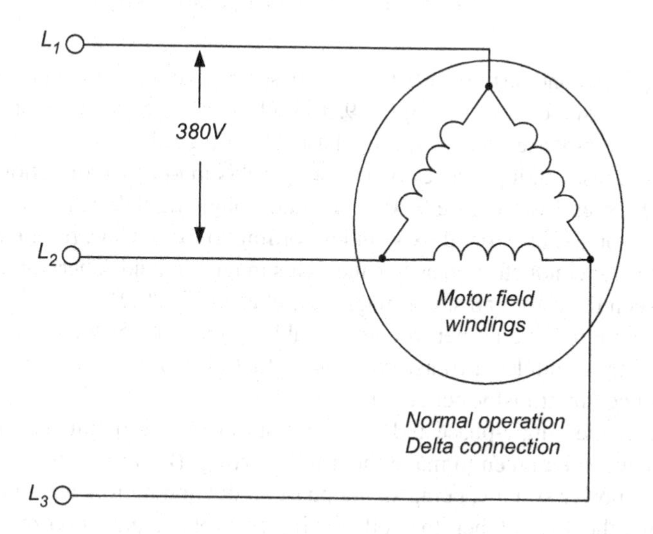

FIGURE 9.19 Star/delta starter after start up.

2. As an alternative way, a resistor or an inductor may be connected to the stator windings.

   For wound rotor-type induction motors, problem can be overcome by simply inserting external resistance to the rotor winding. This would have the extra benefit of increasing the starting torque.

9.3 SINGLE-PHASE INDUCTION MOTOR

In a single-phase induction motor, there is one winding in the stator to which a single-phase AC is applied. The magnetic field intensity in the air gap will be

$$B = Ki(t)\cos\theta$$

If the stator current is

$$i(t) = I_m \cos\omega t$$

Then,

$$B = KI_m \cos\omega t \cos\theta$$

Using the following trigonometric identity,

$$\cos A \cos B = \frac{1}{2}\left[\cos(A-B) + \cos(A+B)\right]$$

the expression for the magnetic field intensity in the air gap becomes

$$B = \frac{1}{2}KI_m \cos(\omega t - \theta) + \frac{1}{2}KI_m \cos(\omega t + \theta) \tag{9.11}$$

First term and second term in equation 9.11 describe two magnetic fields B_{s1} and B_{s2} rotating in opposite directions (Figure 9.20) both with electrical angular velocity ω. Unlike the three-phase source, single-phase AC source does not produce a single rotating field; instead, it produces two rotating fields in opposite directions.

As can be seen from Figure 9.20, total stator magnetic field varies in magnitude but is always on the horizontal axis, and according to Lenz's law, rotor currents will flow in the rotor conductors so as to cancel this magnetic field. Therefore, the torque angle between the rotor and stator magnetic fields will be 180°.

Hence, there will be no net torque available at the start of the motor. In fact at start, machine resembles a transformer since there is no movement, but there is a voltage induced by transformer action.

Therefore, the single-phase induction motors are not self-starting, and special measures have to be taken to make them self-starting. Because of this shortcoming and the first power systems being single phase, induction motors were not available for use until the 1890s when the self-starting poly-phase induction motor became available.

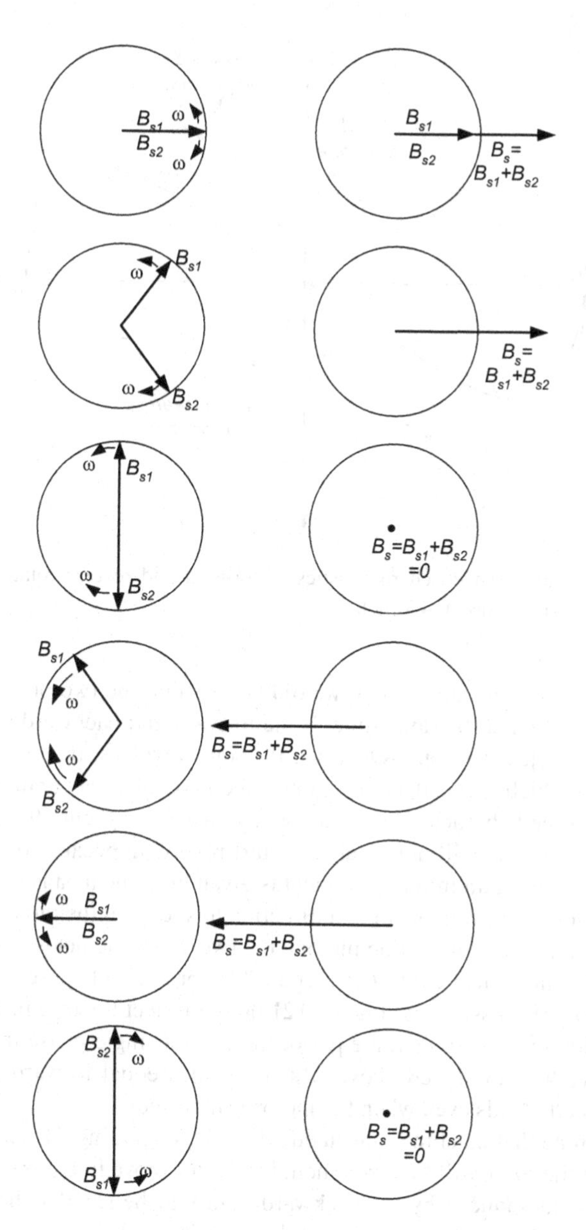

FIGURE 9.20 The addition of two oppositely rotating stator magnetic fields.

However, if an initial movement is provided, the motor will continue to rotate in that direction (see Figure 9.21). Now let us explain how this takes place. If we consider the forward rotating magnetic field only, similar to three-phase induction motors, the torque/speed characteristics would be expected to be as shown in Figure 9.21 (curve f). Conversely, the torque/speed characteristics of a single-phase induction motor due to backward rotating magnetic field only, would be expected to be as shown with curve g in Figure 9.21.

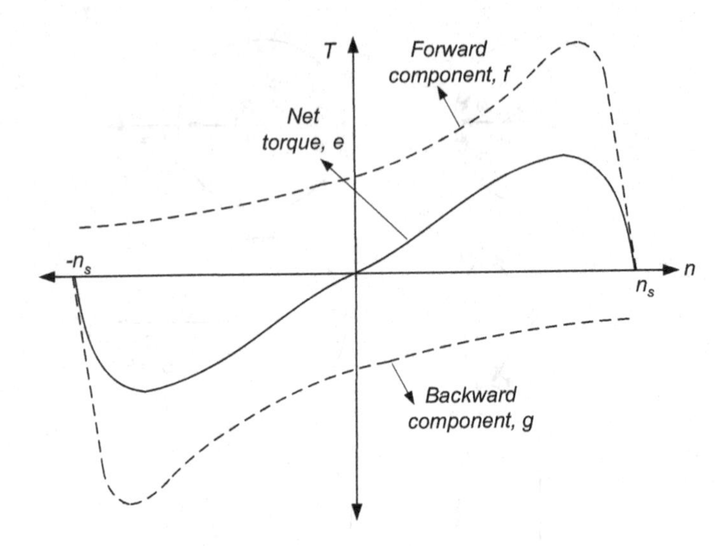

FIGURE 9.21 Torque/speed characteristics of forward and reverse rotating fields added (assumed not to exist at the same time).

Beware that in above discussion, we did not assume the two oppositely rotating magnetic fields exist at the same time. If the forward and backward magnetic fields were to remain equal and opposite when the rotor revolves, the expected net total magnetic field which is the algebraic sum of the two component curves would produce a torque/speed characteristic as shown by curve e in Figure 9.21. This curve e indicates that machine will not be able to start revolving because of zero torque at standstill. However, if an initial movement is given by some means in any direction, a net torque would be available in that direction to speed up the motor.

Since in single-phase induction motors, the two magnetic fields exist at the same time and the same stator current is responsible for each of these stator magnetic fields, the above discussion and Figure 9.21 do not reflect the true picture in single-phase induction motors. In fact, the performance of a single-phase induction motor is much better than predicted above which assumes equal forward and backward rotating magnetic fields even when the motor is in motion.

Let us assume that machine is at standstill at the beginning. Hence, the net starting torque on the rotor will be zero. Should the rotor move in the positive direction, the rotor currents induced by the backward field will be smaller than at standstill with a smaller power factor due to increasing rotor reactance.

This change in rotor current will decrease the backward stator magnetic field and the decrement in backward magnetic field together with smaller rotor current and power factor (meaning smaller torque angle) will decrease the reverse torque produced by the backward magnetic field. Conversely, the rotor currents induced by the forward field would be larger than at standstill with a greater power factor due to decreasing rotor reactance. This rotor current change will increase the forward stator magnetic field. This increment in forward magnetic field together with greater rotor current and power factor (meaning greater torque angle) will increase the forward torque produced by the forward magnetic field. As the speed changes, because the stator voltage is constant, total stator magnetic field can be assumed to be constant (neglecting stator losses and reactances). Therefore, the sum of the two oppositely rotating stator magnetic fields must remain a constant. Hence, with the rotor speed increasing in positive direction, because the two stator fields take place at the same time, the torque produced by the backward stator field gets smaller than the values in Figure 9.21. By the same argument, we can say that when the two stator fields take place at the same time, the torque produced by the forward stator field become higher than the values given in Figure 9.21. In fact, the decrease in magnetic field in reverse direction is equal to the increase in the forward direction so that total magnetic field remains the same. The effect on torque is twofold compared to the values in Figure 9.21. At small slip values near the synchronous speed which happens to be the normal running region, the torque–speed characteristics of the single-phase induction motor become comparable with those of the balanced three-phase induction motor (Figure 9.22).

However in addition to the torques shown in Figure 9.22, double stator frequency torque pulsations will occur as magnetic fields go past each other twice per cycle. Although these pulsations do not effect the average torque, they tend to make the machine noisy. Therefore, elastic mountings would be necessary for these machines. This can be seen from the following equation which is derived from equation (9.11) when ωt is made equal to θ:

$$B = KI_m + \frac{1}{2}KI_m \cos(2\omega t) \tag{9.12}$$

The backward rotating field would cause losses as well as an oscillating double-frequency torque in the rotor structure. This is also true for all unbalanced polyphase AC machines in which rotating field is utilised. Therefore, the operation of AC

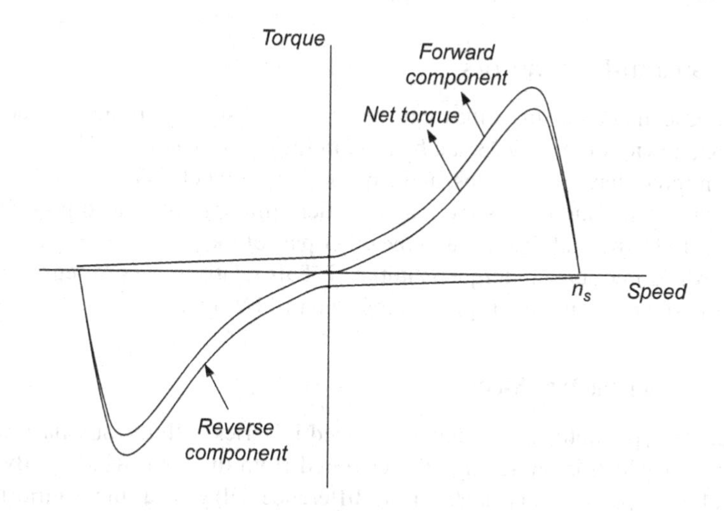

FIGURE 9.22 Torque/speed characteristics of forward and reverse rotating fields added (assumed to exist at the same time).

machines under balanced polyphase conditions is important for efficient operation of these machines. This will eliminate the negative-travelling flux waves as well as some of the losses, while the positive-travelling flux waves are reinforced. Balanced polyphase machines invention led up to smooth power transmission and distribution and hence today's electrical utilities.

Now, we will look into ways of starting single-phase induction motors. There are usually three ways used to start these motors. All of these techniques rely on making one of the rotating magnetic fields stronger than the other one so that motor takes off from standstill. The factors such as the amount of starting torque, running torque, starting current, and cost would be the factors to decide about which one should be chosen. Induction motors are named with their ways of starting:

1. Split-phase motors
2. Capacitor-type motors
3. Shaded pole motors.

9.3.1 Split-Phase Motors

Nearly all single-phase induction motors have an auxiliary winding rotated in space by 90 electrical degrees from the main winding to achieve the start of rotation. To achieve higher resistance, auxiliary winding is wound with a thinner wire. This way, the machine looks like an unbalanced two-phase induction machine which is a self-starting induction motor. Auxiliary winding is usually disconnected with a centrifugal switch which opens at about 75% of the rated speed. Therefore, this winding is only effective during start-up.

Split-phase motors have low starting current. They are used in applications that require moderate starting torque such as fans, blowers, and pumps. They are not very expensive and usually in the power ranges below 1 kW.

9.3.2 Shaded-Pole Motors

Single-phase induction machines can be made self-starting if the "shaded-pole" structure shown in Figure 9.23 can be used in the stator windings. The shaded-pole induction motor has salient poles, and a portion of each pole is encircled by a heavy copper ring. This ring causes the magnetic field through the ringed portion of the pole face, to lag behind that through the other part of the pole face.

This way a low starting torque is initiated. This relatively less expensive method of starting induction motors is preferred for up to 1 kW motors.

9.3.3 Capacitor-Type Motors

In capacitor-type motors, a capacitor is placed in series with the auxiliary winding of the motor which is physically 90° separated from the main winding. By proper choice of the capacitor value, a 90° phase difference will yield a single uniform rotating magnetic field. Therefore, capacitor-type motors behave very much like balanced polyphase motors. In this way, it is possible to enhance the starting torque of these

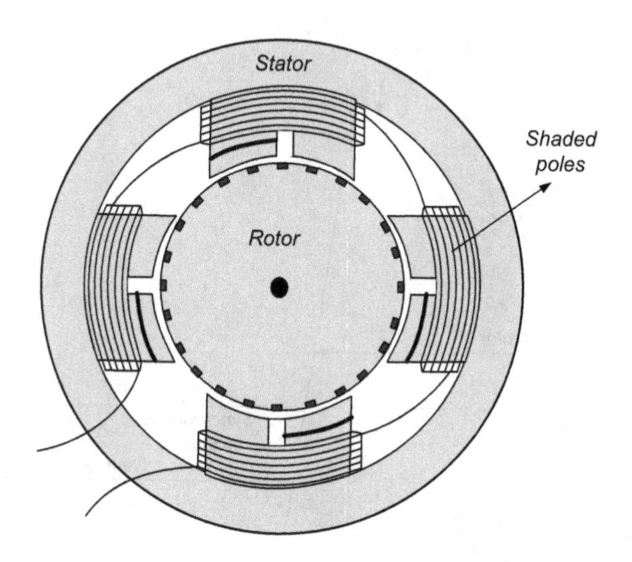

FIGURE 9.23 Shaded pole motor.

machines several times. Capacitor-start type of motors are more expensive but have higher starting torques.

If the capacitor and the auxiliary winding are not taken out after starting, we would have a permanent split-type motor at any load that might be connected. In this type of motors by proper selection, the full performance of the motor can be enhanced. This way, the double-frequency pulsations would be eliminated and the efficiency would be improved by eliminating the backward rotating magnetic field. However, this would be at the expense of decreasing the starting torque of the motor. For this reason, sometimes two capacitors are used to improve both the starting and normal working performance of the motor.

9.4 THE INDUCTION GENERATOR

Due to severe limitations, induction generators are not as widely used as the induction motors. Because of their structure, they cannot produce reactive power but consume continuous reactive power and an external reactive power source must be continuously connected to them. This reactive power controls the terminal voltage of the generator.

Previously, we said that when induction motors are driven beyond the synchronous speed, they become generators. We can see this from the torque–speed characteristics of Figure 9.24. If the external prime mover drives an induction motor at a speed greater than the synchronous speed, the torque will change its direction and the machine will act as a generator (Figure 9.24). For the induction machine to operate as a generator, its stator terminals must be continuously connected to a (polyphase/single phase) voltage source, and it must be driven above the synchronous speed (Figure 9.25). The source connected to the stator terminals fixes the

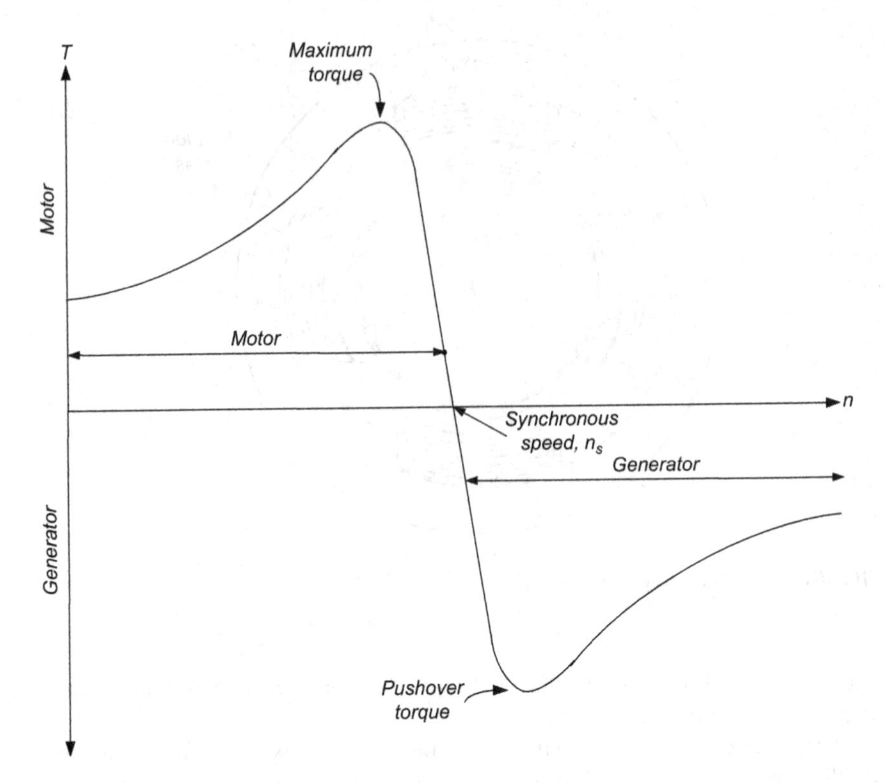

FIGURE 9.24 Induction machine torque–speed characteristics showing generator action.

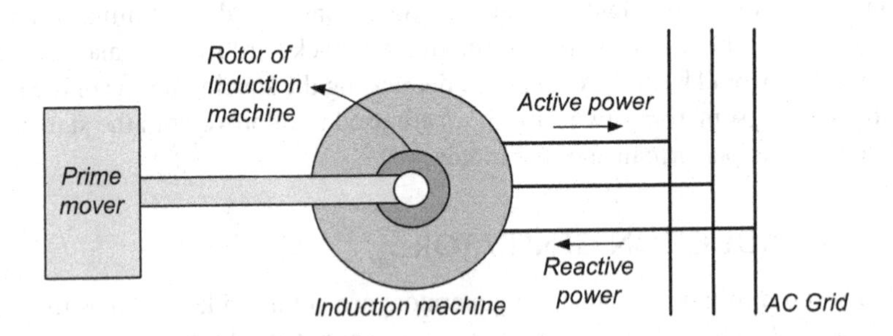

FIGURE 9.25 Induction machine working as a generator.

synchronous speed and also the frequency. Therefore, induction generators are not normally self-excited machines.

If the rotor is made to rotate at a speed more than the synchronous speed, the slip becomes negative. A rotor current is generated in the opposite direction, due to the rotor conductors cutting stator magnetic field. This generated rotor current produces a rotating magnetic field in the rotor which pushes (in opposite way) onto the stator field. This causes a stator voltage which pushes current flowing out of the stator winding against the applied voltage. Thus, the machine is now working as an induction generator (asynchronous generator). The active power supplied back in the line is proportional to slip above the synchronous speed. As the applied torque increases, power delivered to the utilities also increases. A great advantage of induction generators is their speed which does not have to be constant as in synchronous generators. When an induction generator is connected to a grid, it takes reactive power from the grid.

Their structural simplicity is another advantage of them. This actually well suits to applications of electricity generation from wind power in which its stator terminal is connected to a power supply and its rotor is driven by the varying wind (hence varying torque). The generator will overspeed if a torque more than the push over torque is applied to its shaft.

PROBLEMS ON CHAPTER 9

Problem 9.1: If we want a 50-Hz induction motor to drive a load at a speed of 650 rpm, how many poles should the motor have? What is the slip of this motor for the speed of 650 rpm?

Problem 9.2: What would be the synchronous speeds of the induction motors operating at 60 Hz if they have pole numbers of 10 or less? If one of these induction motors is operating at a speed of 750 rpm, how much would be the slip of this motor?

Problem 9.3: A 15-hp four-pole 50-Hz three-phase induction motor runs at 1,450 rpm under full-load conditions. Find the slip and the frequency of the rotor currents at full load. Neglecting the rotor inductance, what would be the speed in rpm of the motor if the load torque is doubled?

Problem 9.4: A certain three-phase 50-Hz induction motor rotates with a speed of 1,400 rpm at a power factor of 0.90 and has the following losses: stator copper loss = 40 W, rotor copper loss = 70 W, stator iron loss = 100 W, and mechanical losses = 5 W. Find the following:

(a) number of poles and slip, (b) air-gap power, (c) output power, (d) input power, (e) efficiency, (f) output torque, and (g) apparent power

Problem 9.5: A 240-V-rms, two-pole, 50-Hz three-phase delta-connected induction motor draws 40 A at a power factor of 85% lagging. The total stator copper and iron losses are 500 W, and the total rotor copper losses are 120 W. The rotational losses are 200 W. Find the power crossing the air gap P_{ag}, the developed power, the output power P_{out}, and the efficiency. What is the slip and speed of rotation in rpm of the motor?

Problem 9.6: (a) In which part of the induction motor the winding responsible for the rotating field is situated? In these motors, does the voltage induced in the rotor increase or decrease as the motor speeds up?

(b) A three-phase 50-Hz induction motor is driving a load, which requires a torque of 20 Nm at a speed of 1,200 rpm. If the efficiency of the motor is 60% and the stator iron and copper losses are 400 W, find the followings: number of poles, slip, frequency of the rotor induced voltage, output power, input power, and rotor mechanical losses.

Problem 9.7: A two-pole, 60-Hz induction motor is running at a speed of 3,500 rpm when driving a load which requires 12 kW of power. If the mechanical losses are negligible, (a) find the slip of the motor, (b) find the developed torque, (c) if the torque is doubled, what would be new operating speed of the motor? (d) What is the new power supplied by the motor when the torque is doubled?

Answers to odd questions:

9.1: 0.13
9.3: 1400 rpm
9.5: 13633.5 W, 13513.5 W, 13313.5 W, 94%, 0.88%, 360 rpm
9.7: 2.77%, 32.74 Nm, 3400 rpm, 11.656 kW.

Bibliography

1. J. R. Cogdell. 1999. *Foundations of Electrical Engineering*. London: Prentice Hall, Inc. ISBN 0-13-092701-5.
2. Alan R. Hambley. 2008. *Electrical Engineering*. New York: Pearson Education, Inc. ISBN 0-13-206692-0
3. Stephen J. Chapman. 2011. *Electric Machinery Fundamentals*. New York: Mcgraw Hill Higher Education. ISBN-10 007108617X.
4. A. E. Fitzgerald, Charles Kingsley, Jr, Stephen D. Umans. 2002. *Electric Machinery*. New York: McGraw-Hill Companies, Inc. ISBN 0-07-366009-4
5. Timothy L. Skvarenina. 2002. *Power Electronics Handbook*. Boca Raton, FL: CRC Press LLC. ISBN 0-8493-7336-0
6. Edward Hughes. 2008. *Electrical and Electronic Technology*. New York: Pearson Education Limited. ISBN 978-0-13-206011-0.
7. Bhag S. Guru; Huseyin R. Hiziroglu. 2000. *Electric Machinery and Transformers*. Oxford: OUP Higher Education Division. ISBN 9780195138900.
8. P. C. Sen. 2013. *Principles of Electrical Machines and Power Electronics*. Hoboken, NJ: John Wiley & Sons. ISBN 111807887X.
9. Soteris A. Kalogirou. 2014. *Solar Energy Engineering and Processes and Systems*. New York: Elsevier, Inc. ISBN-13: 978-0-12-397270-5.
10. Napoleon Enteria, Aliakbar Akbarzadeh. 2014. *Solar Energy Sciences and Engineering Applications*. London: Taylor and Francis Group. ISBN 978-1-138-00013-1.
11. A. E. Fitzgerald, D. E. Higginbotham, A. Grabel. 1967. *Basic Electrical Engineering*. New York: Mcgraw Hill.
12. Luces M. Faulkenberry, Walter Coffer. 1996. *Electrical Power Distribution and Transmission*. London: Prentice Hall. ISBN-10: 0132499479

Index

J

K

L

M

N

O

P

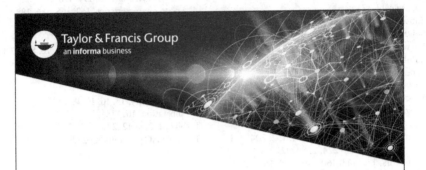

Printed in the United States
by Baker & Taylor Publisher Services

Printed in the United States
by Baker & Taylor Publisher Services